How the Earth Works

Essentials in Earth System Science and Geology

First Edition

By Charlotte Mehrtens

Bassim Hamadeh, CEO and Publisher
Kassie Graves, Director of Acquisitions
Jamie Giganti, Senior Managing Editor
Miguel Macias, Senior Graphic Designer
John Remington, Senior Field Acquisitions Editor
Monika Dziamka, Project Editor
Brian Fahey, Licensing Specialist
Allie Kiekhofer, Associate Editor

Printed in the United States of America

ISBN: 978-1-62661-886-2 (pbk) / 978-1-62661-887-9 (br)

Dr. Barb Tewksbury and the SERC "Cutting Edge Curriculum" workshops provided the inspiration to overhaul my introductory geology class. This book is the result. Thanks to Molly Conroy for research assistance on hydrofracking and figure preparation and to Dr. Phoebe Judge for reading an early draft. Thanks also to Al, Ginny, Marilyn, and Kerry for their support during the writing period. Drs. Tony Fowler, Chris Sinton and Shelly Rayback provided feedback on the preliminary edition.

This book is dedicated to Stockton (Skip) Barnett, who started me down this path.

Contents

PART I INTRODUCTION 1

CHAPTER 1. Science as a Process of Discovery 3

Introduction
How the Scientific Method Works
What is Hydraulic Fracturing?
Potential Environmental Problems
A Scientific Study of Well Water Contamination from Hydrofracking
A Scientific Study on Whether Hydrofracking Triggers Earthquakes
Summary
Review Questions

CHAPTER 2. Introduction to Earth System Science 17

Definition and Characteristics of a "System"
Types of Systems
The Earth System
Why Use a "Systems Approach" to Study the Earth?
Summary
Review Questions

PART II INGREDIENTS AND STRUCTURE 25

CHAPTER 3. The Ingredients of Planet Earth 27

Introduction
Origin of the Elements
Assembling Molecules: Achieving Stability
Chemical Bonds

Isotopes
The Relative Abundance of Elements on Earth
Composition and the Physical Properties of Matter
Differentiation of Planet Earth
Summary
Review Questions

CHAPTER 4. The Earth's Most Important Molecules 41

Introduction
Definition of a Mineral
Mineral versus Crystal
Crystal Structure of Minerals
Determining the Crystal Structure of a Mineral
How and Where Minerals Form
Mineral Composition
The Silicates
The Oxide Minerals
The Carbonate Minerals
The Sulfide and Sulfate Minerals
The Halide Minerals
Native Metals
How Do We Determine the Chemical Composition of Minerals?
Water
The Uniqueness of the Water Molecule
Relationship to Physical/Chemical Behavior to Dipolar Structure of Water
Density
Other Properties of Water
Summary
Review Questions

CHAPTER 5. Introduction to Rocks and the Rock Cycle 63

Introduction
The Three Types of Rocks
Igneous Rocks
Texture of Igneous Rocks
Origin of Magma
Heat within the Earth
Processes That Cause Melting
Decompression Melting

Melting from the Addition of Volatiles
Conduction Melting
Partial Melting
Fractional Crystallization
Volcanoes
Types of Volcanoes
What Controls Which Type of Volcano Forms?
Supervolcanoes
Igneous Rock Bodies
Sedimentary Rocks
Formation of Sedimentary Rocks
Identifying and Classifying Sedimentary Rocks
The Difference between Sediment and Soil
Metamorphic Rocks
The Concept of Metamorphic Grade
Regional versus Contact Metamorphism
Metamorphic Rock Classification
Summary
Review Questions

CHAPTER 6. Telling Time 95

Introduction
Relative Geologic Time
Unconformities
Geologic Time Scale
Numerical Dating
Using Radioactive Decay to Tell Time
Assumptions in the Radiometric Dating Process
Synthesizing Relative and Numerical Ages
Rates of Geologic Processes
Summary
Review Questions

CHAPTER 7. Earthquakes, Seismology, and the Layered Earth 109

Introduction
What Causes an Earthquake?
Why Do Earthquakes Occur?
How Do Geologists Study Waves in the Interior of the Earth?
Seismic Waves: Shock Waves through the Earth

Types of Seismic Waves
Seismic Wave Velocities
Describing the Size of an Earthquake
What Earthquakes Tell Us about the Internal Structure of the Earth
The Lithosphere and Asthenosphere
Origin of the Layered Earth: Differentiation
Where Earthquakes Occur
Summary
Review Questions

CHAPTER 8. Other Characteristics of the Earth's Interior 131

Introduction
Heat Flow within the Earth
Source of the Earth's Internal Heat
How Does Heat Energy Move within the Earth?
Mantle Plumes
Magnetism
Why Does a Compass Point toward North?
Magnetic Polarity Reversals
How is the Orientation of the Earth's Magnetic Field Preserved in Rocks?
How We Sample Rocks for Their Magnetic Record
Magnetic Polarity Reversals over Time
Magnetic Anomalies
Summary
Review Questions

PART III HOW THE SOLID EARTH WORKS 155

CHAPTER 9. Plate Tectonics 157

Introduction
How Plate Tectonics and Continental Drift Differ
What is a Plate?
How Do the Plates Move?
How Fast Do Plates Move?
Plate Boundaries
Plate Tectonics, Volcanism, and Earthquakes
Interpreting Volcano and Earthquake Distributions
Focal Mechanism Studies

Observations of Divergent Plate Boundaries
Summary of Observations of Divergent Plate Boundaries
Processes at Divergent Plate Boundaries
Implications for the Earth System
Observations of Convergent Plate Boundaries
Summary of Observations of Convergent Plate Boundaries
Processes at Convergent Plate Boundaries
Implications for the Earth System
Transform, or Strike-Slip Boundaries
Plate Boundaries Change over Time
Summary
Review Questions

CHAPTER 10. Origin and Evolution of Mountains 191

Introduction
Formation of Mountains at Convergent Plate Boundaries
Volcanism
Fold and Thrust Belt
Crustal Thickening and Isostasy
How Isostasy Is Related to the Uplift of Mountains
Thickening the Crust to Uplift Mountains
Summary
Summary
Review Questions

PART IV WATER, AIR, AND THE EARTH'S CLIMATE 205

CHAPTER 11. The Hydrosphere 208

Introduction
The Production of Water on Earth
The Hydrologic Budget
Gas Exchange with the Hydrosphere
River Transport of Ions
Chemical Weathering
Groundwater
Karst Landscapes
The Effects of Weathering on the Ocean Reservoir

Where do the Ions go When They Enter the Oceans?
Heat Capacity of Water and its Role Thermo-regulating Earth
Surface Circulation in the Ocean
Thermohaline Circulation in the Ocean
Physical Weathering by Water
Waves as Agents of Physical Erosion
Stream Dissection
Glacial Weathering
Soil
The Soil Profile
Soil Types
Weathering Summary
Changes to the Reservoirs in the Hydrosphere
Global Sea Level
Effects of Sea Level Rise on Coastlines
Effects on Precipitation
Changes in Ocean Chemistry: Ocean Acidification
Methane Release from Freshwater Reservoirs
Summary of Future Changes in the Hydrosphere
Summary
Review Questions

CHAPTER 12. The Earth's Atmosphere and Energy Budget 249

Introduction
Origin of the Earth's Atmosphere
Layers of the Atmosphere
Troposphere
Composition of the Troposphere
Stratosphere
Mesosphere
Thermosphere
Heat Transfer and the Earth's Heat Budget
Heat Transfer
The Electromagnetic Spectrum
Heat Budget
Fluctuation in Solar Irradiation
The Carbon Cycle
Greenhouse Gases
What Makes a Gas a Greenhouse Gas?

Volcanic Emissions and the Earth's Atmosphere
Summary

CHAPTER 13. Earth's Climate 275

Introduction
Weather versus Climate
Variation in the Earth's Orbit around the Sun
The Historical Record of Climate Change
Layers in Ice
Climate Proxies
Oxygen Isotopes
Deuterium
Results of Ice Core Study
Glacial Ice Bubble Composition
The Younger Dryas
Holocene Climate Change: The Record of Global Climate Change in the Oceans
Summary of Paleoclimate Data for the Late Pleistocene and Holocene Epochs
The Role of Plate Tectonics in Controlling Climate
Atmospheric Composition and Global Warming
Current and Future Temperature Trends
Missing Heat? Another Example of the Scientific Method
Summa

INDEX 319

Part I

Introduction

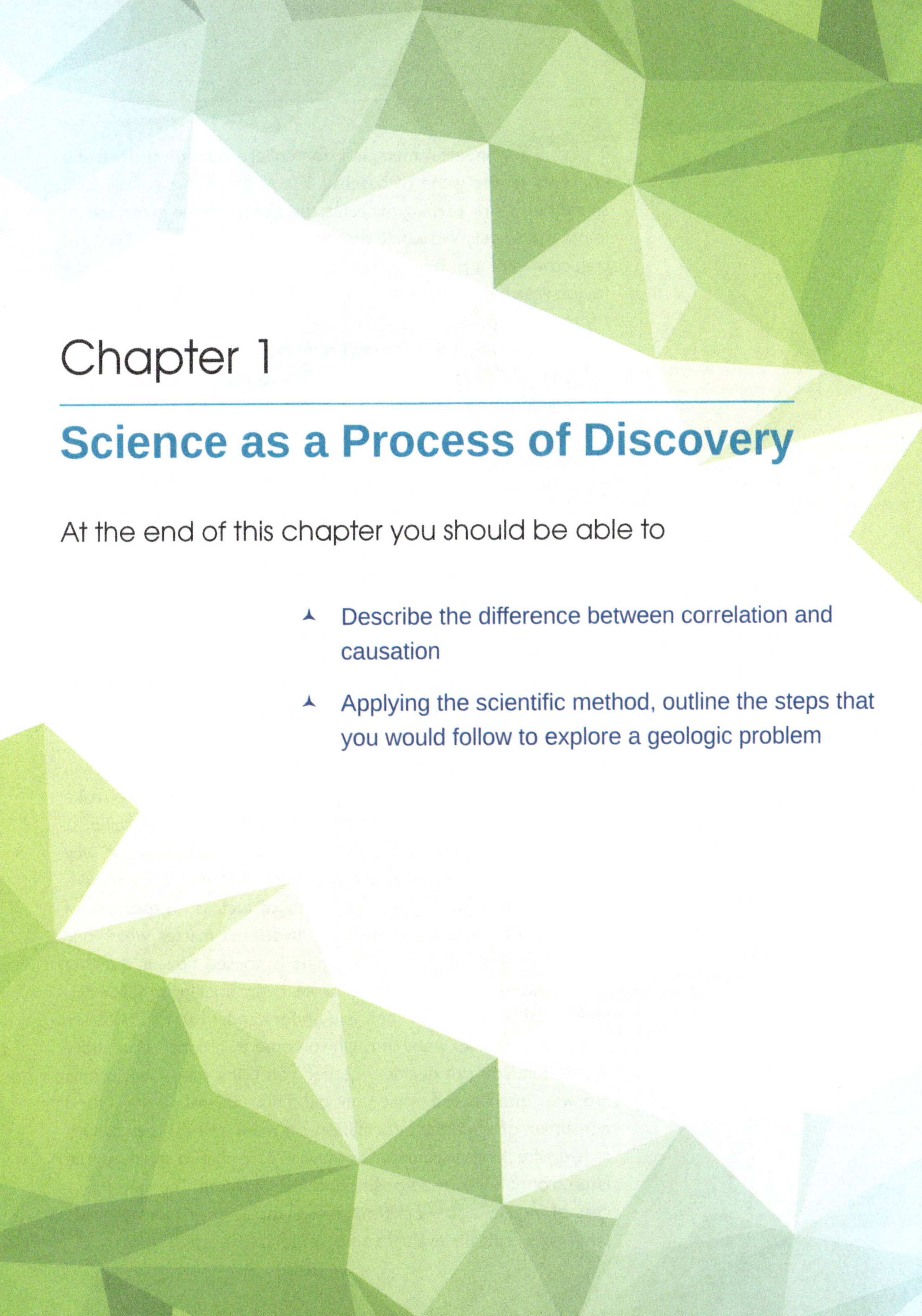

Chapter 1

Science as a Process of Discovery

At the end of this chapter you should be able to

- Describe the difference between correlation and causation
- Applying the scientific method, outline the steps that you would follow to explore a geologic problem

Introduction

There is a wonderful metaphor for college education using the concepts of the game of baseball. If college were baseball, we'd spend two years learning the rules, the history of the game, and its impact on society. We would devote the third year to the theory of trajectories of a thrown ball or swung bat. Then, one week in our fourth year, before we were to graduate from school, we'd actually try to play a game.

Unfortunately, this is an accurate metaphor for how many of us teach and learn science. There are so many terms, concepts, and equations to learn that it can take a huge effort to feel that you know enough to actually apply knowledge to a problem. When you are drowning in definitions and equations, it's hard to remember how you will be able to apply this to anything interesting. Alternatively, we can switch things around and ask, "What are the questions we want to be able to answer?" as a way to identify what we need to know. Science is a process of asking questions and figuring out how to answer them.

Each of the scientific disciplines has its own vocabulary, terms, and equations. They all share, however, the same process of discovery. This process is usually reduced to the phrase "the scientific method," which presents this concept as another easy-to-memorize-and-forget piece of jargon. The method of doing science, however, is fundamentally important to understand if you are going to use the conclusions reached by scientists to make decisions and public policy. Can you "believe" what a scientist says any more than what a talk show host says? Why or why not? This chapter will present you with a modern environmental issue, hydrofracking to retrieve fossil fuel, as an example of how scientists approach a problem. It doesn't matter what your *opinion* on this subject is. The point is to see how a scientist approaches a question such as "Does hydrofracking create environmental problems?" When you understand what a scientist or group of scientists went through to come to a conclusion about a question, you can decide whether you think their conclusions are warranted based what they did. This will make you a good consumer of scientific information on many critical issues, such as hydrofracking or climate change. You'll be able to speak on the issue from an informed position, not an opinion. The same process of inquiry also works outside of the sciences; for example, you could critically evaluate an economic forecast.

How the Scientific Method Works

There is a tremendous amount of debate and discussion around the issue of hydrofracking. Is that what science is, debate and discussion by men and women who have postgraduate degrees? The basis of science is posing a question, developing the methods to answer the question, collecting the data and analyzing them, and then drawing conclusions that interpret the data. The key to this process is in the "analysis" stage, where you test the data to see whether or not the conclusions are false. Much of the testing is done on the methods of collecting the data. Did you collect the data in a way that biases the outcome? Did you collect the data in the correct places or times? Did you collect enough data so that they are statistically reliable? Scientists publish the results of their studies in "peer-reviewed publications," which means that a journal submits the papers to many other scientists for anonymous critique of the methods, the analysis, and the conclusions. A paper that is accepted for publication includes the methods used in the study, so that other scientists can replicate the study and see if they come to the same conclusion. Papers published by scientists do not convey their beliefs or opinions but present observational and measured data that are interpreted in light of existing principles and theories of science. Scientists who would interpret the data differently are able to replicate the study and present their conclusions to the scientific community through the same peer-review process. Other scientists may interpret the data differently and present their conclusions to the scientific community through the same peer-review process.

Does this mean that every scientific publication is "true"? Scientists never use that word because we all realize that our work represents the state of knowledge at a moment in time. As new methods are developed, new instruments constructed, and new data sets assembled, different conclusions might be reached. Published scientific papers represent the state of the discipline at that time.

Let's now apply the scientific method to two example questions. Is there evidence that hydrofracking pollutes drinking water? Is there evidence that hydrofracking triggers earthquakes? The discussion of these issues will necessarily employ terms and concepts, such as "earthquakes," that will be discussed in detail later on in the text.

What is Hydraulic Fracturing?

Some types of rocks have hydrocarbon, or fossil fuel, present between their grains. Sometimes it is easy to extract the hydrocarbon material. Using an oil well is an example of this extraction. Other times the hydrocarbon is tightly bound within the rock, and in order to extract it, geologists "crack" the rock, creating artificial fractures in it, by injecting large volumes of a liquid with additives under pressure (Figure 1-1). This process is termed "hydraulic fracturing" and is often abbreviated to "hydrofracking" or "fracking."

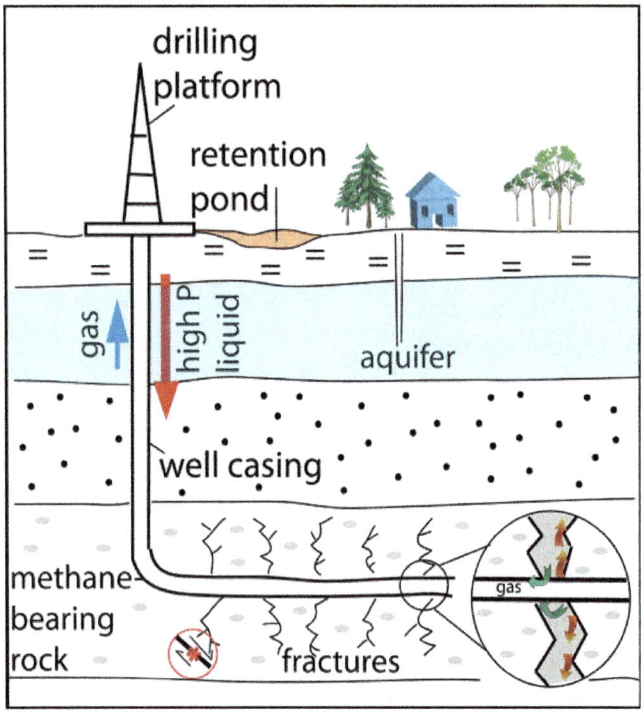

The hydrofracking process has been used to increase the yield of traditional oil wells, and more recently the process has been applied to the retrieval of hydrocarbon from very fine-grained, low-porosity and low-permeability (terms that describe the ability of a rock to store fluid and how well the fluid can move through the rock, respectively) shale rocks. The hydrocarbon present in shale rocks is commonly methane (CH_4, natural gas), which can be refined to various consumable fuels.

Because of the political, social, and economic implications of energy issues, the significance of increased productivity of domestic resources has made the fracking of oil-rich shale a critically important subject for public discussion. The next section of the text will discuss two of the environmental issues that are related to extraction of this natural resource.

Figure 1-1 Components of hydrofracking. There is no vertical or horizontal scale to this diagram. The circular area in the lower right is an enlargement of the flow of liquid (orange) into the fracture and gas (green) out of it. Grey-colored "blebs" are pockets of hydrocarbon. Dots and dashes represent other rock types. Also shown is a layer of rock-containing water (aquifer), which is shown as the source of drinking well water. A small earthquake is shown by the small red circle. P = pressure.

Potential Environmental Problems

Figure 1-1 illustrates the general scenario for a hydrofracking drill site. At some depth below the surface is a fine-grained sedimentary rock containing natural gas (methane, CH_4). At some distance homes and farms sink water wells into layers of rock that serve as drinking water reservoirs, termed *aquifers*. The fracking process injects high-pressure fluids into the drill hole, which serves to create artificial cracks or fractures in the gas-bearing rock. The high-pressure fluids contain chemicals that improve the extraction of the trapped gas, which is then pumped back up to the surface. In Figure 1-1, a retention pond is shown next to the drilling rig. This pond provides water for the chemical mix that is injected into the well. These fluids return to the surface and need to be disposed of, and the most common means of doing so is to pump them back into the rock, filling the space produced by the removal of the fossil fuel material. This fluid is supposed to occupy the same region within the rock that tightly held the natural gas.

An examination of Figure 1-1 might also illustrate to you several potential environmental problems in the hydrofracking process. Your first observation might be that because there are drinking wells nearby, there is the potential pollution of these wells, either from the fluids injected in the well escaping into the adjacent rock layers or from possible leaks out of retention ponds that store drilling fluids. Another potential problem you might have observed involves the triggering of small earthquakes as a result of the injection of injection of waste water from the drilling process back into the rock. There is currently a great deal of debate about whether these environmental problems have occurred or are occurring. There are obvious monetary reasons to pursue fossil fuel extraction; however, are there environmental costs to this extraction? If so, what are they? You only have to listen to cable news stations to hear the opinions on this from media personalities. How would a scientist approach these questions?

A Scientific Study of Well Water Contamination from Hydrofracking

A scientist might start with the question "Is there any reason to assume that fracking contaminates well water?" In other words, is there a *theoretical reason* why this might be a problem? Depending on the partial pressures of the gas and liquid, gases are soluble in water. We also know, from basic chemistry, that methane is highly combustible. Because disclosure of fracking chemicals is voluntary for many energy companies and only a few states have laws requiring this, it may not be possible to ascertain what chemicals are being used in a natural gas extraction. A congressional panel on hydrofracking and water pollution assembled a partial list of the chemistry of the fluids injected into the wells (Table 1-1), many of which are highly toxic. Basic groundwater geophysics tells us that

Table 1-1 Chemical Components of Concern: Carcinogens, SDWA-Regulated Chemicals, and Hazardous Air Pollutants

Chemical Component	Chemical Category	No. of Products
Mathanol (Methyl alcohol)	HAP	342
Ethylene glycol (1,2-ethanediol)	HAP	119
Diesel[19]	Carcinogen, SDWA, HAP	51
Naphthalene	Carcinogen, HAP	44
Xylene	SDWA, HAP	44
Hydrogen chloride (Hydrochloric acid)	HAP	42
Toluene	SDWA, HAP	29
Ethylbenzene	SDWA, HAP	28
Diethanolamine (2,2-iminodiethanol)	HAP	14
Formaldehyde	Carcinogen, HAP	12
Sulfuric acid	Carcinogen	9
Thiorurea	Carcinogen	9
Benzyl chloride	Carcinogen, HAP	8
Cumene	HAP	6
Nitrilotriacetic acid	Carcinogen	6
Dimethyl formamide	HAP	5
Phenol	HAP	5
Benzene	Carcinogen, SDWA, HAP	3
Di (2-ethylhexyl) phthalate	Carcinogen, SDWA, HAP	3
Acrylamide	Carcinogen, SDWA, HAP	2
Hydrogen fluoride (Hydrofluoric acid)	HAP	2
Phthalic anhydride	HAP	2
Acetaldehyde	Carcinogen, HAP	1
Acetophenone	HAP	1
Copper	SDWA	1
Ethylene oxide	Carcinogen, HAP	1
Lead	Carcinogen, SDWA, HAP	1
Propylene oxide	Carcinogen, HAP	1
p-Xylene	HAP	1
Number of Products Containing a Component of Concern		**652**

Table 1-1 Data on hydrofracking chemicals. From US House of Representatives Committee on Energy and Commerce Staff Report, 2011. HAP = hazardous air pollution; SWDA = potential candidate for regulation under Safe Water Drinking Act.

the flow of water from drinking water aquifers is a function of pressure gradients (water will flow from a region of high pressure to low). Because well fluids are injected under high pressure, if these fluids are not contained, physics tells us that there is reason for concern that they would migrate up into aquifers. In summary, there are theoretical reasons to believe that we should investigate the potential for contamination. This does not mean that there *is* contamination, just that there is a rationale to investigate this. You could also choose to rephrase this concept to be more conservative: "it would be irresponsible to *not* explore the potential for drinking water contamination by natural gas extraction."

If we believe that there is theoretical reason to be concerned about water pollution, a scientist might then pose the question "Is there any evidence that water wells have been contaminated from fracking?" Daily news reports present anecdotal evidence that this is the case; however, geoscientists would need to establish a methodology for a research project that explores this question in an organized fashion. At the start of a research project a geoscientist would, for example, define where she would go to collect data, how would she collect data, and discuss how she would know that the sites chosen for study are representative fracking settings and not geologically or hydrologically unusual. She should also know what chemicals are going to be tested for, as well as have information about the subsurface geology. In other words, a good research project is constructed so that by its conclusion, you will be able to answer the original question: "Is there any evidence that water wells have been contaminated from fracking?"

Figure 1-2 shows the data simplified from a more complete data set in Osborne et al. (2011) on the relationship between methane in drinking water wells and distance to a natural gas well.

This data are plotted on a *bivariate graph*. Examine the axes of the graph so that you know what is being plotted.

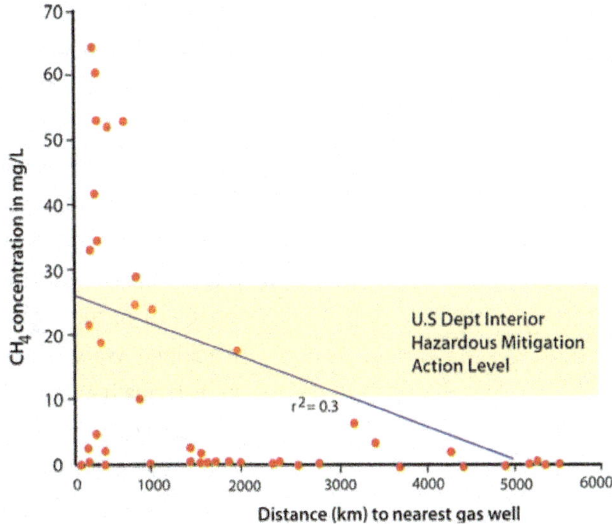

Figure 1-2 Simplified graph of methane concentrations (CH_4 mg/l) as a function of distance to the nearest gas well (active and inactive). The distance estimate is an upper limit and which does not take into account the direction or extent of horizontal drilling underground (this would decrease the estimated distances to some extraction activities). Also shown on the graph is the Department of the Interior range of values for CH_4 concentration levels that trigger hazard mitigation activities (yellow band). Original data in Osborn, et al. (2011). The variables show a positive correlation.

Read the caption of the graph for information on the units of measurement and other details. The authors of this study interpret the graph for their readers in the text of their paper. When you examine the graph reproduced here, would you say that there was reason to assume that there is a correlation between the distance of a drinking water well to a natural gas extraction well? Generally speaking, what is the difference in methane levels from drinking water wells close to natural gas wells and those far away from natural gas wells?

If you looked at these data and said, "There is a correlation between the levels of methane present in drinking water and the distance of a water well from a natural gas extraction well," you would be stating that you see a *correlation* of the data.

Correlation, however, does not mean that there is a causal relationship, it could, in fact, just be coincidence. The next step in the scientific process would be to pursue research that would demonstrate *why* this relationship exists and, furthermore, *can exclude* other causes for the relationship not related to the hydrofracking process.

The scientific paper from which Figure 1-2 was taken also included extensive chemical analysis of the drinking water, which those authors interpreted to conclusively point to contamination of this water from methane released from the drilling process. Other geoscientists reading this published article disagreed with the authors' conclusions, and published their concerns (Davies, 2011; Saba and Orzechowski, 2011; Schon, 2011) in the following issue of the journal. These authors' concerns included the possibility that the chemistry of the drinking water wells was tainted *prior to* the onset of natural gas extraction. Because the study lacked "before fracking" water quality data to compare to the "after fracking" water quality data, Osborne and his colleagues could not exclude the possibility that the water was already contaminated, and that the contamination was unrelated to drilling (in other words, they lacked the so-called "smoking gun"). Another criticism of Osborne's study was the nonrandom selection of wells in the original study and the possibility that the results that the researchers saw were extremely localized to a unique geology of the site and not related to the fracking process in general. In other words, Osborne and his colleagues "overgeneralized" the results of their study. Without coming to any conclusion regarding whether Osborne and his co-workers were "right," this series of published exchanges illustrates both the rigor (pointing out the shortcomings of the research

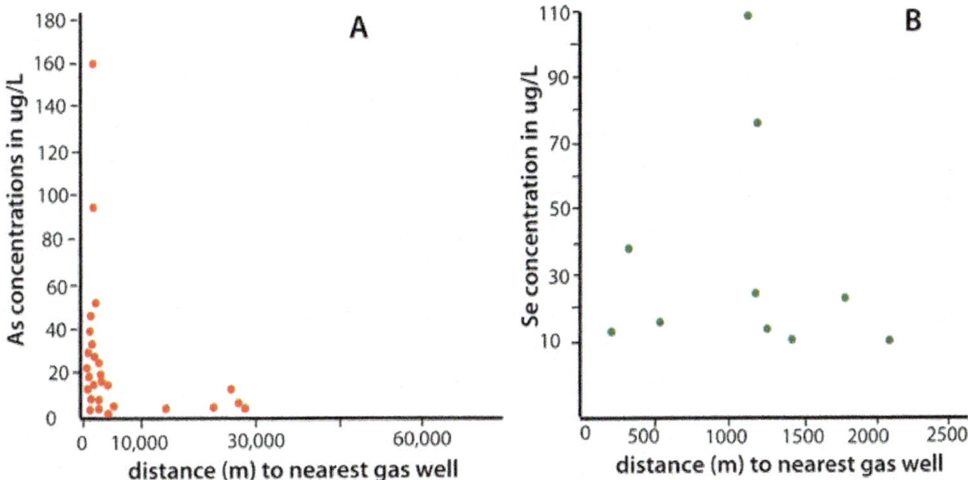

Figure 1-3 Simplified graphs of data on concentration of chemical elements as a function of distance from natural gas extraction wells, active and inactive. A = values for arsenic; B = values for selenium. Note the difference in distances from wells in the two data sets. Original data in Fontenot, et al. (2013).

methodology) and openness of the scientific process (the nature of the criticisms and the identity of the scientists were public).

Another study of the water quality in drinking water wells adjacent to natural gas extraction wells by Fontenot and coworkers (2013) examined concentrations of several toxic elements, including arsenic and selenium, in drinking water. Figure 1-3 shows a simplified set of results from this study.

Examine these two graphs and draw at least one conclusion about the relationship of variables for each graph. In this study, does there appear to be a correlation between the distance from a drinking water well to a natural gas extraction well and concentrations of both arsenic and selenium?

These data were chosen for our discussion because they demonstrate a result (one apparent correlation and one with no correlation) that suggests to a geoscientist that the contamination of drinking water wells by fracking is not as simple as "it gets contaminated" or "it doesn't get contaminated." In this same study, other metals and total dissolved solids are present in higher concentrations closer to operating gas wells as we might predict; however, there was geographic variation in the data—not all water wells close to gas wells are contaminated, only some of them. Clearly, the contamination is not as simple as "the gas wells did it."

Now we can ask the question "What else is going on at natural gas wells that might regularly contaminate groundwater, but not always?" The study by Fontenot and his coworkers suggests that other processes that might be occurring. For example, localized fractures in the well casing of the gas extraction wells could produce localized contamination. Another possibility is that localized lowering of the groundwater table around natural gas wells could change the water chemistry of the aquifer, triggering a variety of chemical reactions. In fact, Fontenot and his coworkers came to the conclusion that localized processes such as mechanical failure (well case cracks, fluid spills, etc.) around active natural gas extraction wells best explain their data. Their study suggests that problems in drinking well contamination can exist adjacent to natural gas extraction wells, but may not occur in all wells. Some individuals consider the contamination of water from the mechanical failure of wells associated with drilling for gas extraction to be the same as "fracking contaminates water," but others do not. A recent EPA study that determined that fracking was not responsible for widespread groundwater contamination was careful to limit their conclusion to situations where mechanical failure within a well did not occur.

It is worth noting that at the end of their paper, Osborne and his colleagues stated that they hoped that their work would provide an impetus to the creation of a federal (or statewide) establishment of baseline water quality data prior to the initiation of drilling, followed by long-term monitoring of data. Because hydrofracking fluids are currently not regulated as a hazardous waste under the Resource Conservation and Recovery Act, and fractured wells are not covered under the Safe Drinking Water Act, there are little baseline data available to assess the degree to which contamination might occur. Similarly, the Fontenot research group also recommends the creation of a coordinated large study to develop groundwater quality data before, during, and after natural gas extraction.

Do the two studies discussed here answer the question "Does fracking contaminate drinking water?" If you respond, "No," the next step is to ask, "What study would I design so that I could conclusively determine whether fracking does or does not contaminate wells?" You have just taken another step in the scientific process. Research is *iterative*, meaning that you establish a research question and collect the data, and it's entirely possible that the results, like those of Fontenot, don't answer the question but lead you to ask other, new questions. Science is an ongoing process. If you are a "question-asker" and an explorer of the natural world, science is a great field.

Our discussion of fracking and water contamination is not intended to tell you the "answer" to the question "Does hydrofracking contaminate groundwater?" You should be able to explain why the question is not this simple. As you'll see throughout this course, there are no simple answers to complex questions. When you study the Earth and how it works, you will realize that it is a complex "machine," so we should not expect simple answers when questions arise about how it "works"!

A Scientific Study on Whether Hydrofracking Triggers Earthquakes

An earthquake is the release of stored (potential) energy when a rock breaks. Large earthquakes are associated with geologic settings on Earth where large forces related to the movement of tectonic plates are occurring. An example of this would be the famous San Andreas Fault system of California. Other regions experience smaller, less frequent earthquakes that are related to the recent geologic history of a region, as rocks are able to store stresses over millions of years. A recent earthquake in 2011 in Virginia is an example of this. However, over the past several years (2010–2013) some regions of the United States that have rarely experienced earthquakes have seen an increase in the number of these events. For example, Ohio, Kansas, and Oklahoma have seen the number of small earthquakes recorded by the US Geological Survey increase to more than one hundred, from a preceding average over the same time period (three years) of twenty-one. Many of these earthquakes are occurring in regions where hydrofracking for natural gas retrieval drilling has begun. Is there a relationship between earthquakes and fracking? To make this possible relationship more visually clear, Figure 1-4 shows the correlation.

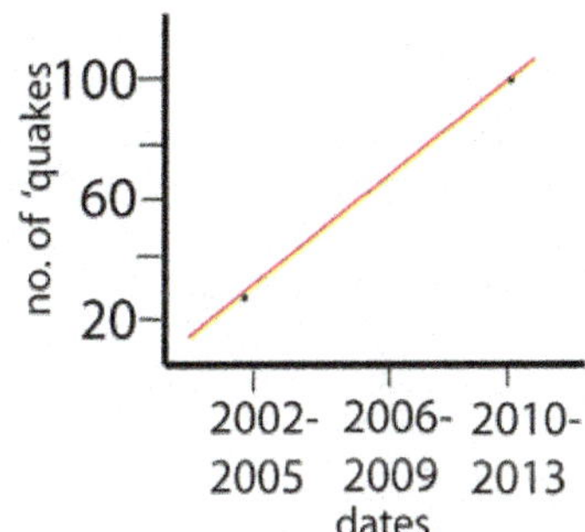

Figure 1-4 With two data points it would be tempting to assume a direct correlation between the number of earthquakes and when they occurred. Two data points are not sufficient for demonstrating a strong statistical correlation, but the apparent relationship does suggest that further data collection is warranted.

Geologists are drawn to the question "Does fracking cause earthquakes?" because the answer expands our knowledge about how earthquakes in general operate. This information could be applied to earthquake-active areas such as California. However, we are also interested in this question because of the risks to life and property from earthquakes. Are facilities in regions that historically have not been concerned about earthquakes now at risk? If earthquakes are triggered by fracking, the implications of this relationship should be a matter of discussion in the development of regulatory policy. So, in order to answer the question of whether fracking

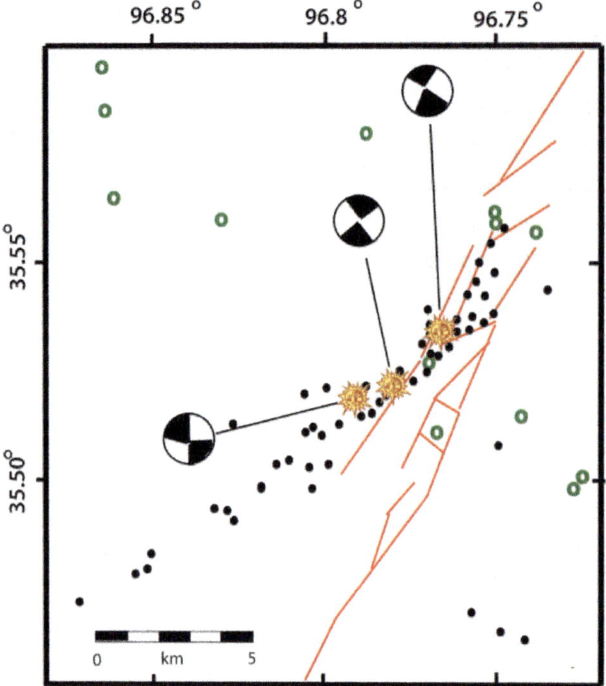

Figure 1-5 A simplified map of a portion of central Oklahoma around the Wilzetta fault system (red lines). Fluid disposal wells are shown in green circles. The large black and white circles are a type of graph that shows the orientation of stresses within rocks involved in the earthquake. In Chapter 9 we will discuss how these graphs are interpreted. In this example the graphs indicate to geologists that the movement is primarily sideways slip. Original data from Sumy, et al. (2014).

does or does not cause earthquakes, geologists analyzed data collected on a swarm of earthquakes in Oklahoma.

Chapter 7 will discuss earthquakes, why they occur, and how we describe their size. In brief, the data analyzed by geologists are obtained by interpreting the record of the movement of rocks interacting with waves of energy released by an earthquake. These records are termed *seismograms*. From the way the rocks move, geologists can determine the nature of the stress the rocks were under. They represent these data on a type of graph that looks like a beach ball, although the technical name is a "focal mechanism solution" (Figure 1-5). Geologists chose to study the earthquakes that occurred along one fault in Oklahoma (a *fault* is a fracture in a rock where movement has occurred). Although a fault was present in these rocks, earthquakes were not present until wells were installed and hydrofracking began. There is theoretical reason to suggest that the fluids identified as triggering earthquakes are associated with the disposal of the fracking liquids, because fluid would lubricate the fault and reduce the friction between the rocks, making it easier for them to slide by one another, triggering an earthquake. A good analogy for this process would be an ice skater's moving across the ice. The pressure of the skate blade on the ice melts a very thin film of water that reduces the friction between the solid ice and skate blade. The film of water reduces friction and allows sliding. The same process occurs in rocks: the introduction of a fluid along a fracture in the rock reduces the friction present and allows the rocks to slide by one another.

The analysis of more than one hundred seismograms was able to document that the rocks along the fault slipped in a direction and motion that would be consistent with changes induced as a result of fluid injection. It is important to understand why this would be the case; in other words, what is the explanation for this linkage? The study described here focused on identifying what effect the fluid injection had on the physical state of the surrounding rocks. The authors attribute the slippage along faults as the result of lowering the confining pressure that was holding the rocks in place.

This study confirmed other research conducted in Texas and Colorado that established linkages between earthquake frequency and fluid injection; however, it was the first to identify specific earthquakes with movement of fault segments where fluid was injected. Scientists now feel confident that there is a causal relationship here, which does support their theoretical reasoning for why such a relationship would exist.

From the perspective of the scientific method, we had initial observations (lots of earthquakes since we started drilling), the posing of a question (are these related to one another?), identifying a location where the geological setting was well understood (one well-studied fault system in Oklahoma), a statistically large sample set (hundreds of data sets), rigorous analysis to see if there were any other ways to explain the data, and a conclusion that was very specific to faults in this region. Should the conclusions be expanded to other fracking sites; in other words, did this study prove that fracking always causes earthquakes? No, it did not. It showed that it did cause earthquakes in this location, and, as a result, it suggests to the scientific community that the potential exists. Other earthquake-prone fracking areas should undertake the same types of analysis.

Summary

In this chapter we've posed questions to explore two potential environmental issues with hydrofracking. There are other issues related to hydrofracking, such as the potential for increased global climate change from the release of methane (see Chapter 13), that we won't discuss because our focus is to explore how scientists ask questions and design research to explore these questions. In order to answer the question "Does hydrofracking pose environmental hazards to water contamination and trigger earthquakes?" what geology concepts do we need to understand? You might respond, "I need to understand how fluids move through rocks," or "I need to understand why earthquakes happen." Once you frame these questions, you can begin the knowledge acquisition process (reading, conducting experiments, making observations) that will help you develop the expertise needed to critically evaluate media coverage of this issue and express an informed opinion. In the meantime, asking questions ("Do we know enough about X in order to proceed with this?") is where it all starts.

One final note about the scientific process. There are several places in the preceding text where the sources of information were cited, for example, "Sumy et al. (2014)" ("et al." is Latin for "and others," or the list of other author names). You will hear numerous times in college that you need to "reference your sources," or "cite your sources." While you might feel that these are required as a means of checking for violation of intellectual property (plagiarism), the real reason that citations are required is that they allow readers to follow up on the original source if they wish to pursue a subject further on their own. Think of citations or references as "a trail of breadcrumbs" that help curious readers explore a subject further.

Chapter 1 Review Questions

- Using the hydrofracking example, summarize the steps that geoscientists would take in order to answer a question such as, "does hydrofracking contaminate drinking water?"
- What is the difference between "correlation" and "causation"?
- Examine Figure 1-3A and B. Write a caption for each graph that describes what data the graph plots as well as the major conclusion that the author wants the reader to reach.
- Scientists use the term "theory" very differently than the lay public. Describe this difference.

Chapter 2

Introduction to Earth System Science

Nothing exists in isolation.

—Elizabeth Viggers, 2011

When one tugs at a single thing in nature, he finds it attached to the rest of the world.

—John Muir

At the conclusion of this chapter you should be able to

- Define what a system is and describe its characteristics
- Explain why the Earth is a system
- Describe what type of system the Earth is, what its reservoirs are, and many of its fluxes
- Identify the energy sources that run the Earth system

Definition and Characteristics of a "System"

A *system* can be thought of as a product that results from the interactions of multiple parts. A watch is an example of a system. So is a car. So is Planet Earth. All systems share several characteristics. Matter and/or energy recycles from one component, or sphere, to another. Energy is needed to run a system, and, following the laws of thermodynamics, energy can't be created or destroyed but can change from one form to another. A change in one component of a system induces a change in another so that a system is self-regulating. This means that you can't increase the amount of material in one reservoir indefinitely; at some point it will increase the outflow to another reservoir. A system is more than the sum of its parts.

The Earth is a complex system. In order to understand how it works, we need to understand what the parts of the Earth system are, how they function, how they are powered, and how a change in one component will impact the others and the overall functioning of the system.

Types of Systems

Based on differences in energy and material flow, we recognize three types of systems: open, closed, and isolated (Figure 2-1). In an open system, both energy and matter enter and leave. In an isolated system, no energy or matter enters or leaves. In a closed system, energy enters and leaves, but matter does not. Which type

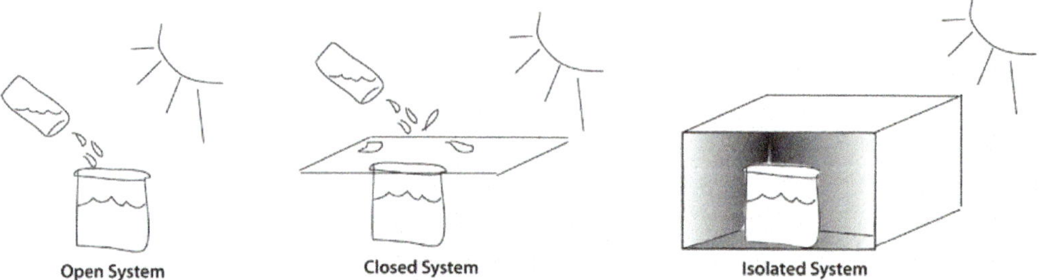

Figure 2-1 A beaker filled with water represents our system. Sunlight represents energy, and another beaker of water represents input of material. The diagram shows three systems: open (energy and matter enter and leave), closed (energy enters and leaves but not matter), and isolated (neither energy nor matter enter or leave). In the third system, the beaker of water is isolated from both energy and matter by being placed in a box.

of system is the Earth? In order to answer this question, we need to understand the ways in which matter and energy can enter and leave Earth. In order to address this, let's look at the components of the Earth system.

The Earth System

One way to graphically illustrate a system is shown in Figure 2-2 as "spheres." The primary topic of this text is the *geosphere*, the matter comprising the Earth, the processes that move matter around on the Earth, and the energy that drives this movement. Humans and all other living organisms are part of the *biosphere*, the second component of the Earth system. The gaseous envelope around the Earth comprises another sphere, the *atmosphere*. Finally, the *hydrosphere* describes where this fluid, upon which all life forms depend, resides and how this form of matter cycles.

We can think of each of the spheres in Figure 2-2 as representing the total amount of matter in that sphere, for example, all the water on Earth. Within the Earth, where does water reside? In systems terminology, *reservoirs* represent storage sites for material. The oceans, for example, are a reservoir in the hydrosphere. In the geosphere, fossil fuels are reservoirs for organic carbon. The processes that move matter from one reservoir to another are termed *fluxes*. Evaporation is a flux that moves water from the ocean reservoir to the atmosphere reservoir (clouds). Burial

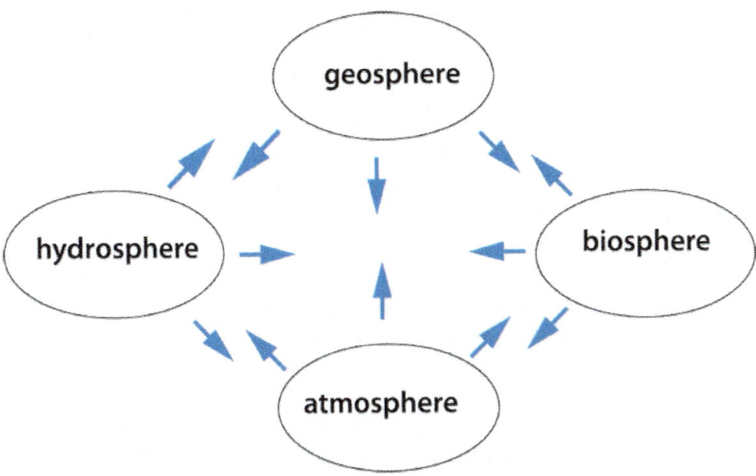

Figure 2-2 Reservoirs in the Earth system.

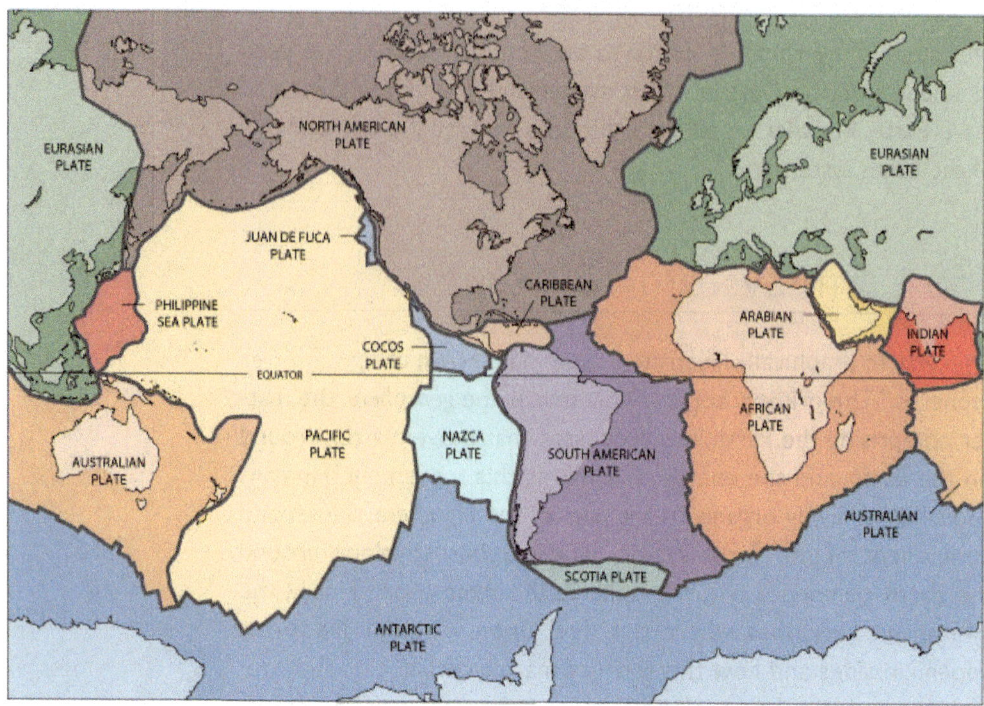

Figure 2-3 Today's major tectonic plates.

Image from http://pubs.usgs.gov/gip/dynamic/slabs.html.

of dead organic material is a flux that moves carbon from the biosphere to the geosphere (coal and petroleum fossil fuel). Fluxes are shown as arrows in Figure 2-2.

The self-regulation attribute of systems operates by feedback loops, which prevents an unlimited increase in the amount of matter in any reservoir. As we will see in Chapter 13, as the amount of carbon dioxide increases in the atmosphere, it triggers an increase in the flow, or flux, of CO_2 into the hydrosphere (the oceans). The level of dissolved CO_2 gas in the oceans will increase for a period of time. We are currently documenting this increase as we monitor "ocean acidification." This flux cannot continue indefinitely, but will be regulated by the ultimate equilibration of CO_2 flux between the atmosphere and hydrosphere. These responses are not instantaneous; there usually is a lag time before self-regulation "kicks in."

We can use the "flow-chart" approach to visually illustrate the reservoirs and fluxes within any cycle. The first step, of course, lies in understanding what all the reservoirs are in each of the four Earth system spheres. Then we need to identify all the fluxes that move

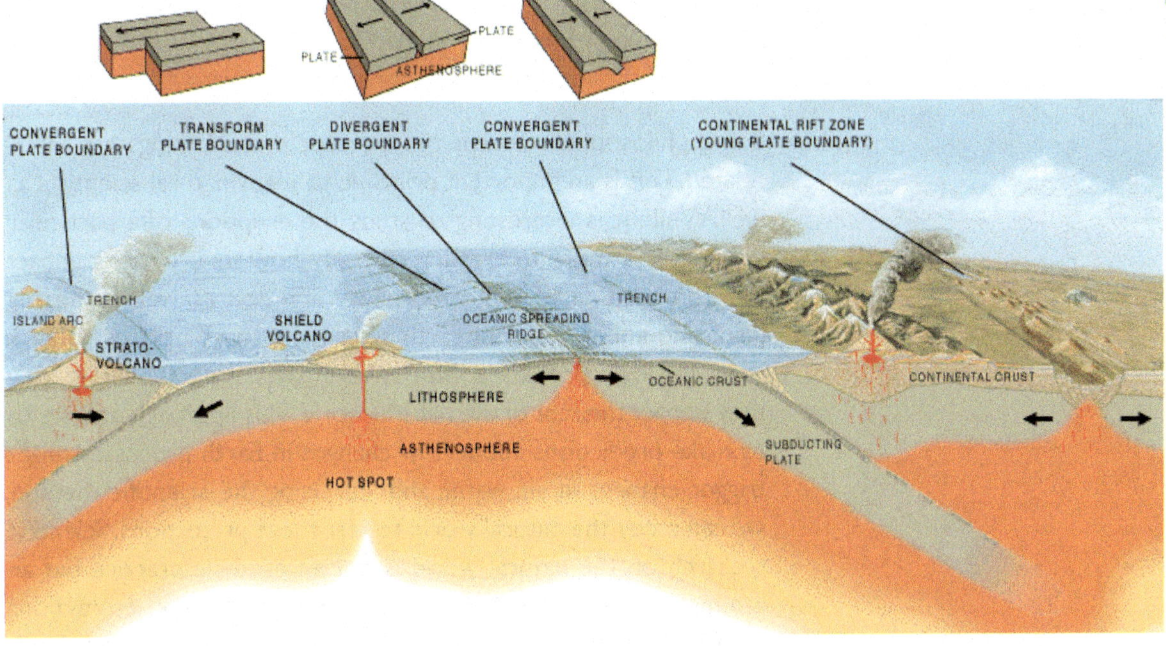

Figure 2-4 Where tectonic plates meet, one of three types of boundaries can form depending on the relative motions of the two plates. A wide range of geologic processes occurs at plate boundaries, including volcanism and earthquakes, and material in the geosphere cycles from one reservoir to another.

Image from http://en.wikipedia.org/wiki/Plate_tectonics#mediaviewer/File:Tectonic_plate_boundaries.png.

matter around from one reservoir to another. Finally, we need to identify the energy sources that fuel this movement. This text will focus on doing this for the geosphere. We will not be able to study all the other spheres in the same detail, but we will examine several ways in which the geosphere interacts with the biosphere, atmosphere, and hydrosphere by examining a critical set of interactions that control the Earth's climate.

As later chapters in the text will discuss more fully, the interrelated processes that run Planet Earth are the components of plate tectonics, the idea that the outermost Earth is divided into 100 to 150 kilometer–thick slabs of rock that move relative to one another (Figures 2-3 and 2-4). The energy sources that drive plate tectonics are the Earth's internal heat and solar radiation. The chemical and physical processes that result from plate movements impact all other spheres of Planet Earth.

So, based on this information, is the Earth currently an open, closed, or isolated system? What does this imply about the supply of resources?

Why Use a "Systems Approach" to Study the Earth?

Some philosophers accept the concept that all things are interrelated. This is an important principle to many natural scientists as well. While it is interesting to study the eruptions of a particular volcano, it is more interesting to study how its eruptions impact the composition of the atmosphere and soil as well as nearby plant and animal life. The Earth system approach asks us to look at all geologic processes with an eye toward documenting how they impact the rest of the natural world. This approach enables us to make predictions about how changes in Earth processes might trigger changes in air, water, and life. Using the scientific method, we can study the natural world to test these predictions. Scientific research on the natural world is not an esoteric practice but an attempt to understand the nature of the complex interactions within and between "spheres." As passengers on *Spaceship Earth*, it's important to understand how our "vessel" operates.

Summary

One way to describe how the various components of the Earth are related to one another is to view the Earth as a complex system. Systems are composed of components that interact with one another and the result of those interactions are synergistic, in other words, the whole is more than the sum of its parts. Energy is needed to drive interactions between components. Systems can be open, in which case both energy and matter enter and leave, isolated, in which neither energy nor matter enter or leave and closed, in which case only energy enters and leaves, but not matter. The Earth is currently a closed system (although very early in its formation during intense meteor bombardment it could have been considered an open system). In systems terminology the components are termed reservoirs, temporary storage sites for matter. The reservoirs in the Earth system include the geosphere (solid Earth), biosphere, atmosphere and hydrosphere. The flow of matter between reservoirs is termed a flux, and in the Earth system, fluxes are the physical and chemical processes that occur on Earth, such as evaporation or respiration. Many of the geologic processes that drive fluxes within the geosphere are parts of a unifying theory in geology termed plate tectonics. Plate tectonics says that the outermost later of the Earth is divided into a series of rigid slabs that move and interact with one another in response to heat flow from the Earth's interior. Processes in the geosphere that are driven by plate tectonics also influence other spheres, for example, volcanic eruptions represent the interaction of the geosphere and atmosphere. This text focuses on processes in the geosphere that influence the atmosphere and hydrosphere, and to some extent, the biosphere.

Chapter 2 Review Questions

- Using the language of "systems," define what is meant by a "reservoir" and a "flux"
- List the reservoirs in the Earth system and at least one flux between each.
- What are the energy sources that "run" the Earth system?
- The "paradigm" that describes how the geosphere works is termed "plate tectonics." What does this paradigm say?
- There is abundant evidence that Mars once had water on its surface. If Mars is a "closed system," what does this change imply about where the water went? If Mars is not a closed system, what does this imply about where the water went? If Mars is not a closed system, what type of system is it?

Part II

Ingredients and Structure

Chapter 3

The Ingredients of Planet Earth

We are stardust; We are golden.

—Joni Mitchell

At the end of this chapter you should be able to

- Summarize the process that creates matter and where this is occurring
- Compare and contrast covalent, ionic, and hydrogen bonding
- Define what atoms, ions, isotopes, and molecules are
- List the three most abundant elements in the whole Earth and the outermost layer, the crust
- Explain why Planet Earth is layered
- Extrapolate your explanation for the origin of the layered Earth to other planets in our solar system. Would you predict that they are layered as well? Why or why not?

Introduction

Because *Spaceship Earth* is constructed of chemical compounds in solid, liquid, and gaseous states of matter, it is important to review the nature of matter. The elements found on Earth form and interact as a function of the organization of electrons in orbit around individual atomic nuclei, so we will review basic atomic structure and how this influences the properties and behavior of the Earth's ingredients.

Origin of the Elements

The material that makes up Planet Earth is thought to have originated in a brief flash of time at the formation of the universe in the big bang. Chemical elements are thought to originate through the process of *nucleosynthesis*, which says that larger atomic particles are created from collision of preexisting protons and neutrons (see Sidebar 3-1 for a review of atomic structure). In the nucleosynthesis model, the big bang created hydrogen and helium, and a series of nuclear reactions impacted these atoms, progressively generating the larger and heavier elements. The energy for this process comes from the explosion of large dying stars, or supernovae. The explosion of these massive stars, or supernovae, generates atomic particles, including protons and neutrons, at extremely high temperatures. When the speeding neutrons slam into other nuclei, they create new elements and set off a chain reaction of ejected electrons (beta particles), which form yet more elements through a process termed *neutron capture*. Following the explosion of a giant star, the process of neutron capture and subsequent beta decay repeats over and over again, forming progressively heavier elements. Thus, it is not an exaggeration to say that you are composed of "stardust," as all the elements in our solar system were generated through repeated star formation and supernovae explosions. Some elements were produced in greater abundance because of their stability after nuclide collisions. Different nuclide stabilities control the nature of chemical and physical reactions throughout the universe as well as physical properties (density and volatility) of materials.

The physical and chemical processes that occurred at the birth of our solar system would have also been able to occur in other solar systems in the universe, so the same elemental synthesis that

Sidebar 3-1 Review of the Building Blocks of Earth's Materials

Atoms are the smallest individual particles that retain all the properties of a given element and its components. Although oversimplified, the basic representation of an atom's structure proposed by Niels Bohr will serve our purpose.

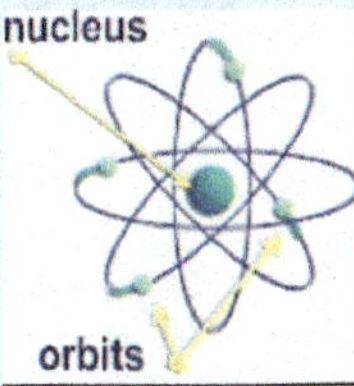

SB 3-1: EPA / Copyright in the Public Domain.

Protons and neutrons are found in an atom's nucleus. Electrons orbiting the nucleus are confined to specific shells that are arranged at specific distances from the nucleus. Each shell has a specific amount of energy associated with it, and there are a maximum number of electrons in each shell. The illustration above implies that the electrons are in fixed positions as they orbit the nucleus; however, this is a schematic representation—there exists a probability that electrons are in these regions. The nucleus, considered as a solid mass, is termed a *nuclide*.

An *element* is the most fundamental substance into which matter can be separated by chemical means. NaCl (halite, or salt) is not an element because the bonds between the Na and Cl can be broken, separating the two elements of Na (sodium) and Cl (chlorine). Na and Cl are elements because neither can be further subdivided. Each element has a unique atomic structure.

An atom that loses an electron loses a negative charge, leaving it with an excess positive charge; an atom that gains an electron gains a negative charge, giving it a new negative charge. In both cases, an ion is created.

Na^+ is a *cation*; Cl^- is an *anion*. They combine to form NaCl (the mineral halite, or salt).

Compounds and *molecules* form when elements combine, or bond:

$$Li^+ + F^- = LiF, \text{ the molecule lithium fluoride.}$$

A *molecule* is the smallest unit of bonded atoms that retains all its properties. A *compound* is a molecule made up of at least two different elements.

built Planet Earth is also able to occur elsewhere (Figure 3-1). The vast majority (approximately 99%) of the helium and hydrogen produced at the big bang at the formation of the universe are contained in stars that are slowly converting hydrogen to helium. Our sun has been doing this for billions of years.

Within the supernovae explosions that are the death throes of large stars, nuclear reactions produce neutrons that blast into space. Neutron capture describes the collision and absorption of neutrons into elements, increasing their atomic mass. Some of these neutrons decay to form protons and electrons, which are also absorbed into elements. As a result of this process occurring over and over, the larger and heavier elements were assembled in the first seconds

Figure 3-1 The HUDF (Hubble Ultra Deep Field) image in visible and near-infrared light shows the evolving universe within the southern-hemisphere constellation Fornax. This image shows more than ten thousand galaxies and many stars in the process of formation. Image credit: NASA, ESA, H. Teplitz and M. Rafelski (IPAC/Caltech), A. Koekemoer (STScI), R. Windhorst (Arizona State University), and Z. Levay (STScI).

after the big bang. So, it is safe to say that the elements present on Planet Earth represent all the matter available to recycle in the Earth system; in other words, the Earth is a *closed system*.

In the late nineteenth century, scientist Dmitri Mendeleev recognized that there is an underlying order or organization to all the elements known on Earth. Mendeleev's accomplishments in generating what is now called the periodic table of the elements (Figure 3-2) represents true scientific genius. First, he grouped elements by what we now know as increasing atomic weights; Mendeleev accomplished this before atomic structure was understood. Second, he left gaps in his table, predicting that at some point these elements would be discovered. His predictions have been borne out, with the current table listing elements by increasing atomic number and many of Mendeleev's "missing elements" inserted into their correct places.

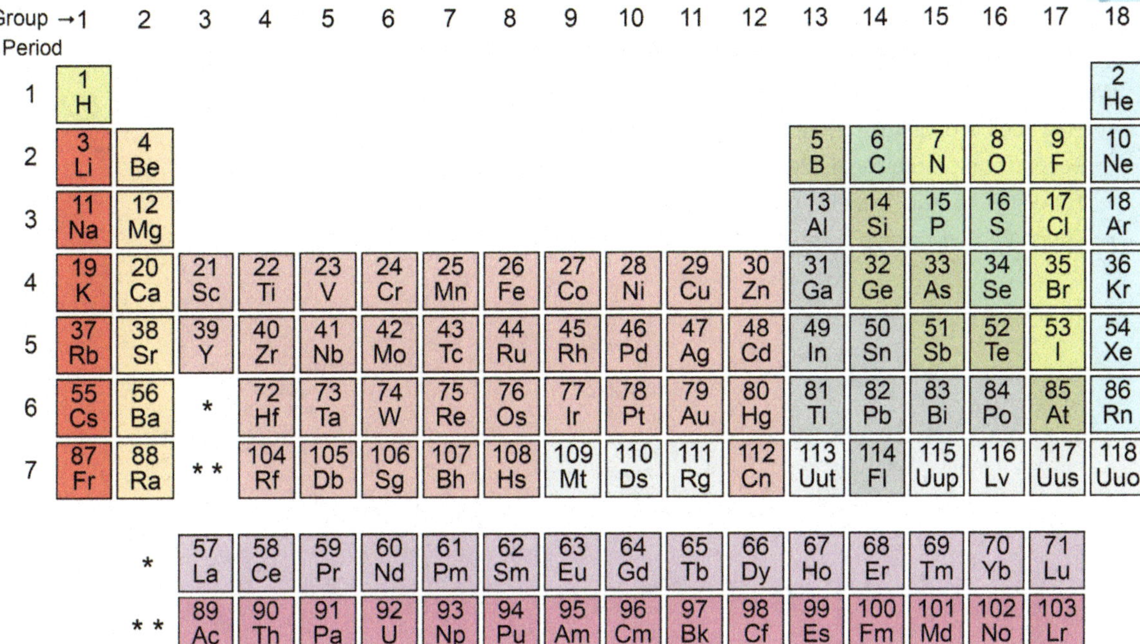

Figure 3-2 Columns run vertically and rows run horizontally. A new row (period) forms when an electron is added to a new shell. Elements in a column (group) have the same number of electrons in a subshell, and for this reason they have similar properties to other elements in that group. The first row, with only H and He, can have its electron shells filled with up to two electrons. The second row, from Lithium (Li) to Neon (Ne), corresponds to the eight electrons necessary to progressively fill both the innermost and second orbital shells. The third row, from sodium (Na) to Argon (Ar), represents the progressive addition of electrons to reach the ten that fill the third shell of orbitals. The column on the extreme right (Group VIII) represents the elements termed the *noble*, or inert, gases, which are characterized by a filled outermost shell; thus, they are highly stable compared, for example, to the alkalides.

One drawback to the periodic table's organization around an element's atomic number is that it can easily give us the impression that the nucleus containing protons is the most important part of the atom. In fact, while the number of protons in the nucleus of an atom does control how many electrons are needed in orbit to balance the positive charge of the nucleus, because the electron clouds around the nucleus take up many, many orders of magnitude more space than the nucleus, interactions between atoms involves interactions between electron clouds. Arguably, electrons and their positions are the most important atomic particles. The orbital areas of electrons (orbital shells) represent the regions where bonds are formed between atoms, forming molecules. So, it is important to understand how these orbital shells are organized (Sidebar 3-2).

Sidebar 3-2 Review of the Structure of Electron Shells

Within each shell there are subshells:

- First shell: s orbit holds 2 electrons (total = 2)
- Second shell: s orbit holds 2e; p orbit holds 6e (total = 8)
- Third shell: s orbit holds 2e; p holds 6e; d holds 10e (total = 18)
- Fourth shell: s = 2e, p = 6e, d = 10e, f = 14e (total = 32)

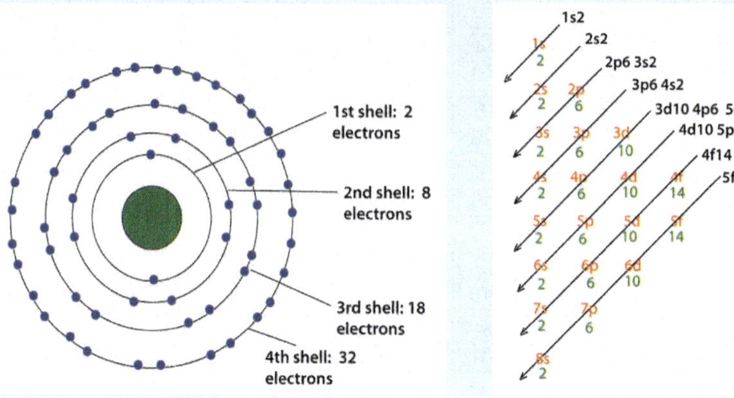

The periodic table is important to the study of Earth materials because the atomic structure on which it is based controls the behavior of the elements, specifically, which elements will bond together and how they will bond. Thinking in terms of the Earth as a system, the chemical elements are the ingredients of life on Earth. Nitrogen (N), phosphorus (P), oxygen (O), and a host of other elements are necessary to support the biosphere. Many of these elements exist in their "free" or elemental (unbonded) state; for example, we can mine pure gold (Au), but most other elements bond together to form molecules, for example NaCl (halite, or salt).

Assembling Molecules: Achieving Stability

Elements will be most chemically stable when the electrons present in electron orbitals are stable, in other words, when the outermost or valence shell is completely filled with electrons. This is the state of the inert gases, such as neon. Because most of the elements, however, are not the stable, inert gases, most will engage in chemical reactions to form molecules. For these elements, stability is achieved when atoms share and donate or receive electrons from other atoms, forming bonds (Figure 3-3). As a result of these interactions between electron orbits, atoms combine to form *molecules* (e.g., H_2O) and *compounds* (e.g., O_2). Atoms that achieve stability by losing electrons from their outermost orbital shell or by gaining extra electrons to fill their outermost shell are termed *ions*. A *cation* is an atom that has lost an electron to another atom and thus has a $+1$ charge; an *anion* is an atom that has gained an electron and has a charge of -1.

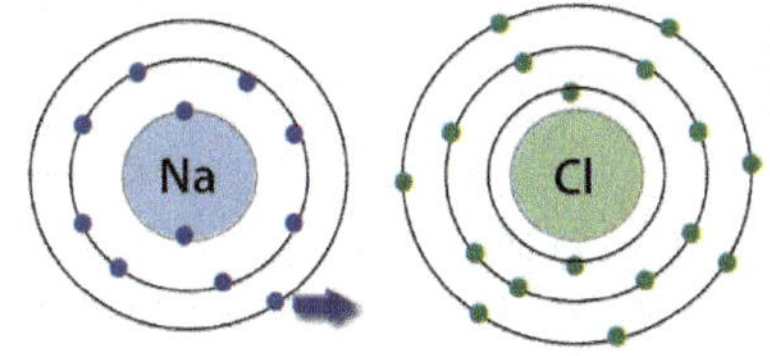

Figure 3-3 Achieving stability through bonding. The sodium atom has one electron in its valence shell, that could hold up to 8 electrons. The chlorine atom has 7 electrons in its valence shell, so it accepts an electron from sodium. This makes the ion Na^{-1} and the cation Cl^{+1}. They combine to form NaCl (the mineral halite).

Chemical Bonds

The instability of unfilled valence shells results in elements combining, through gain, loss, or sharing of electrons, to form compounds held together by bonds. There are three main types of bonds for us to study: ionic, covalent, and weak (hydrogen) bonds.

Elements that form *ionic bonds* are those that have gained or lost electrons in their valence shell, and they combine to produce a compound with no charge. Figure 3-4 shows how the ionic bond forms between lithium and fluorine to form lithium fluoride.

A *complex ion* is a compound with an ionic charge that combines with another ion to form a compound. A geologically important example of a complex ion is calcium carbonate, $CaCO_3$.

Covalent bonds occur when there is a *sharing* of electrons between elements. Covalent bonds form when elements are able to hold on to their electrons (the nucleus is able to exert force so that electrons don't leave their orbits for other atoms), so they deal with unfilled outermost electron orbits (or valence shells) by sharing. For example, in Figure 3-5, carbon has six electrons in three orbits: 1s2, 2s2, 2p2. This last shell can have six electrons in it to be most stable, but it only has two electrons. To get to its most stable configuration, it would like four more electrons. It gets these by sharing them with oxygen (O). O has eight electrons distributed in shells: 1s2, 2s2, 2p4. Its outermost shell would like to be filled with six electrons, so it shares two electrons with carbon to form covalently bonded CO (carbon monoxide).

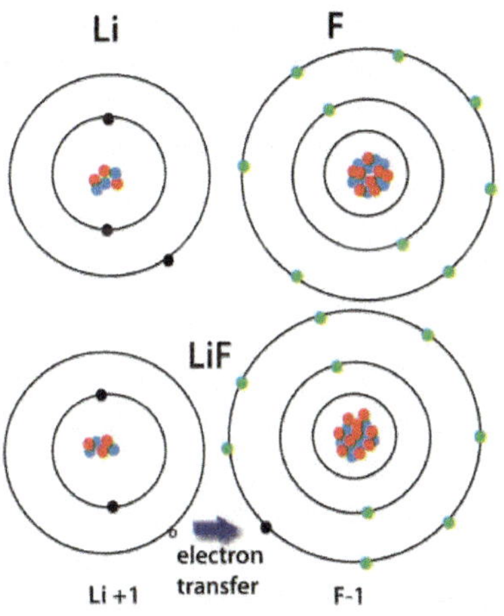

Figure 3-4 Ionic bonding. Lithium (Li) has 3 electrons in orbit around the nucleus with only 1 electron in the valence shell. Fluoride (F) has seven valence electrons, so it accepts one electron from lithium. $Li^{+1} + Fl^{-1} =$ LiF lithium fluoride.

What controls whether an ionic or covalent bond holds a compound together? It depends on an atom's (or ion's) *electronegativity*: a measure of how well a nucleus holds on to its own electrons compared to those in adjacent atoms. This is a function of the relative size of an element's nucleus and electron orbits. In the case

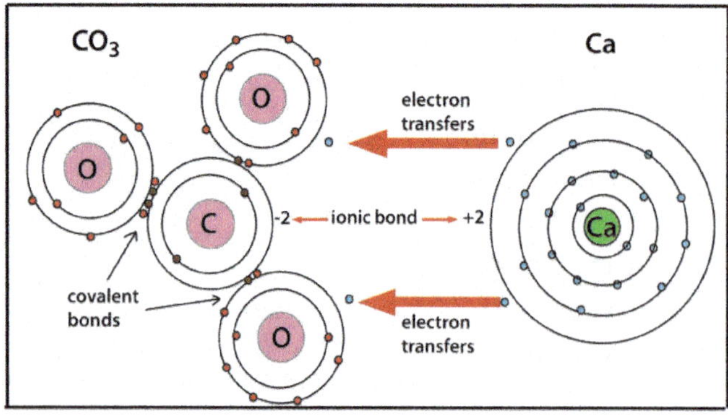

Figure 3-5 The bonding between carbon, oxygen and calcium to form $CaCO_3$. The compound CO_3 accepts two electrons from Ca, so: $CO_3^{-2} + Ca^{+2} = CaCO_3$

of chlorine (Cl), it takes less energy to take an electron from Na than it would to share an electron. Elements that behave like this include the alkali metals that bond with nonmetals, in other words, elements on opposite ends of one another on the periodic table. They have large differences in their electronegativity, so they usually form ions and ionic bonds. A molecule that is so important to the Earth system is carbon dioxide (CO_2), which forms from a covalent bond; carbon and oxygen are near one another on the periodic table, so there are small differences in their electronegativity. It takes less energy to share electrons than to give or take one from the other element. In nature, there really aren't "pure" covalent and "pure" ionic bonds; rather, they exist in a continuum of characteristics of these end-member bond types. The important thing to remember is that the different properties of the bonds impart different physical properties to the molecules that form from them.

A *hydrogen bond* is the attractive force between a hydrogen atom in one molecule to an ion in a different molecule. The water molecule is our best example of a compound held together by hydrogen bonds. In the case of the water molecule, H_2O, the hydrogen atom attached to oxygen in one molecule of water is attracted to the oxygen in another molecule of water. This occurs because the location of where the two hydrogen atoms are bonded on the oxygen atoms leaves this molecule with asymmetric positive charges on one side (where the two hydrogen atoms are) and a more "negative side," opposite where the hydrogen atoms are (Figure 3-6). This asymmetry results in the attraction of the positive side of the H_2O molecule to the more negative side of another H_2O molecule. Another important aspect of the water molecule is the physical effect that the very different sizes of the atoms have on the shape of the molecule. The oxygen atom is much bigger than the hydrogen atom, so the most efficient "fit" of the atoms is for the H_2O molecule to be shaped like a tripod: O is on one side, and the two H atoms are opposite "legs" (Figure 3-6). This means that there are more electrons on one side of the molecule than the other, so we say that water is a dipolar molecule. This term means that even though the electrostatic charges of the atoms are fulfilled, there is a physical imbalance of charges across the molecule. The effect of this is that water is attracted to other ions even though it has a neutral charge; this is why water is such a good solvent.

Compared to ionic and covalent bond strengths, hydrogen bonds are much weaker. They are, however, extremely important bonds on Planet Earth, as the interactions between hydrogen atoms in the water molecule explain this substance's physical properties, including how it can exist in all three states of matter (solid, liquid, and gas) on the surface of our planet.

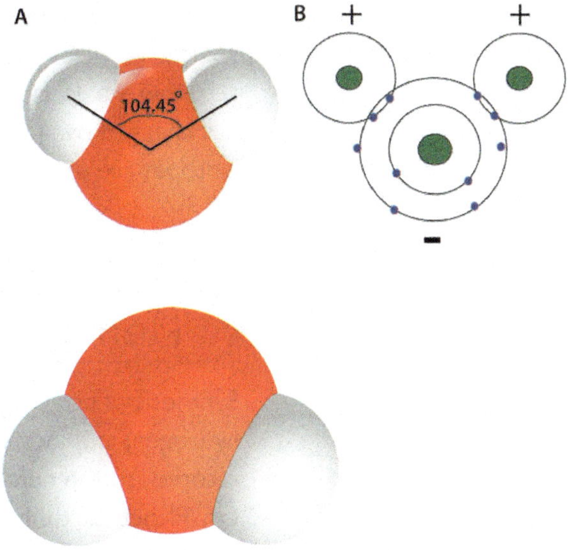

Figure 3-6 The structure of water. A. illustrates the"tripod" shape of the covalently bonded H_2O molecule B. illustrates the hydrogen bonds that link the positive (hydrogen) side of the molecule to the negative (oxygen) side of the molecule.

Fig. 3-6B: Sakurambo / Wikimedia Commons / Copyright in the Public Domain.

Isotopes

An *isotope* is an element with variable atomic mass (number of protons + number of neutrons) but the same atomic number (number of protons). For example, the element uranium (U) has 92 protons in its nucleus. Most U has 146 neutrons, for an atomic weight of 238 ($^{238}U_{238}$). Another isotope of uranium also has 92 protons and 143 neutrons, for an atomic weight of 235 ($^{235}U_{92}$). The element oxygen exists in three isotopes: $^{16}O_8$, $^{17}O_8$, and $^{18}O_8$. The atomic number is the same for each of these; however, they vary by the number of neutrons in the nucleus. Of these isotopes of oxygen, ^{16}O comprises more than 99% of all oxygen on Earth; less than 1% is ^{18}O, and only a minute amount is ^{17}O. These isotopes of oxygen are stable isotopes, meaning that once formed, they maintain their atomic numbers and weights over time. ^{16}O was formed as part of the nucleosynthesis process described at the beginning of this chapter. The remaining isotopes of oxygen form from the alteration of other elements; $^{14}N_6$ captures a $^{4}He_2$ nucleus in stars to produce $^{18}O_8$, for example. The isotopes of uranium are unstable isotopes, which means that over time they transform, with the release of energy, into other elements. ^{235}U and ^{238}U are, like $^{16}O_8$, other "primordial isotopes"; in other words, they were formed in the immediate stages of the formation of matter through nucleosynthesis after the big bang.

There are many isotopes of uranium that form from the capture of nuclear particles by the primordial element. In Chapter 13 we will see how stable isotopes are used for documenting how climate has changed over time, and in Chapter 6 we will see that unstable isotopes are used as "geologic clocks" for telling time.

The Relative Abundance of Elements on Earth

Synthesis of both organic molecules (involving carbon, oxygen, and hydrogen) and inorganic molecules is occurring every day in laboratories, so the number of possible molecules is virtually limitless. On Planet Earth, however, we see that about a dozen elements are most abundant in the solid earth (hydrogen, carbon, oxygen, magnesium, silicon, sulfur, calcium, nitrogen, potassium, and iron; Figure 3-7), and the molecules they form construct approximately 98% of our solid, liquid, and gaseous world.

Only twelve elements occur near the Earth's surface at levels greater than or equal to 0.1%! Oxygen and silica make up more than 70% of the outermost Earth. In the next section of the text, we will see that this is not true of the composition of the entire Earth. Why do we think that the composition of the outermost layer, termed the *crust*, is not characteristic of the entire Earth? Why would Earth be differentiated, or layered? These are important questions to ask, as they help us understand how the Earth formed and how it may, or may not be, unique within our solar system.

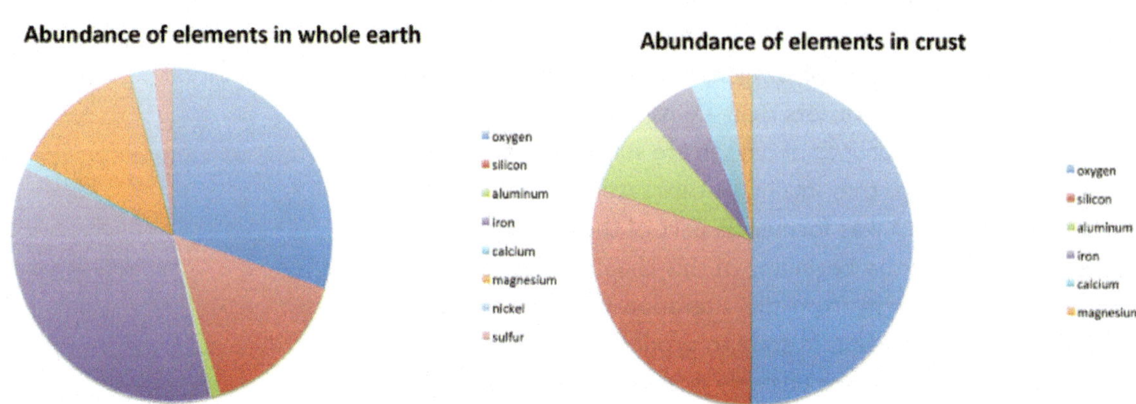

Figure 3-7 The overall abundance of elements in the Earth by mass. Oxygen and silica make up more than 70% of the rocks at the Earth's surface (the crust). Data are shown in Table 3-1.

Table 3-1 Comparison of the Abundance, by Mass, of elements in the Crust and Whole Earth, excluding the atmosphere. N/A = Amounts well below 1%.

Source: NASA / Copyright in the Public Domain.

Element	Abundance in Crust	Abundance in Whole Earth
Oxygen	47.00%	30.0
Silicon	28.00%	15.0
Aluminum	8.10%	1.1
Iron	5.00%	35.0
Calcium	3.60%	1.1
Sodium	2.80%	N/A
Potassium	2.60%	N/A
Magnesium	2.10%	13.0
Others	0.80%	<1.0
Nickel	N/A	2.4
Sulfur	N/A	1.9

Composition and the Physical Properties of Matter

Volatility is the property that describes whether an element or molecule is in the solid, liquid, or gaseous state at a particular temperature and pressure. Volatility is an important concept to discuss because this property, along with the nucleosynthesis processes that produced the elements after the big bang, determined the chemical composition of all the planets. Highly volatile substances will change states from solid to liquid (*melting point*) or liquid to gaseous (*boiling point*) at relatively low temperatures. Materials that we call "gases" are characterized as highly volatile elements, such as hydrogen and helium. On the other hand, substantially less volatile substances, also called refractory elements, require extremely high temperatures to change states of matter and include elements such as calcium and magnesium. Most molecules on Earth lie somewhere between these extremes.

The Earth is one of the "inner" planets (Mercury, Venus, Earth, and Mars), all of which are characterized as relatively high density and "rocky" compared to the "outer" planets (Jupiter, Saturn, Uranus, and Neptune), which are lower density and gaseous. These differences reflect differences in the abundance of volatile and refractory elements comprising them. One reason that Earth can be considered a "Goldilocks" planet is that it has, as rocky

Table 3.2 Densities of some Important Earth Materials

Substance	Chemical formula	Density (g/cm^3)
water	H_2O	1
gypsum	$CaSO_42(H_2O)$	2.32
calcite	$CaCO_3$	2.71
olivine (forsterite)	Mg_2SiO_4	3.27
magnetite	Fe_2O_3	5.17
iron	Fe	6.98
gold	Au	17.1
uranium	U	18.9

planets go, a greater than expected abundance of volatile elements, such as carbon dioxide. The reason for this enrichment is still unknown but might be related to material brought to our planet from meteorite and comet impacts early in Earth's history.

Density is a physical property that describes the amount of solid matter in a given volume of space (g/cm^3). We know that most atoms are empty space of the electron orbitals and most of the weight is in the nucleus. It makes sense, therefore, that the higher the atomic weight of atoms in a molecule, the more dense the molecule will be.

Density is hugely important in understanding the distribution of material within the Earth, and the density change as a function of temperature change is important in explaining the vertical movement of liquids, solids, and gases on Earth.

Temperature is a measure of the motion of atoms. In materials at a high temperature the atoms are moving rapidly. It is difficult to maintain the bonds that keep material in the solid state, so high-temperature matter often melts. When you burn your finger on the stove, you have (painfully) broken bonds between organic carbon molecules. At very low temperatures the slow movement of atoms makes it easier for elements to feel the effects of electrostatic attraction, so bonds form more readily.

Differentiation of Planet Earth

After accretion from nebular dust formed the planets, the coalescing material separated or differentiated internally as a function of its density. More dense materials sank toward the Earth's interior, and less dense materials rose toward the surface. This includes the lowest-density materials of all, the gases, which accumulated in a thin layer above the Earth's surface, forming our atmosphere. The Earth's gravitational field kept this thin, gaseous envelope in place. In Chapter 8 we will review the evidence for the internal layered structure of the Earth, and in Chapter 12 we will examine the structure and composition of the atmosphere.

Are the other planets internally layered? Because the origin of the layered Earth is due to the distribution of the chemical elements vertically as a function of their density, this same process would have occurred following the accretion of the other planets. What would differ, however, is the relative abundance of volatile and refractive elements and their densities. In other words, the layers themselves would be expected to differ. Based on satellite data, Jupiter, for example, is thought to have a rocky outer layer and an inner metallic hydrogen layer (liquid hydrogen that exists at extremely high pressures).

Summary

Because the Earth is a closed system, all of the material on Earth recycles from one reservoir to another, but where did this material originally come from? The process of nucleosynthesis describes the generation of all material except hydrogen and helium through a process termed neutron capture or nucleosynthesis. This process describes how larger atoms are formed from collisions of subatomic particles. The energy and matter needed to do this comes from exploding dying stars or supernova. The elements created through nucleosynthesis can be organized into the periodic table of the elements, which groups Earth materials into a scheme based on the number and distribution of subatomic particles, notably protons and electrons. The distribution of electrons is especially important because of its influence over the tendency of atoms of an element to combine by bonding into molecules. The types of bonds which form are important in controlling the chemical behavior of the molecule. Density is a physical property that reflects the mass of atomic particles in a given amount of space and the different densities of elements, and the molecules they form, controls their distribution within the Earth. The planets accreted from dust and gases created at the big bang, nearly 15 billion years ago. As this material clumped together gravitational forces generated enough heat to produce a melted Earth. The distribution of material of different densities in this melted Earth produced an internally layered planet with the lowest density materials forming the atmosphere above. The solid Earth is composed primarily of twelve elements, of which iron, oxygen, silicon, and magnesium are the most abundant. In the crust, oxygen is most abundant, followed by silicon aluminum, and iron. If we consider the atmosphere in the "whole Earth composition," various gases, such as nitrogen, could also be included in the discussion. In the next chapter we examine where these elements are stored on our planet.

Chapter 3 Review Questions

- Summarize the process of nucleosynthesis and where it occurs.
- Compare and contrast covalent, ionic and hydrogen bonding
- Define the terms: atoms, ions, isotopes and molecules
- List the three most abundant elements in the whole Earth. Do the same for the outermost layer, the crust.
- Explain why the interior of the Earth is layered. Does this imply that other planets would be internally layered as well? Why or why not?
- What is the valence shell of an atom and why is it important?
- Which physical property is most responsible for producing a layered earth?
- If we hypothesize that Mars is internally layered, what event can we infer happened sometime in the history of the formation of Mars?

Chapter 4

The Earth's Most Important Molecules

At the end of this chapter you should be able to

- Define what a mineral is and how it differs from a rock or from a crystal
- List and define the physical properties that we use to identify a mineral
- Explain what determines the physical properties of a mineral and discuss at least one mineral example
- Identify and describe the fundamental characteristic of the most abundant group of minerals on Earth
- Explain why water is a unique molecule on Earth and the origin of its unusual physical properties

- Describe the properties of water that are important in heat transfer on Earth
- Describe why water is such a good catalyst for chemical reactions
- Discuss whether or not you think that the minerals on Planet Earth are the same as they are on the Moon or Mars or Venus
- Solve problems such as the following:
  If the density of water is 1,000 kilograms per cubic meter, how may kilograms of water will there be in a glass container having a volume of 5 m^3?

Introduction

As we've said earlier in the text (Chapter 2), the Earth is a closed system, which means that no new material is entering, although energy enters and leaves. The first part of this chapter examines the reservoirs of solid Earth materials that occur in molecules that we call minerals. *Minerals* are defined as naturally occurring inorganic solids of specific chemical composition and characteristic crystalline structure. When we discussed cycles in Chapter 2, we described the "storage areas" for matter as *reservoirs* and the processes that move material from one reservoir to another as *fluxes*. We can think of various minerals as "mini-reservoirs" that temporarily "store" elements such as silicon and sodium. Minerals combine together to comprise rocks (Chapter 5). To use a common analogy, "if rocks are analogous to *books*, minerals are analogous to *words*." In other words, minerals comprise rocks, just as words comprise books.

The concept of minerals as reservoirs is not an abstract one. Many of the elements that are used in modern electronics are contained in minute quantities within minerals. Called the *rare Earth elements*, or *REEs*, their global distribution and the environmental impacts related to their mining and processing will be political and economic issues for the remainder of this century. *Ores* are economically valued minerals, such as those that contain significant amounts of iron, nickel, copper, and gold. Like REEs, the mining and processing of these minerals also have significant environmental impacts. Some of the nations with the richest ore mineral reserves are also politically less stable. What will be the environmental, economic, social, and political costs of their exploitation?

Because different minerals form under different physical and chemical conditions, the study of minerals sheds light on processes occurring within the Earth. There are more than 3,500 known minerals; however, not all of them are equally important recorders of Earth processes. We already defined ore minerals. Some other minerals have artistic or aesthetic value and are termed *gems*. While more than three thousand minerals sounds like a lot, when you think of the previous chapter and the almost limitless ways in which chemical elements can combine, rather than make the comment, "So many minerals!" we should really question, "Why so few minerals?" If you look at Figure 3-7, you might hypothesize that the answer to this question is related to the abundance of only a few elements on Earth: Si, O, Fe, Mg, Al, S, Ca, and K. Of all the minerals that these seven elements can make, we will focus on a small subset that is critical to explaining how the Earth works.

There are two different approaches that we can take to the study of minerals. We can study their composition, exploring the different elements present and their abundance. We can also examine the crystalline structure of minerals. How are the atoms arranged and linked? Both approaches are important because composition and crystalline structure determine a mineral's physical properties. The physical properties of a mineral allow us to identify them (see Sidebar 4-1), but they also influence how a mineral will react with other solids, liquids, and gases. Because minerals combine to form rocks, the minerals' physical properties also explain why rocks have the properties they do. If the physical properties of the rocks are different on and within the Earth, it is because the composition and crystalline arrangement of minerals is different. The physical properties of minerals are determined by their composition and atomic structure, which is determined by the types of bonds and arrangement of atoms. Minerals really are the building blocks of our planet!

The second part of this chapter discusses the most important molecule on Planet Earth: H_2O, or water. Water is a necessary ingredient for life as we understand it, and as we explore other planets in our solar system, the search for water is of paramount importance. While the water reservoir in the Earth system warrants its own chapter (Chapter 11, on the hydrosphere), we will

Sidebar 4-1 Crystal Habit Examples

Acicular (needle-like). Example shown: natrolite

Bladed. Example shown: actinolite

Cubic: Example shown: halite

Tetrahedral. Example shown: magnetite

introduce the water molecule now so that we can examine how it interacts with minerals and rocks.

Definition of a Mineral

A *mineral* is defined as a naturally occurring, inorganic crystalline solid with a fixed chemical composition. Let's look at the components of this definition:

- Naturally occurring: laboratory synthesized (synthetic) crystalline solids are not minerals (ex, cubic zirconia).
- Inorganic: minerals are not composed of the carbon-hydrogen bonds that define "organic" molecules (for example, sugar: $C_{12}H_{22}O_{11}$).
- Crystalline solid: minerals occur in the solid state, which means that they can retain their shape indefinitely (a liquid, for example, conforms to the shape of its container), and the atoms that comprise them occur in fixed, ordered positions within a lattice. A mineral may have more than one crystalline shape. For example, the minerals graphite and diamond have the same chemical composition but very different crystalline structures. We say that these two minerals are polymorphs of one another.
- Fixed chemical composition: this refers to our ability to write a chemical formula for a mineral, showing the elements that comprise the mineral.

Based on this definition, ice is a mineral: it is naturally occurring; it is inorganic; it is solid at temperatures below 0°C; the atoms occur in fixed, ordered positions; and the chemical composition is H_2O.

Mineral versus Crystal

What is the difference between a mineral and a crystal? Are all crystals minerals? Are all minerals crystals? We just reviewed the definition of a mineral. A *crystal* is defined as a solid in which the atoms or molecules form a series of continuous, flat, three-dimensional surfaces, which are termed *crystal faces*. You are familiar with snowflake (ice) crystals and the wide variety of shapes that they can assume. Note that the definition of a crystal does not

Figure 4-1 Examples of crystals (clockwise from top left): quartz; halite; and gypsum crystals, the largest crystals ever found, in Naica Cave in Mexico. These forms represent the symmetry of their crystal faces, which reflects their atomic structure.

include the modifiers "naturally occurring" or "inorganic"; thus, you can examine sugar crystals or synthetic crystals (fake gems!). All minerals, however, have a crystalline form. Most of the time the crystals are too small for you to see. Occasionally crystals grow in conditions where they can become quite large. In these cases it is possible to observe and measure the orientations of the crystal faces, which represent the macroscopic expression of "fixed ordered positions within a lattice."

Crystal Structure of Minerals

All minerals are crystalline, which means that the elements that comprise them are organized into regular geometric patterns. A mineral's *crystal habit* is the characteristic form it takes when it grows. Recognizing a mineral's crystal habit can be helpful in identifying the mineral. A few of the more distinct crystal habits are shown in Sidebar 4-1.

The crystal structure of each mineral reflects the arrangement of the atoms and the bonds between them. Figure 4-2 shows the atomic structure of the mineral halite (NaCl), an ionically bonded cubic mineral.

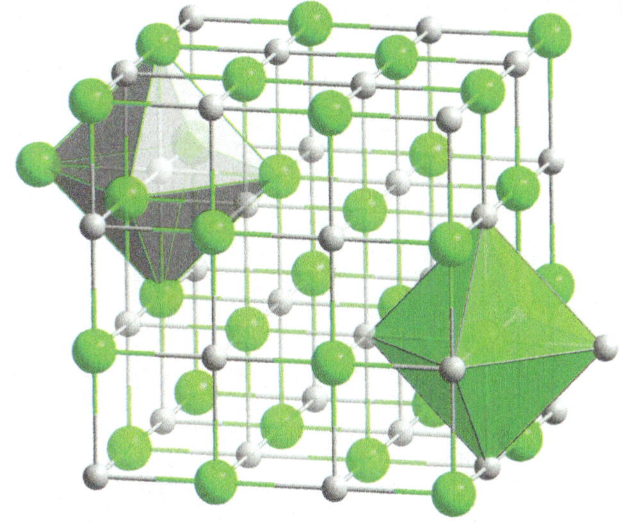

Figure 4-2 The structure of halite. A "ball and stick" model of this mineral is shown here ("balls" = atoms and "sticks" = bonds) green ball = Na; grey ball = Cl.The cubic crystal habit of halite owes its origin to the equal spacing of Na and Cl ions.

Determining the Crystal Structure of a Mineral

Chemical analyses of minerals tell us which elements are present but not the arrangement of atoms. How do we "see" atoms and where they are? "Seeing" atomic structure is done by bombarding a mineral with a high energy stream of electrons and recording where the electrons are reflected, or bounced back, and where they pass through (Figure 4-3).

Because not all crystal arrangements are equally "easy" to construct in nature, it's important to know the crystalline structure of a mineral. $CaCO_3$ is the chemical composition of two different minerals: calcite and aragonite. These minerals differ in their crystalline structure. The shape of the calcite crystal is rhombohedral, while the shape of the aragonite crystal is a longer, skinnier orthorhombic shape. As a result, calcite and aragonite form in different chemical conditions.

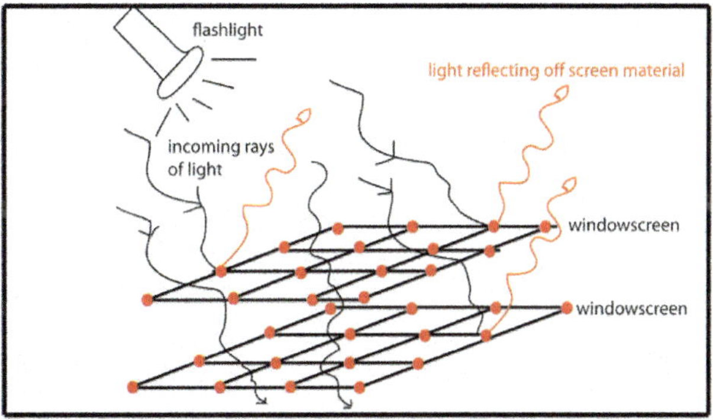

Figure 4-3 Cartoon of how X-ray diffraction is used to determine crystalline structure. Black lines with arrows represent the stream of electrons fired at a sample, whose atoms are shown as red spheres. The windowscreen represents bond spacing between atoms or a crystal lattice plane. Some electrons can pass between atoms while others encounter the atom and are reflected (orange arrows). A collector on the other side of the sample counts the number of electrons in an area and use the spacings between received electrons to determine the sample's crystalline lattice structure.

How and Where Minerals Form

Crystals of minerals can form in a variety of ways, but the two most common

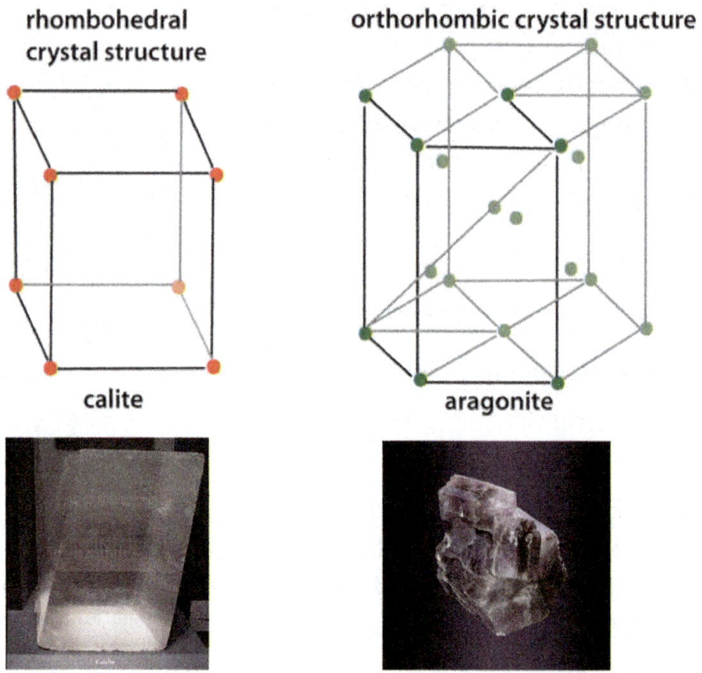

Figure 4-4 The minerals calcite (left) and aragonite (right), have the same $CaCO_3$ composition but different arrangements of the atoms in their respective crystalline structures. Ball and stick models of crystal structures above photographs of crystals.

Fig. 4-4B: Sergei Kazantsev / Copyright in the Public Domain.
Fig. 4-4C: Copyright © 2010 by Didier Descouens / (CC by 3.0) at http://en.wikipedia.org/wiki/File:Aragonite_2_Enguidanos.jpg.

processes are when a liquid freezes and precipitation from a solution (other processes include precipitation within living tissue, or biomineralization). *Solidification*, or the crystal formation that accompanies the cooling of a liquid, occurs in nature every time a snowflake forms. It also occurs when melted rock (magma) cools. *Precipitation from a solution* describes the initial condition where atoms or molecules are dissolved in a solution, such as water, and they begin to bond with one another, forming solids and separating themselves chemically from the solution when the electrostatic attractions between atoms is greater than the vibrational energy of the atoms. Under what conditions does this happen? When a solution is supersaturated, you have so many atoms in a solution that they bump into one another and electrostatic charges create bonds regardless of how fast the atoms are vibrating. Crystals will also form through cooling because heat provides thermal energy to keep atoms in motion. Removing heat (cooling) means that

Sidebar 4-2 Physical Properties of Minerals

The physical properties of minerals are determined by their composition and atomic structure, which is determined by the types of bonds and arrangement of atoms.

Color: the color of any substance is the result of the interaction of visible light with a material. A substance, in this case, a mineral, has a color that represents the wavelength of light that it has reflected back (not absorbed). Many minerals come in a variety of colors, so this property is not definitive in identifying minerals.

Streak: the color of the powdered mineral when you rub it on an unglazed ceramic plate. While the color of a mineral may not be diagnostic, the color of its powder is quite useful in mineral identification; however, if a mineral is harder than the ceramic plate, it will not produce a streak.

Luster: this describes the way in which a substance scatters light. The two important types of luster are "metallic" and "nonmetallic," to distinguish whether a mineral looks like metal. Other terms describing luster include "resinous" and "glassy."

Hardness: a measure of the relative ability of a mineral to resist scratching. Moh's scale of hardness has established the relative hardness of ten minerals to which an unknown sample can be compared (Table 4-1). The hardness of a mineral reflects the strength of its bonds—their resistance to being physically broken. Quartz (with its strong covalent bonds) is harder than calcite (with its weaker ionic bonds).

Fracture and cleavage: these terms describe the way in which a mineral breaks. Cleavage describes the tendency to break along preferred planes, which represent planes of weakness in a bond direction. A *cleavage plane* represents the flat (planar) surface that is the macroscopic representation of the weaker bond direction. Calcite will break along preferred planes of weakness because the bonds that Ca^+ and CO_3^- are weaker (ionic) than the bonds that hold the C and O together (covalent). Fracture describes the absence of cleavage and represents minerals where the bonds are all equally strong in all directions.

Specific gravity: this describes the density of a mineral. If you "heft" two mineral samples of approximately the same size, the more dense mineral will feel "heavier."

Crystal habit: this describes the shape of a single crystal with well-formed crystal faces. Sidebar 4-1 illustrates some examples of these terms.

Special properties: some minerals have distinctive properties. Magnetite, for example, is magnetic; halite tastes salty; calcite effervesces when a drop of weak acid is placed on it.

you remove the energy, that prompting atoms to move around. As the atoms move more slowly, electrostatic charges between atoms trigger bonds to form. The process of forming a crystal through either cooling or precipitation is initiated when a "seed" is present that can serve as a nucleation site for further growth. In the atmosphere, raindrops or snowflakes usually nucleate around a tiny dust particle in the air. This solid surface provides just enough of a "template" for atoms to bond, even temporarily, before they being the process of forming bonds with one another. Mineral formation from cooling will be an important process when discussing the origin of igneous rocks (Chapter 5), which form from the cooling of melted rocks.

If you can make a mineral by precipitation of solidification, you can destroy minerals by melting or dissolving them. What melting represents is the addition of heat to a substance so that the atoms

Table 4-1 Moh's Scale of Relative Hardness		
Moh's Relative Hardness	**Mineral**	**Scratch Test**
1	Talc	Scratched by fingernail
2	Gypsum	Scratched by fingernail
3	Calcite	Scratched by penny
4	Fluorite	Easily scratched by knife
5	Apatite	Scratched by knife
6	Orthoclase	Barely scratched by knife
7	Quartz	Scratches glass
8	Topaz	Scratches quartz
9	Corundum	Scratches topaz
10	Diamond	Scratches corundum

or molecules begin vibrating. If enough heat is added to a material, the vibration breaks the bonds of the crystal and the atoms and molecules move freely again. Dissolution of a crystal represents a similar process, but in this case the addition of the dipolar water molecule breaks bonds that are holding the crystal together. The dissolution of minerals will be an important process when we discuss how rocks break down, a subject covered in more detail in Chapters 5 and 11.

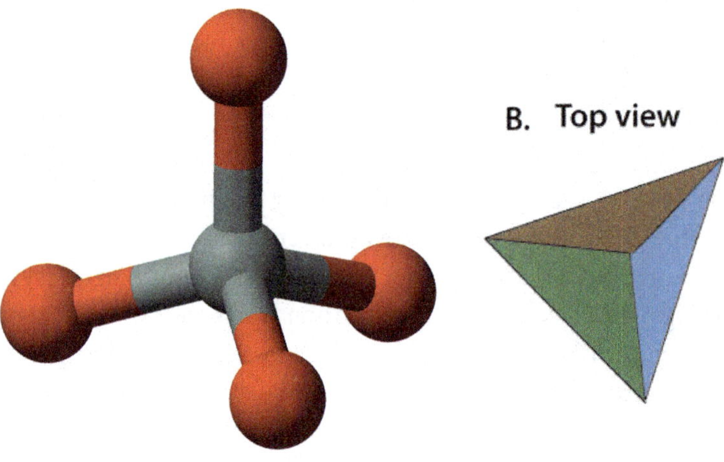

Figure 4-5 The structure of a silicate tetrahedron: A smaller silicon atom, with only four electrons in its valence shell forms a covalent bond with one electron, each from four oxygen atoms. Because of the size difference between the silicon and oxygen atoms, the most physically stable configuration is a tetrahedron with the tiny silicon atom in the center. While the silicon atom is filled, each oxygen has one "unsatisfied" electron in its valence shell, so the molecule has a −4 charge overall.

Mineral Composition

As mentioned earlier, only a few elements are very abundant on Earth, so it is not surprising that when we examine the composition of minerals, we group them according to which of these elements—specifically, the cations—are present. Thus, geologists have classified minerals into broad groups, such as the silicates, carbonates, sulfides, and so on (see Sidebar 4-3).

Of all the classes of minerals, those comprising the silicates are the most abundant, making up more than 95% of the rocks in the rocks of the outermost layer of Earth. Thus, we will discuss them in more detail than we will for the other groups of minerals (the second most abundant minerals are the oxides—for example, Fe_2O_3, iron oxide, or the mineral hematite).

The Silicates

As their name implies, the silicates are composed of the cation silicon (Si). The silicon is bonded to oxygen, forming a molecule that we call the silicate tetrahedron: SiO_4 (Figure 4-5). This molecule is an anion, as it has a −4 charge (SiO_4^{-4}).

Because the silicate tetrahedron is an anion, this molecule will bond with other elements to form a variety of different minerals (see Sidebar 4-3), so it really is a "building block" molecule. If a silicate tetrahedron bonds with a cation, such as Fe or Mg, we call this an *isolated* or *independent tetrahedron*. The mineral fayalite in the olivine group (Fe_2SiO_4) is an example of this group of minerals. If the silicate tetrahedron covalently bonds two of the oxygen atoms with two oxygen atoms in another tetrahedron, we call the resulting mineral a "single-chain tetrahedron." The mineral pyroxene ([Mg,Fe] SiO_3)is an example of this type of mineral. A double-chain tetrahedron can form if two or three atoms are shared between tetrahedrons, for example, the amphibole group of minerals. The sheet silicates are two layers of linked tetrahedra sharing three oxygen atoms. The layers, usually containing cations between them, produce a well-developed cleavage in the minerals of this group, as the mineral breaks along the weaker bonds between sheets. The mica group of minerals are an example of this structure. The framework silicates form from bonding of all four oxygen atoms to other tetrahedrons. The minerals quartz (SiO_2) and potassium feldspar ($KAlSi_3O_8$) are examples of this group.

Sidebar 4-3 Silicate Structures

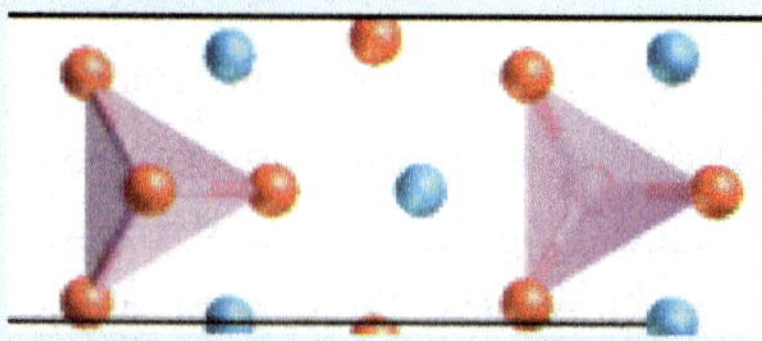

A. Isolated, or independent tetrahedron: ex, olivine $[Mg,Fe]_2SiO_4$

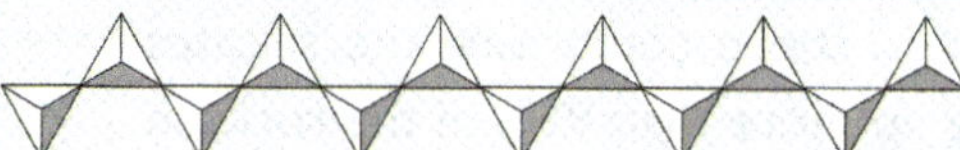

B. Single chains of tetrahedral linked by shared oxygen. Ex: pyroxene $[Mg,Fe]SiO_3$

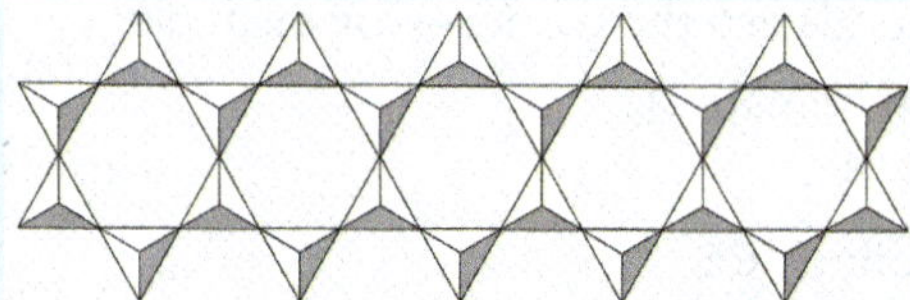

C. Double Chains: two oxygen shared from each tetrahedra. Ex: hornblende $Ca_2(Mg,Fe,Al)_5(Al,Si)_8O_{22}(OH)_2$

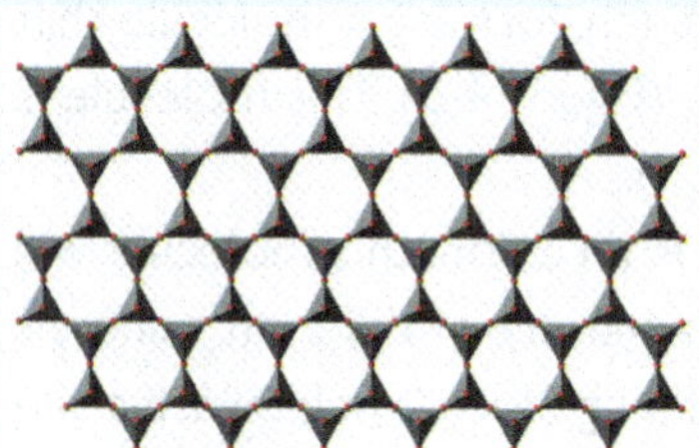

D. Sheets: 3 oxygen at corners all shared, producing flat sheets. Ex: Muscovite $KAl_2(AlSi_3O_{10})(F,O\underline{H})_2$

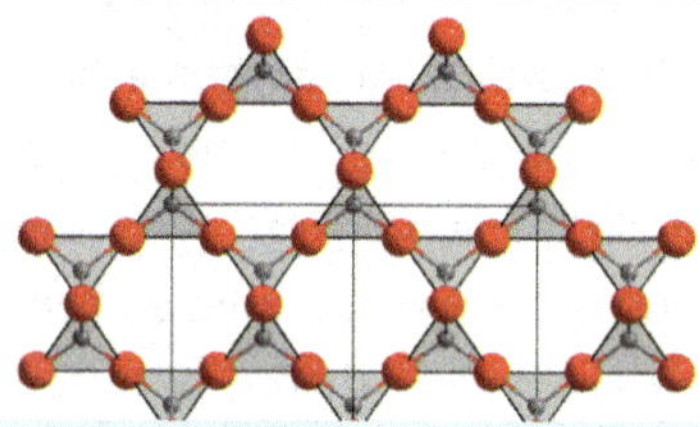

E. Framework: All 4 oxygen shared in a 3D framework. Ex. Quartz SiO_2

One of the most abundant minerals in most rocks is quartz (silicon dioxide, or SiO_2). You may note that this mineral doesn't appear to be constructed of the SiO_4^{-4} anion. Why would SiO_2 form instead of SiO_4? The silica tetrahedron forms when there are other cations present (such as Mg, Fe, Al, K, etc.) because the tetrahedral structure takes less energy to form and bond with adjacent tetrahedrons than is required for the SiO_2 structure, which forms only when there are no competing ions present.

In addition to the effect of competing ions, temperature is a factor. The stable pyramid, or tetrahedron, is, from a kinetic point of view, the easiest physical construction to build when you have two atoms of such different sizes. The SiO_2 molecule, on the other hand, represents one small silicon atom that has to evenly space itself between two larger oxygen atoms, sharing two electrons from each. Thus, while SiO_2 seems to be the most simple molecule to construct, from an "ingredient" point of view, it is very difficult to form the bonds that have to be very carefully "balanced" or spaced apart. The SiO_4 bonds form readily, while the atoms are still energetically vibrating, because the physical tetrahedron structure is an easy one to "build." The SiO_2 bond forms when the atoms are vibrating much more slowly because of the difficult balancing act this molecule requires. A substance's temperature is really a measurement of the vibration level of atoms, so the covalent bonds to build SiO_4 happen at higher temperatures than do the bonds for SiO_2.

The Oxide Minerals

As discussed previously, the oxygen atom needs two electrons for its valence shell to be full, so oxygen is a highly active element, bonding with cations at every opportunity. Oxygen forms the anion O^{-2}, and the oxides represent the second largest group of minerals, forming from the bonding of the metals (everything to the left of Group 3 on the periodic table) to oxygen. A common oxide is hematite: Fe_2O_3.

The Carbonate Minerals

The CO_3^{-2} molecule is the building block for this group of minerals. The carbon atom shares one electron with each of two oxygen atoms and two electrons with a third oxygen. Most commonly, the cations Ca or Mg bond to it, forming minerals such as calcite ($CaCO_3$) and dolomite ($[CaMg]_2CO_3$).

The Sulfide and Sulfate Minerals

The sulfide anion has a –2 charge (S^{-2}), and metals commonly bond to this, for example, FeS_2, the mineral pyrite, or "fool's gold." The sulfates consist of the SO_4^{-2} anion, and cations commonly bond to this, forming mineral such as gypsum (the constituent of the building material sheetrock): ($CaSO_4 \cdot 2H_2O$).

The Halide Minerals

This group of minerals is characterized by elements from Column 17 on the periodic table (Figure 3-2), called the *halogens*. These are elements such as Cl^- and F^-. These anions will bond with cations from columns on the left side of the periodic table, such as Na^+ and Ca^{+2}, to form minerals such as halite (NaCl).

Native Metals

Occurrences of a single metal, such as copper (Cu), silver (Ag), or gold (Au), are grouped in this class of minerals.

How Do We Determine the Chemical Composition of Minerals?

The crystal structure of a mineral is determined through by passing X-rays through a crystal and measuring where electrons are reflected (Figure 4-3). The chemical composition of a mineral can be determined through a variety of elemental analysis techniques involving the identification of elements based on the nature of their interaction with high-energy electrons, an analytical technique called *spectrometry*. Various spectrophotometers involve different detectors, but essentially all can detect the presence of elements at extremely small quantities (parts per billion). Spectrophotometers can do this by exposing a sample to a gas composed of ionized (charged) elements. Atoms in a sample interact with this ionized gas, and, because of their different charge to atomic mass ratio, they separate themselves within the instrument. Based on the directions they scatter, a detector is then able to identify the different atoms present and their abundance.

Water

As of 2014, Planet Earth is the only planet within our solar system that has been identified as having abundant water in all three states of matter. While Mars appears to have ice, much of the water on Earth is in the liquid state and available for incorporation in the biosphere. All life on Earth depends on water; in autotrophs, such as plants, water is used in photosynthesis and nutrient transport in leaves and stems. In heterotrophs, water is the main component of cells and is a catalyst in the chemical reactions of metabolism.

Besides the biosphere, water plays a significant role in all the other spheres in the Earth system, to the extent that we recognize the *hydrologic cycle*, or hydrosphere, as one of the major "spheres" within the Earth system. This text will not be able to discuss all reservoirs within the hydrosphere in detail, but Chapter 11 will examine the ways in which it intersects with the geosphere and atmosphere.

Given the importance of water to Planet Earth, you would think that there would be a lot of it here, but our entire planet contains $< 0.1\%$ H_2O! If we examine the crust of the solid Earth, the atmosphere, and the oceans, this value rises to approximately 7% by weight, still a small amount. The differences in these values indicate that water was concentrated at or near the Earth's surface as the interior of our planet evolved. Very recent discoveries in geology suggest that much more water is contained within the crystal lattices of minerals than previously thought. We normally don't think of minerals as reservoirs of water, but this is the case. As the internal processes within the Earth bring these minerals to the surface (Chapters 9 and 10), this water is released and becomes part of our available water cycle.

The Uniqueness of the Water Molecule

Dipolar Structure

The two atoms of hydrogen and one oxygen in water are bonded by covalent bonds. Each hydrogen atom, with one electron in its valence shell, shares it with the oxygen atom. This covalent bond fills the outer shells of both atoms and results in the formation of the water molecule with no net electrostatic charge (Figure 4-6).

The remaining electrons in the oxygen atom are repelled by the additional electrons from hydrogen, and they arrange their

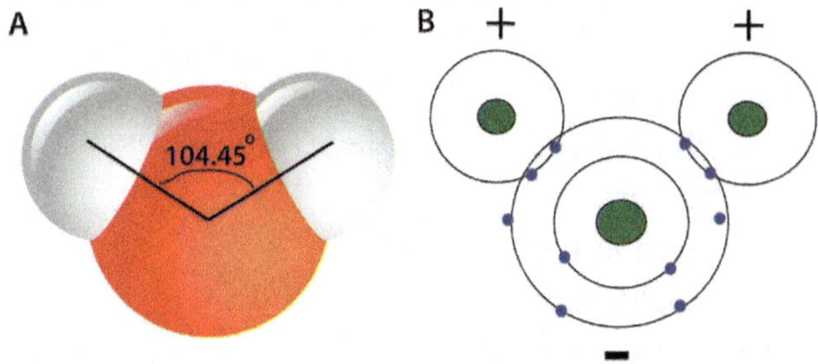

Figure 4-6 The structure of water molecule. A shows larger oxygen atom bonded to two smaller hydrogen atoms. An angle of 104.45° separates the two hydrogen atoms. B shows that the polarity of the water molecule arises from more electrons on one side of the molecule and more protons on the opposite side.

Image A from http://commons.wikimedia.org/wiki/File:Water_molecule.svg

orbits to move as far away from them as possible. This repulsion is countered by the positive attraction of protons in the oxygen's nucleus, so the compromise position of electron pairs places them *almost* at a perfect tetrahedral angle of 104.5° from one another (see Figure 4-6). The "almost" is critical, because the angle between electron pairs achieved by the compromise is 104.5°, a slightly distorted tetrahedron. The resulting molecule has a "more oxygen-dominated" side with unbounded electrons and a corresponding "more hydrogen-dominated" side dominated by the protons of the hydrogen nucleus. Even though the water molecule has no electrostatic charge, the physical alignment of positive and negative charges makes water a *dipole* (in physics, the separation of positive and negative charges is defined as a dipole).

The dipolar nature of the water molecule means that it readily forms hydrogen bonds with other water molecules. The "positive side" of the molecule dominated by the proximity of the hydrogen nucleus's protons is attracted to the electron-dominated "negative side" of the unshared oxygen electrons. These hydrogen bonds are much weaker (up to 50%) than the covalent and ionic bonds we've examined so far, but they are critical to understanding the physical and chemical properties of water and water's behavior on the Earth's surface.

Relationship to Physical/Chemical Behavior to Dipolar Structure of Water

The fact that water exists in all three states of matter on Earth but nowhere else in our solar system should suggest to you that its unusual chemical structure imparts unusual properties. Table 4-2 lists some of the most significant properties of water, most of which will explain why water is so important in the Earth system.

Table 4-2 The Physical and Chemical Properties of the Water Molecule.
From: http://en.wikipedia.org/wiki/properties_of_water

Property	Comparison with Other Molecules	Importance in the Earth system
density (ρ)	maximum at 4°C, not at its freezing point. Expands upon freezing.	This behavior prevents water bodies from freezing from the bottom upwards. Also causes stratification of water
melting and boiling points	abnormally high due to strength of H bonds	Water can exist at the Earth's surface in all 3 states
heat capacity	Due to strong H bonds, it is the highest of any liquid except ammonia (NH_3)	The storage of heat helps regulate and transfer energy
heat of vaporization	extremely high due to strong H bonds	Important in heat transfer between atmosphere and oceans and regulates temperature
absorption of radiation	significant for some wavelengths and frequencies	H_2O is a greenhouse gas in vapor state
solvent	excellent (universal solvent)	Important in transfer of material from solid to dissolved states (ex, weathering)

Density

Recall that *density* describes the mass of material in a given volume of space. Examine the following illustration (Figure 4-7) of the density of water as a function of temperature.

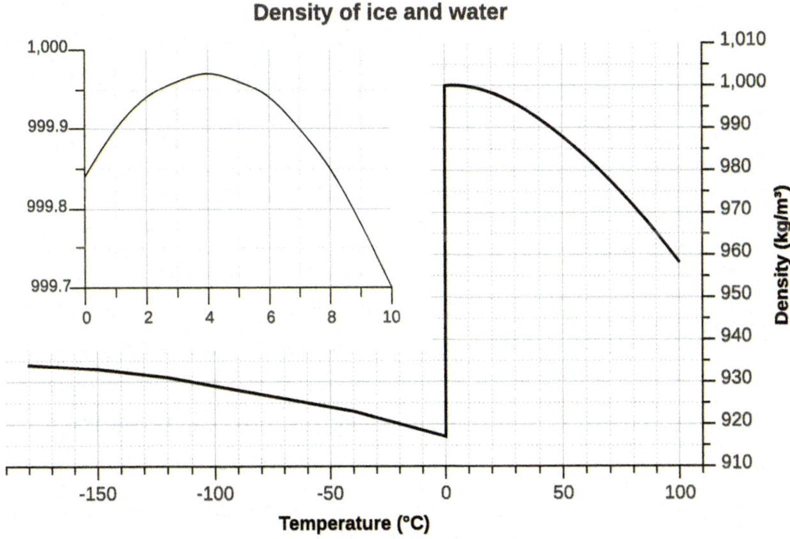

Figure 4-7 The density of water as a function of temperature at 1 atm of pressure. Insert in graph shows detail of density-temperature relationship over 1 to 10 degrees Celsius. Most substances are denser in the solid state than the liquid, however water behaves differently.

From this graph you should note the following:

- Water at 0°C, or ice, has its lowest density, significantly lower than water in its liquid state.
- Water reaches its maximum density at 4°C.
- The density of water decreases above 4°C.

It may not be readily apparent to you that these density changes are unusual, but most other substances become more dense as they cool. Why? Temperature is a measure of the activity, or motion, of atoms. At high temperatures, atoms vibrate wildly, moving apart from one another from collisions and electron repulsions. As temperatures drop, atoms vibrate more slowly and can get closer to one another. Thus, the drop in density of water in its solid state is atypical. The hydrogen bonding between water molecules explains this behavior. As molecular vibrations slow with cooling temperatures, the dipolar molecules can arrange themselves in hydrogen-bonded tetrahedra (Figure 4-8). Hydrogen bonds are longer than the covalent bonds holding the molecule together, so as these bonds form, so does wider spacing between molecules.

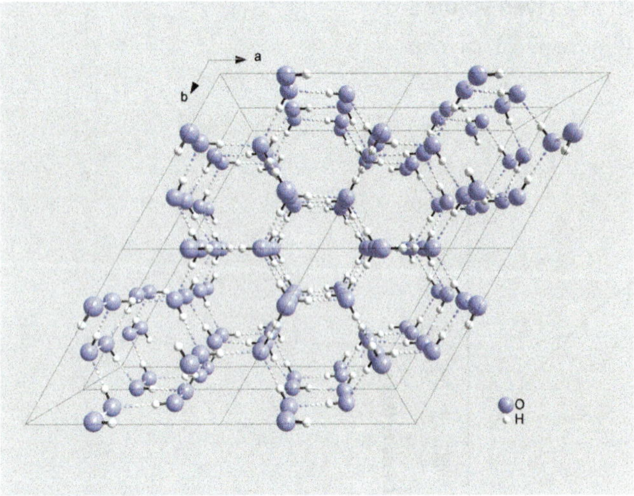

Figure 4-8 The hexagonal structure of ice.

This also explains why pipes burst when water freezes: the water molecules are forced apart as the hydrogen bonds form. The expansion of water at its freezing point is a huge factor in the geosphere as well. The freezing and expansion of water is a process in physical weathering: the physical breakdown of rocks when water seeps into cracks in rocks at the Earth's surface and freezes and expands.

Because ice is less dense than water, ice floats. In addition to explaining why ice cubes float in your soda, it also explains why when lakes, ponds, and rivers freeze in the winter, they freeze from the top down. If solid water were more dense than liquid, ice would sink and freeze a body of water from the bottom up. Once ice forms on a lake or pond, it serves as insulation from the cold air temperatures. As any winter angler knows, this allows fish and other organisms to survive through the winter in such environments.

Between 0° and 4°C, hydrogen bonds begin to break, and the water molecules can begin to pack more closely together, so the density of water increases. As you increase temperature above 4°C, the vibrational activity of the molecules increases as a function of temperature (which is normal behavior for substances), which results in decreasing density with heating. Water reaches its boiling point at 100°C, becoming a gas (water vapor).

Other Properties of Water

Melting and Boiling Points

The *melting point* of a substance is the temperature when it changes state from solid to liquid at atmospheric pressure. The *boiling point* of a substance is the temperature where the vapor pressure of the gas escaping from the liquid is equal to the atmospheric pressure. Water is unusual in both attributes. Examine Figure 4-9, which plots the temperature of the boiling and melting points of water and several other hydrogen-bearing molecules as a function of their molecular weight (the weight of one molecule). From experiments, chemists know that the attraction between molecules, termed the Van der Waals force, holds them together, and the magnitude of this attraction between molecules is a function of their molecular weight. From this relationship we would expect water to have a much lower boiling point: −80°C instead of 100°C! The same dramatic difference can be seen in the values for the predicted and observed melting point of water: nearly −100°C instead of 0°C. These anomalous values are the result of the extreme dipolar nature of the water molecule, creating enhanced Van der Waals forces. These forces must be overcome with additional heat to

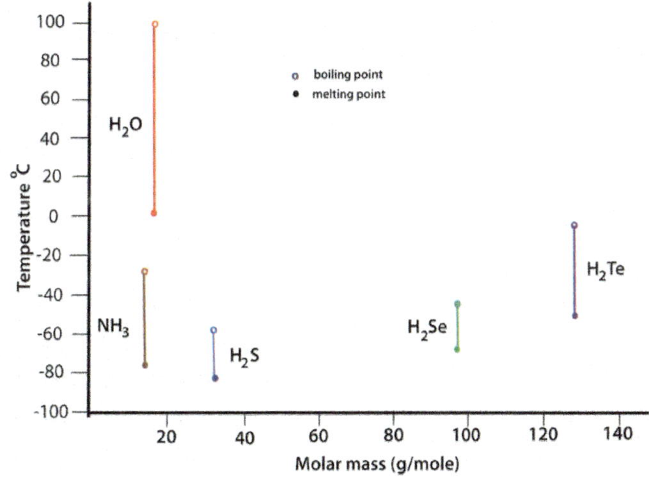

Figure 4-9 Comparison of the melting and boiling points of hydrogen-bonded molecules: water (H_2O), ammonia (NH_3), hydrogen sulfide (H_2S), hydrogen selenide (H_2Se) and hydrogen telluride (H_2Te). Another common molecule bonded to hydrogen is methane (CH_4) but its boiling and melting points fall well off the graph (−161.5°C and -182°C, respectively). Note that water is anomalous for the wide temperature range of boiling and melting points as well as its high values relative to its molecular mass.

vaporize. These extra attractions make it possible to form crystal lattices at higher temps (=vibrating atoms) of freezing.

Heat Capacity

A substance's *heat capacity* is the amount of energy required to change the temperature of the substance by 1°C. Only ammonia (NH_3) has a higher heat capacity than water. Why is this? The hydrogen bonds that hold water together absorb heat energy, energy that could be increasing the vibrational motion of atoms. This energy is stored in water's bonds, and is released once the bonds are broken.

Latent Heat of Vaporization

The energy required to transform a liquid to a gas at a given pressure is termed *latent heat*. Changing the state of matter from liquid to gas requires the input of energy to break the hydrogen bonds holding the molecule together. The value for water is 540 cal/cm^3 at 100°C. This energy is released again when water vapor condenses back into a liquid. Latent heat is one of the most important energy sources in the hydrosphere (see Chapter 11). Latent heat stores solar energy and moves it around the Earth's surface. Imagine water vapor evaporated (liquid to vapor) at the equator. The water vapor rises into the atmosphere and spreads toward the poles, ultimately cooling and condensing (vapor to liquid). The stored latent heat of vaporization is transferred from the equator and moved poleward. The changing states of water provide a "heat pump" for Earth.

Solvent

Water is such a good solvent because of its dipolar structure (Figure 4-6), with one side of the molecule having a positive change while the other has more of a negative charge. If you add an ionic substance to the other—for example, salt (NaCl)—these ions (Na^+ and Cl^-) will be more attracted to the dipolar water molecule than to each other. The bonds between NaCl will break, which we recognize as the salt having "dissolved." Water also dissolves other dipolar molecules, such as sugar, that readily form hydrogen bonds with the H_2O. We describe water as a *catalyst*, a substance that promotes chemical reactions, and it does so because of its dipolar structure and ease of forming hydrogen bonds with other atoms and molecules. We will see examples of water as a catalyst for breaking down minerals and rocks in Chapter 11.

Summary

A mineral is a naturally occurring inorganic solid with a fixed chemical composition and specific crystalline structure. Crystals are three dimensional solids characterized by flat, planar faces. All minerals are crystalline, but not all crystals are minerals. Minerals form in nature from precipitation from solution, solidifying out of a cooling liquid, or through biomineralization (synthesis within the tissues of organic material). We determine the crystalline structure of a mineral by examining the three dimensional arrangement of atoms using lab instrumentation. We also determine the chemical composition of minerals through lab analyses. It is far more useful for us, however, to observe that the physical properties of minerals reflect their chemical composition and physical structure. The properties of minerals include their hardness, luster, color, cleavage or fracture, streak, and specific gravity. Some minerals have additional special properties such as magnetism.

In the thousands of types of minerals that exist, one group is particularly important. The silicate minerals are composed of a basic "building block" of covalently bonded silicon and oxygen atoms. These basic structures are then joined together by ionic bonds into a variety of combinations, or silicate structures. Other important groups of minerals include the carbonates, which involve the covalent bonding of carbon and oxygen to one another and the ionic bonding of this molecule to calcium and other cations.

Water is one of the most important molecules on Earth and it is unusual in that it exists in all three states of matter at surface temperatures and pressures. Water can exist in this fashion because of its dipolar structure and the resulting hydrogen bonds that form. The covalently bonded oxygen and hydrogen atoms produce a molecule where the distribution of electrons is not uniform across the structure, instead one side of the molecule appears more positively charged because of the location of the hydrogen atoms. Weak hydrogen bonds cause the dipolar molecule to bond to many other molecules, which is why water is such a powerful solvent. Hydrogen bonds also explain why the density of water does not change uniformly with temperature, and why solid water (ice) is less dense that liquid water. Hydrogen bonds can also absorb heat energy, so water has a very high heat capacity. The attraction of hydrogen bonds between water molecules make them more cohesive when heated, thus water has a higher boiling point than its molecular weight would suggest. The additional energy (termed latent heat) that has to be added to water to change its state from liquid to gas is stored in the molecular bonds. This heat is released into the environment when water condenses back into the liquid state. Latent heat plays a critical role in moving heat around the Earth's surface.

Chapter 4 Review Questions

- There are three occurrences or processes in nature where the crystallization of minerals from a liquid can occur. What are these?
- What is the definition of a mineral?
- Are all minerals also crystals? Are all crystals also minerals?
- What are the physical properties we use to identify a mineral?
- What determines the physical properties of a mineral?
- Complete the following sentence: Using "systems terminology," minerals are ...
- What is the fundamental characteristic of the most abundant group of minerals on Earth?
- How do we determine the chemical composition and the crystalline structure of a mineral?
- Describe the composition and structure of the silicate tetrahedron
- What is the defining characteristic of a dipolar molecule?
- Water is arguably the most important molecule on Earth. What are water's unique properties?
- What physical property of water makes it important in heat transfer on the Earth's surface?
- Would you predict that the minerals comprising rocks on Mars would be the same or different from those on Earth? Why?
- Calculate the density of a mineral (in g/cm3) given a mass of 0.05kg in one cubic meter.

Chapter 5

Introduction to Rocks and the Rock Cycle

At the end of this chapter you should be able to

- Draw and label the rock cycle
- Identify a hand sample as either an igneous, metamorphic, or sedimentary rock
- Distinguish between rocks and minerals in a selection of samples
- Describe the textures of a hand sample of igneous, metamorphic, and sedimentary rocks
- Calculate the geotherm over a distance within the Earth
- Describe the conditions that cause rock to melt

- Explain the role that water plays in the melting point of rocks
- Discuss how the diversity of processes in melting and cooling explain the origin of the many different intrusive igneous rock compositions
- Explain how Bowen's reactions series could produce a felsic rock
- Summarize what controls the viscosity of a magma
- Describe the different materials erupted from a volcano
- Identify different types of volcanoes in a photograph
- Explain the significance of finding volcanoes on other planets
- Describe the types of environments where sedimentary rocks form
- Describe the differences between sediment and soil
- Summarize the conditions that produce metamorphic rocks

Introduction

As we've described earlier, minerals are the reservoirs of Earth materials, such as iron, silicon, and oxygen. In the Earth system, these materials cycle from one reservoir (mineral) to another. Because rocks are composed of minerals, this cycling of matter takes the form of cycling from one type of rock to another, a phenomenon we call the *rock cycle* (Figure 5-1). The three types of rocks are igneous, formed from the cooling of magma; sedimentary,

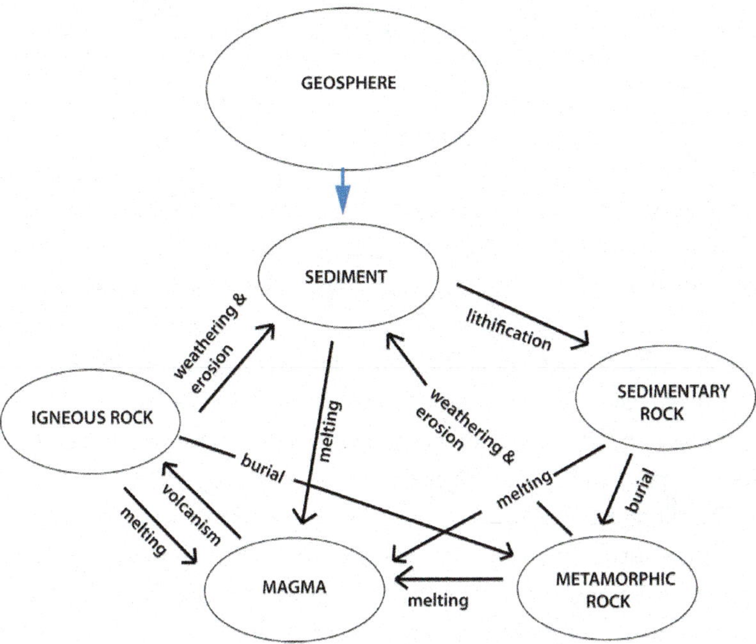

Figure 5-1 The reservoirs of the rock cycle are shown, along with fluxes (geologic processes) that move material from one reservoir to another. The rock cycle represents the "inner workings" of the geosphere portion of the Earth system (Figure 2-2).

formed from the cementation of sand, clay, and other particles; and metamorphic, rocks altered from heat and pressure. There are temporary reservoirs between these end-member rock types; for example, magma and lava represent rock material that has melted but not yet crystallized to form an igneous rock. Sediment is the reservoir of detritus (e.g., sand and clay grains) on the Earth's surface before it is buried and *lithified*, or cemented, into rock. A range of geologic processes cycle material from one reservoir in the rock cycle to another. Chapter 9 will examine the mechanisms that are part of plate tectonics driving this cycling.

Because it follows naturally from the material discussed in Chapter 5, we will start our examination of rocks by looking at igneous rocks that all form from various minerals that have crystallized from cooling *magma*, or molten (melted) rock. We will then examine how these igneous rocks are altered by Earth processes to form other types of rocks and, in so doing, Earth materials are recycled. Recall that because rocks are composed of minerals, we are really talking about changing the mineral reservoirs of Si, O, Fe, Ca, Al, and other elements on Earth.

Igneous Rocks

Figure 5-2 Photograph of a coarsely crystalline igneous rock (granite). The interlocking crystals that are characteristic of igneous rocks are clearly visible. The pink mineral is potassium feldspar. The white mineral is quartz, and the black mineral is amphibole. Penny shown for scale.

Igneous rocks are formed by minerals that crystallize from the cooling of *magma*, or melted rock. Igneous rocks can be distinguished from other rock types by the interlocking crystals that compose them (Figure 5-2). They also might feel more dense than other types of rocks because many igneous rocks are rich in iron and magnesium. These tend to be dark in color as well; however, color is not the best criteria for identifying rocks—the fabric, or arrangement of crystals or grains, is much more reliable.

Recall that temperature is a measurement of the vibrational energy of a substance. When the temperature drops, vibrating atoms slow down and electrostatic charges between atoms begin to form bonds, constructing crystal lattices of minerals. Which minerals form depends on the composition of the starting liquid. Thus, the composition of an igneous rock depends greatly on the composition of the magma. The other important control of composition is the magma's cooling history.

Igneous rocks are identified and classified based on two attributes: the minerals that are present, and the texture of the rock (the size, shape, and arrangement of constituent minerals). See Sidebar 5-1 for examples of terms used to describe igneous rock textures.

Texture of Igneous Rocks

Before we go on to discuss how we describe the texture of an igneous rock, we need to clarify some terms. *Lava* is magma that has reached the Earth's surface before cooling and forming minerals. Lava produces *extrusive* igneous rocks. Lava is what we see flowing down the sides of Hawaiian volcanoes or erupting out of the top of Mt. Vesuvius, along with gases and ash. When lava reaches the Earth's surface, it cools very quickly in the presence of air or water, from temperatures greater than 1000°C in the Earth's interior (above the melting points of the minerals) to surface temperatures. Because of rapid cooling, the crystals form very rapidly and are thus very small in size. Extrusive igneous rocks are characteristically very fine-grained (the term "grain" describes crystal size). *Magma* produces *intrusive* igneous rocks, as magma has never reached the Earth's surface before cooling. If the magma cools intrusively, it cools more slowly (and may cool *very* slowly over many millions of years!). Intrusive igneous rocks are

Sidebar 5-1 Igneous Rock Textures

(A) Aphanitic: fine-grained texture (below left: basalt).

(B) Phaneritic: crystals large enough to see with naked eye (below: gabbro).

(C) Porphyritic: two different crystal sizes present.

Porphyritic andesite.

(D) Glassy: a glass lacks crystals.

Obsidian.

(E) Vesicular: gas escape vacuoles.

Vesicular basalt.

characteristically coarse-grained (i.e., with large crystal size). Thus, the texture of an igneous rock tells us whether the rock formed in an intrusive or extrusive setting.

Magma begins moving toward the Earth's surface because temperature controls a substance's density; as the vibrational energy of atoms increases, density decreases as the atoms spread out. The pressure from surrounding rocks also helps push magma toward the surface. Rising magma will begin cooling and crystallizing minerals. The rate at which magma cools is variable, depending on how rapidly heat is lost to surrounding rock or air or water. Due to the insulation of the surrounding rock, this cooling could be slow. Slow cooling produces large crystals because the atoms have time to organize themselves into well-formed crystal lattices. Fast cooling, which happens when magma approaches or reaches the Earth's surface, produces rapid cooling and igneous rocks with small or very small crystals. If cooling happens instantaneously, a glass forms (this is the rock obsidian). The porphyritic igneous rock texture records a change in the cooling history of magma. Examine the photograph of the porphyritic andesite in Sidebar 5-1 and suggest what cooling history would produce such a texture.

Origin of Magma

As seen in Figure 5-1, magma is another reservoir in the rock cycle. Many of the major elements that make up Planet Earth are present in magma and in the igneous rocks that it forms. Physically, where within the Earth does magma form? Because magma forms from melting rocks, we need to look at where the temperatures are high enough to reach the melting points of different minerals within the rocks. As we'll see in Chapter 7, all but one layer within the Earth, the outer core, is dominated by solid rock. Solid rock is possibly not what your image of the interior of the Earth is like, yet earthquake wave analysis indicates that this is the case. Does magma come all the way to the surface from the outer core, or are there processes that produce localized melting?

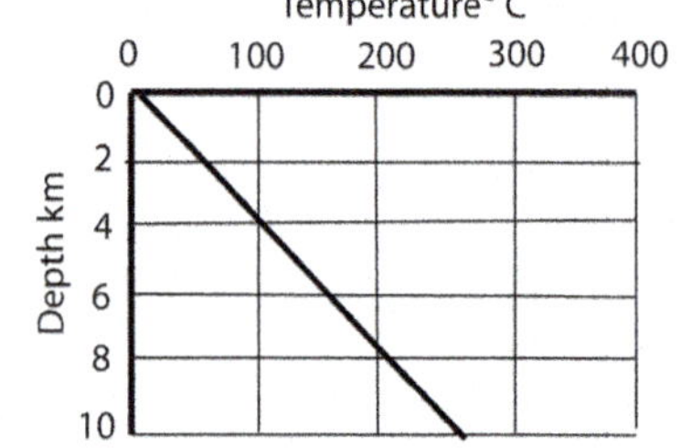

Figure 5-3 Shallow geothermal gradient.

Heat within the Earth

The temperature gradient within the Earth is called the *geotherm* (Figure 5-3). Our direct measurements of temperatures within the Earth are confined to the deepest mines (3.9 kilometers) and drill holes (2.1 kilometers). At the TauTona mine in South Africa, temperatures reach 55° to 60°C at its nearly 4-kilometer depth. If you assume an average surface temperature of 20°C, this represents a geothermal gradient of 25-30°C/ kilometer; the shallow (surface

Sidebar 5-2 Geothermal Energy

The heat of the Earth's interior has become an energy source for humans. How do we access this energy? Wells drilled down to the water table come in contact with groundwater. If that water has come in contact with hot rock, it can be heated (by conduction). If the temperature of the hot water is extremely high, as it is in places such as Iceland, steam can be used to generate electricity.

Krafla geothermal energy complex, Iceland. The Krafla volcano itself is covered with clouds in this photo. Steam plumes mark the location of pressure release vents. The magma chamber responsible for heating groundwater is approximately 2,000 meters below the surface.

In Iceland, near-surface magma heats overlying rocks through which groundwater circulates. We say that the geotherm is "elevated" here because it is hotter at this depth than the normal geotherm predicts. In other areas of the world, very localized "elevated geotherm" conditions can form if there is a heat source in rocks at depth, most commonly from the presence of a rock type that contains minerals undergoing radioactive decay (Chapter 6), releasing heat.

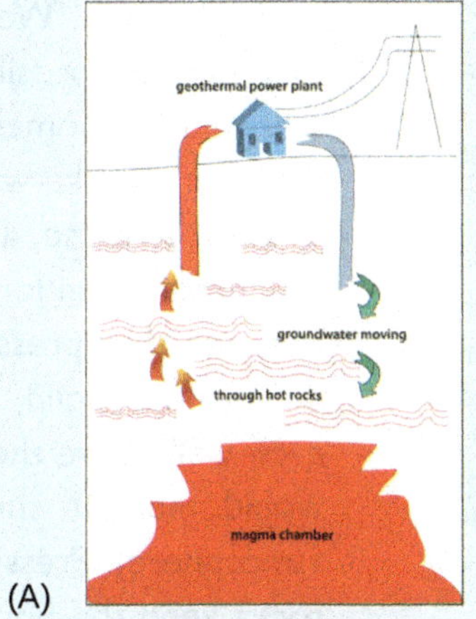

(A)

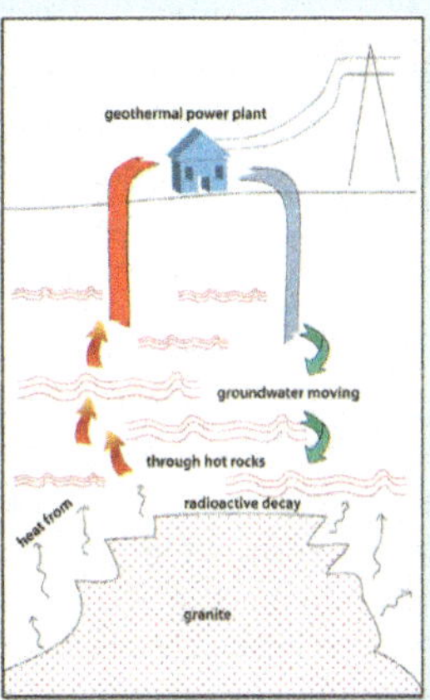

(B)

down to ~10 kilometers) geothermal gradient is usually rounded off to 30°C/ kilometer. If you were to extrapolate this value to a depth of 3,000 kilometers within the Earth, the temperature would be an absurd 30,000°C! Even given the great pressure within the Earth, all materials would melt well before we reached this value, so clearly the shallow geothermal gradient is not maintained to greater depths. Figure 5-4 illustrates the whole Earth geotherm.

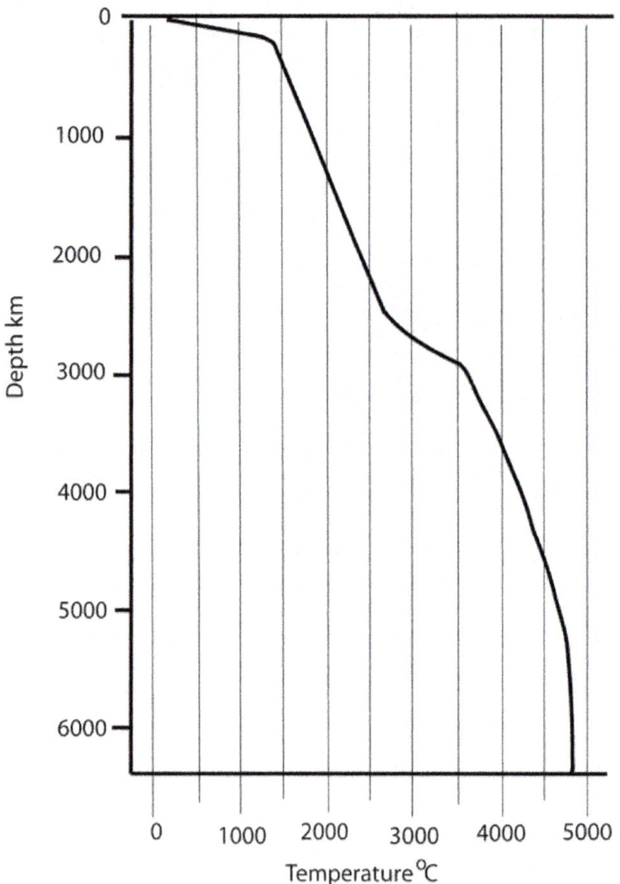

Figure 5-4 Whole Earth geothermal gradient. Figure 5-3 is detail of the gradient in the very top of this figure. Note the change in slope of the temperature curve (i.e., change in gradient). Values for temperatures at different depths are not obtained by direct measurement but are estimated based on the velocities observed for earthquake waves (Chapter 7) as well as from some experimental studies on the composition and stability of rocks.

Examining the temperature estimates of the entire Earth reveals several important features. Note how steep the gradient is near the Earth's surface. We already noted that this averages approximately 10°C/kilometer. Try to interpolate values of depth and temperature at other depths within the Earth from Figure 5-4 so that you can calculate the geothermal gradient, for example, for the interval between 1,000 and 2,000 kilometers, or 5,000 and 6,000 kilometers.

The "melting point" of a substance is the temperature at which it changes state from solid to liquid. The melting points for most minerals, and thus most rocks, are between 600° and 1,200°C. At what depth in the Earth are those temperatures attained? You probably said, "Well under 500 kilometers," or possibly even, "Around 100 to 200 kilometers." Yet we know from earthquake waves that the Earth is solid at these, and greater, depths. The explanation for this discrepancy is the role that pressure plays in maintaining the solid state of minerals even well above the temperature they would melt at surface pressures (1 atmosphere). Pressure from overlying rocks keep the bonds between atoms from breaking.

Processes that Cause Melting

What causes rocks to melt and form magma? There are three processes that can cause melting: decompression melting, melting by adding volatiles, such as H_2O and CO_2, and conduction.

Decompression Melting

If pressure keeps bonds together at temperatures where the vibrational energy would break them, and if the pressure exerted on a rock decreases, the rock could melt (Figure 5-5). The pressure from the weight of overlying rocks increases with depth within the Earth, so if a rock physically moves from deep in the Earth's interior toward the surface, the pressure drop plus elevated temperatures in the interior will allow melting to occur. The composition of magma produced from decompression melting is more commonly very rich in iron (Fe) and Magnesium (Mg), as these elements are more abundant in the Earth's interior layer, the mantle. We call a magma rich in iron and magnesium a *mafic magma* (ma = magnesium and f = iron). How does a rock move from deep within the Earth's interior toward the surface? As we will discuss in detail in Chapter 10, the forces that drive the movement of the tectonic plates horizontally across the Earth's surface also have a vertical component. Forming a structure similar to a convection cell in boiling water, hot rock can rise toward Earth's surface. This happens beneath divergent plate boundaries.

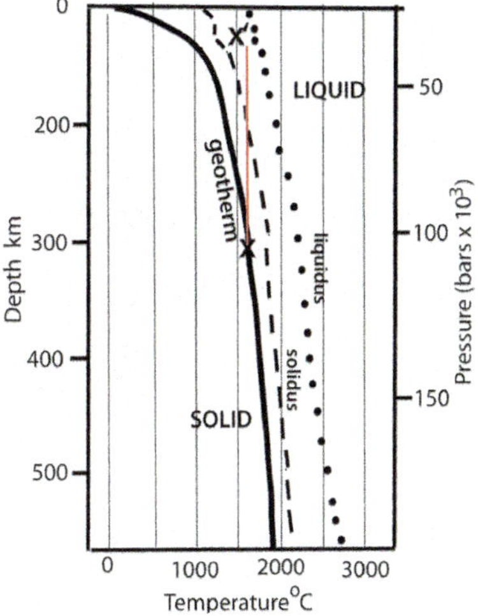

Figure 5-5 Graphical representation of why decompression melting produces magma with no temperature change. A rock at point X rises toward the Earth's surface (red line) and as pressure drops, the rock crosses the solidus (the line representing the average start of melting) and begins to melt. Partial melting of the rock has occurred as the rock rises to point X'. If the rock crossed the liquidus (dotted line), it would melt entirely.

Melting from the Addition of Volatiles

Volatiles are substances with low boiling points, such as water and carbon dioxide. When these molecules are added to magma at high temperatures, they act as catalysts that help break the bonds between atoms in minerals. Thus, geologists describe two melting points for rocks, termed "dry" and "wet" (Figure 5-6). Where and how do we add volatiles to hot rock? One of the easiest ways to do so involves melting minerals that have these molecules bound up in their lattices. As the bonds break, this material is released, free to interact with other minerals as a catalyst.

Conduction Melting

When hot rock from the Earth's interior rises toward the surface, it transfers heat to adjacent rocks. Because the interior of the Earth is layered (Chapter 7), the hot rock rising toward the surface will be composed of different minerals than those comprising the crust. The crustal rocks melt at lower temperatures than the rocks stable deep within the Earth, so as they feel the heat from magma (formed from decompression melting), they themselves melt (Figure 5-7). Magmas produced from conduction melting can have a wide range of compositions that depend on what rock melted.

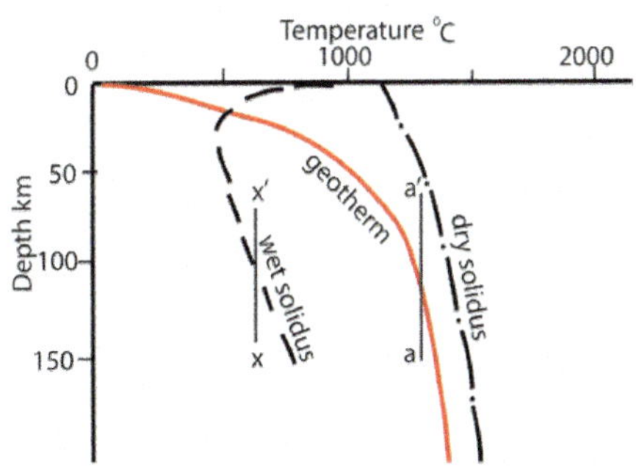

Figure 5-6 The effect of adding volatiles on the melting point of rocks. Rock x starts at approximately 150 kilometers and rises to a depth of approximately 50 kilometers, beginning to melt at approximately 100 kilometers as it rises to x'. Rock a follows the same path toward a'; however, the absence of volatiles keeps the melting point temperature higher. Rock a won't melt until it crosses the dry solidus at less than 50 kilometers.

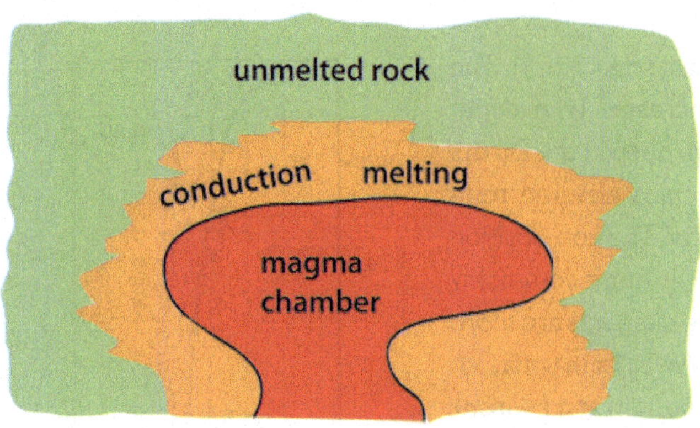

Figure 5-7 Conduction melting. The cartoon shows the melting of host rock that occurs when hot magma rises toward the Earth's surface.

Partial Melting

It's important to remember that because of their composition and the nature of bonds between them, not all minerals are stable at the same pressures and temperatures; thus, rocks won't all have the same melting points, either. This means that when a rock is heated, different minerals will begin to melt at different temperatures and pressures. In other words, the composition of mineral assemblages influences the melting point of a rock. We can predict the sequence of melting that would be expected to occur in a rock that is heated. Recall that the crystal lattice of a mineral forms when the electrostatic attractions between atoms is not great enough to overcome the atoms' vibrational energy. When a mineral melts, it is because the process operates in reverse: as atoms begin to vibrate with more and more energy, the electrostatic charges won't be able to hold the bonds together, and the crystal lattice will break apart. This is what melting is: the destruction of bonds forming the solid structure crystal lattice. As we heat a rock, the first minerals to melt will be those that formed at the lowest temperatures, because it will take less energy to break apart the bonds within minerals.

What minerals have lower melting points? The minerals that crystallize at the highest temperatures include silicates such as olivine and pyroxene. Temperatures would have to get really high (>1,000°C) to melt these minerals. On the other hand, quartz, potassium feldspar, and muscovite crystallize at a temperature of 600°C, so it takes much less heat to break the bonds in these minerals. Thus, the addition of heat from conduction melting can produce a magma rich in Si, K, Al, and O. Magma rich in these elements is termed a *felsic magma* ("felsic" is derived from "fel," for feldspar, and si, for silica, or quartz).

As mentioned above, conduction melting will produce magmas of a wide range of compositions, depending on what rocks melt. Partial melting can also produce a wide range of magma compositions as a wide range of low temperature minerals melt. We call magmas that are intermediate in composition (neither mafic nor felsic) *intermediate magma*.

Partial melting is a very important process in varying the composition of magma and, as a result, varying the composition of the

Sidebar 5-3 The M&M's Analogy for Partial Melting

Suppose we have 100 M&M's of different colors:

20 blue = 20% of the total M&M's

20 red = 20%

10 green = 10%

40 brown = 40%

10 yellow = 10%

(Total = 100%)

Now we remove 50 M&M's, so that we now have a total of 50 M&M's.

If we removed 10 of the blue, 5 of the red, all the green (10), 20 of the brown, and 5 of the yellow, we would have:

20% blue (there were 20% originally)

30% red (there were 20% originally)

0% green (there were 10% originally)

40% brown (there were 40% originally)

10% yellow (there were 10% originally)

(Total = 100%) (Note: We also started with 100%.)

So,

- The relative percentages of remaining M&M's colors changed as some colors were removed.
- We can say that the remaining M&M's became enriched in red M&M's relative to the starting percentages. Other M&M's colors, such as the green, became depleted.

What does the M&M's analogy have to do with magma composition?

- If we melt a rock of composition of 10% Fe, 10% Mg, 50% SiO_4, 20% Ca, and 10% Al,
- And we remove 5% of the Fe and 5% of the Mg and 10% of the SiO_4,
- The magma becomes 6.25% Fe, 6.25% Mg, 50% SiO_4, 25%Ca, and 12.5% Al. In other words, the new magma is enriched in silica, calcium, and aluminum (i.e., more felsic).

igneous rocks that form from the cooling of this magma. Sidebar 5-3 gives an example of how much the relative abundance of different elements can change as a result of partial melting.

Why we should care about how all of the different types of igneous rocks are produced? Recall that rocks are composed of minerals, and minerals are reservoirs (storage sites) for different elements. Because different types of rocks are composed of different minerals, this means that different elements are available on or near the Earth's surface for us to use. Understanding the genesis of rocks helps guide mineral exploration and recovery.

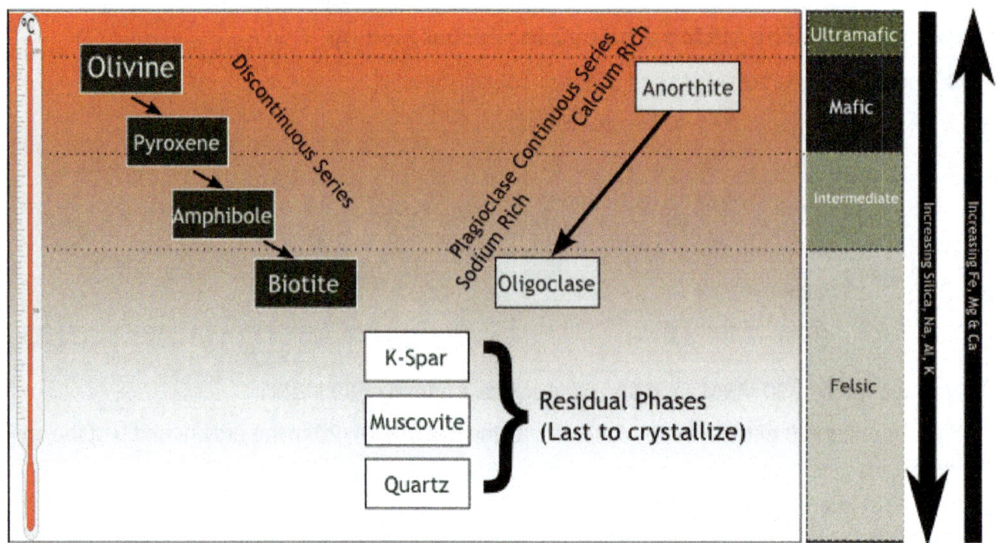

Figure 5-8 Bowen's reaction series. Temperature decreases from the top of the illustration to the bottom. The upward pointing black arrow represents increasing amounts of Fe, Mg and Ca in the early-forming minerals. The downward pointing arrow indicates increasing amounts of Si, Na, Al and K in later-forming minerals.

Fractional Crystallization

Partial melting is one way we can produce a wide range of magma compositions and resulting igneous rock types. *Fractional crystallization* is the name we give to the second process that produces variation in igneous rock composition from a magma that starts out mafic. Fractional crystallization describes the sequential crystallization of minerals from a cooling mafic magma, such that it becomes progressively more felsic. As the magma cools and the freezing point of a mineral is reached, its crystals will settle out of the liquid magma (because of their slightly higher density in the solid than the liquid phase), falling through the magma chamber. The magma composition progressively changes as elements comprising the early-forming minerals go out of solution. This sequence of mineral formation was experimentally tested by Neil Bowen, who started his experiments with a rock whose composition represented that of the mantle (peridotite). He then heated this rock to approximately 1,200°C and allowed it to slowly cool. Bowen described both the sequence of mineral formation and the evolving composition of the magma melt, so the sequence is named *Bowen's reaction series* (Figure 5-8).

Starting with melted peridotite, the first mineral to crystallize is the iron-magnesium-rich silicate, olivine, in its isolated tetrahedral

Sidebar 5-4 Bowen's Reaction Series

(A) A magma chamber (a pluton in the making), within temperature of the Earth begins to cool....

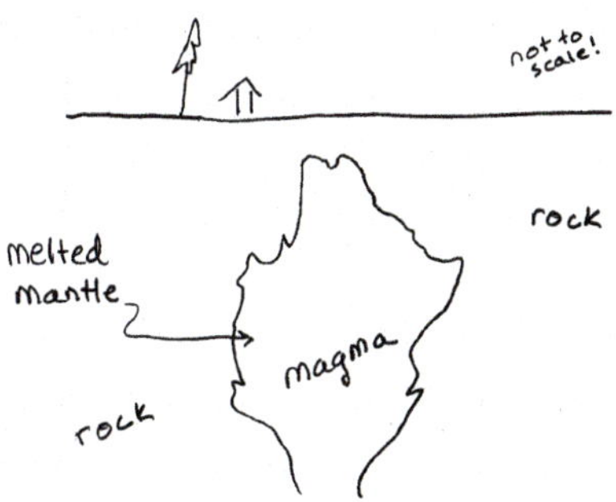

(B) As the magma cools, it hits the crystalline *olivine* (~1225°). The removal of the Fe and Mg to make olivine leaves the remaining magma proportionally richer in other elements.

(C) At the same time the olivine begins to form, so does Ca-rich *plagioclase feldspar*. Some olivine settles to the bottom of the magma chamber, and some reacts with the evolved magma to form pyroxene. The remaining magma becomes enriched in other elements.

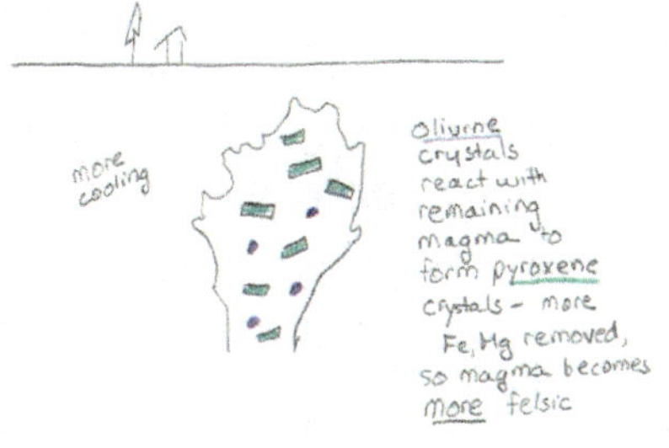

(D) With further cooling, almost all Fe and Mg are used up to make *amphibole*. The plagioclase continues to crystallize, incorporating more Na into the lattice then Ca. The remaining magma has very little Fe and Mg and is mostly other cations (K, Al) and Si and O.

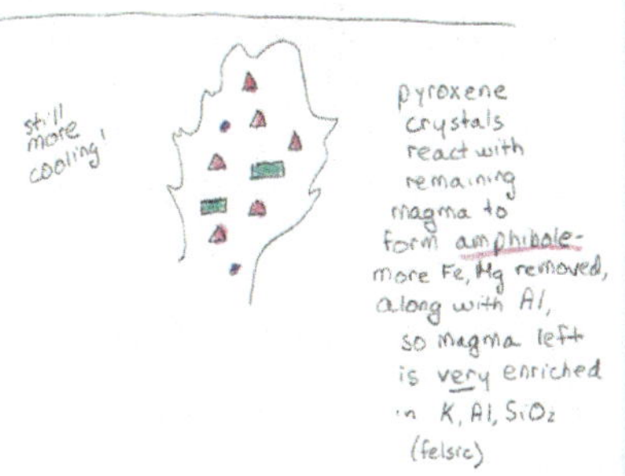

(E) With further cooling, K and Al form muscovite and then potassium feldspar and finally quartz. The rock type formed from this magma is *granite*. crystallize from it have less Fe and Mg thus are more felsic.

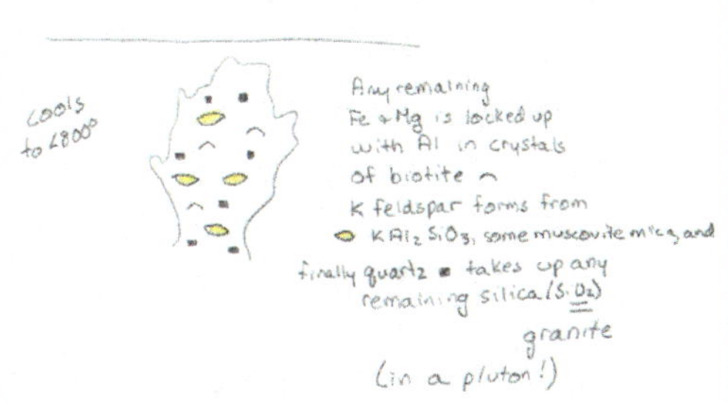

(F) What if we go through this cooling sequence but remove magma while cooling? As crystallization begins in the magma chamber, the olivine, pyroxene, and plagioclase feldspar minerals form. The remaining magma becomes proportionally richer in the other elements. If *this* magma is removed, the minerals that can

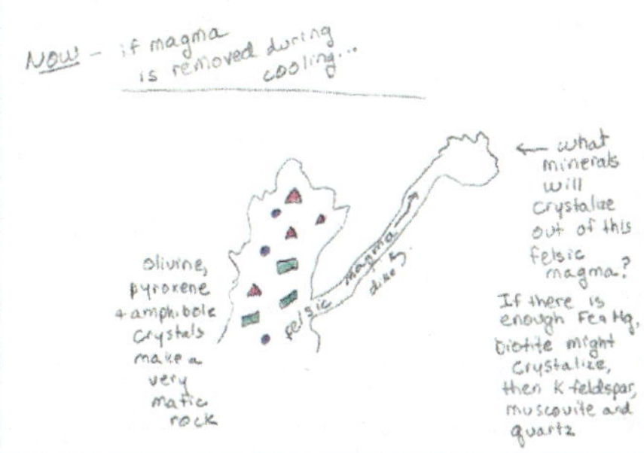

structure. With the removal of Fe and Mg atoms, the magma is slightly less enriched in these elements. With further cooling, crystallization will continue. Some of the early formed olivine crystals settle out of the magma, but some react with the magma to form another iron-magnesium silicate, pyroxene, in its single-chain structure. Some pyroxene crystals will also settle to the bottom of the magma chamber, but others will react with the more iron and magnesium-depleted magma to form the double-chain silicate mineral amphibole. As cooling continues, so do these processes: some amphibole crystals settle out, but others react with the remaining much less mafic magma to form the sheet silicate biotite. Because each mineral in this part of Bowen's reaction series forms discretely, either settling out of solution or reacting with the magma, this sequence is termed the *discontinuous series*.

Mantle rock has more elements present than just iron, magnesium, and the silicate tetrahedron, so Bowen noted that while the olivine-to-biotite mineral sequence was crystallizing, the calcium and sodium atoms were bonding with silicon and oxygen to form the framework silicate mineral plagioclase feldspar. Initially, the cation present in the crystal lattice is Ca, but with cooling, Na enters the lattice in the calcium's place. Because the plagioclase feldspar (usually shortened to just "plagioclase") crystals form with a continuous substitution of Na for Ca as the magma cools, this part of Bowen's reaction series is called the *continuous series*.

By the time biotite and sodium plagioclase have formed, the magma is now felsic. The remaining minerals that can crystallize include the sheet silicate mineral muscovite, which incorporates aluminum and potassium. With continued cooling orthoclase feldspar forms, a potassium-rich framework silicate. At the lowest temperatures, and with all cations removed from the magma, quartz (SiO_2) crystallizes (Sidebar 5-4).

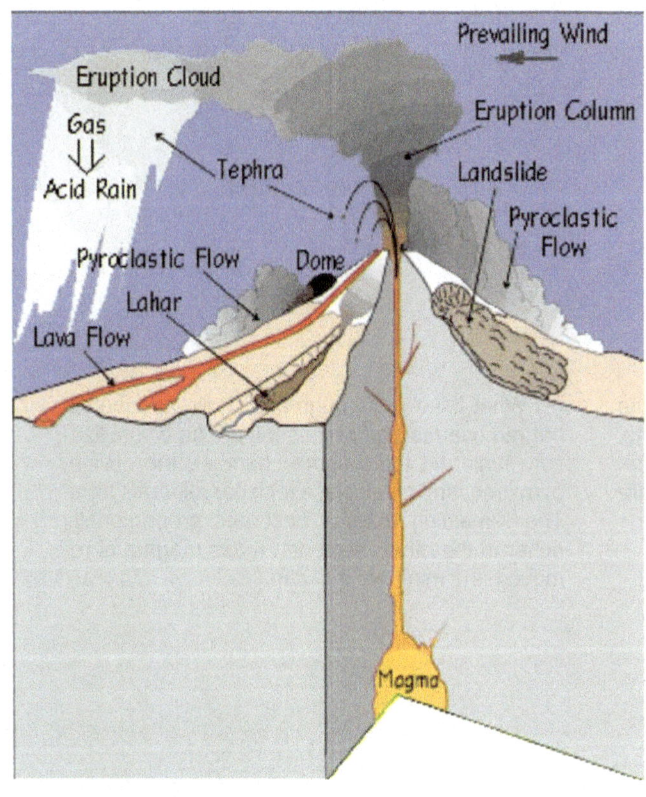

Figure 5-9 Types of volcanic hazards.

Why is Bowen's reaction series important? It explains how you can generate many different types of igneous rocks from melting and cooling mantle rock. As we will discuss in Chapter 11, it also describes, in reverse, mineral stabilities in the presence of surface conditions. In his study of soil formation, Goldrich noted that the chemical durability of minerals at the Earth's surface generally followed Bowen's sequence of formation, only in reverse. Minerals that form early, at elevated temperatures, break down more readily than do minerals that form late in the sequence. Thus, quartz is more stable than olivine or plagioclase is. We will discuss the interactions between rocks, minerals, and water in more detail in Chapter 10.

Volcanoes

Bowen's reaction series describes the sequence of crystallization of minerals within a magma chamber. Often these magma chambers lie beneath volcanoes. The eruptions of volcanoes play a significant role in the Earth system because they represent "burps" of material from the Earth's interior onto its surface and atmosphere. Thus, eruptions are a flux in the Earth system, transferring material from the Earth's interior, some of which impacts Earth's climate. The 1991 eruption of Mt. Pinatubo in the Philippines illustrates the role that volcanic eruptions play in cycling material from the Earth's interior to the surface. Geologists estimate that during this multi-day eruption approximately 8.8×10^5 metric tons of zinc, 6×10^5 metric tons of copper, 5.5×10^5 metric tons of chromium, 3×10^5 metric tons of nickel, 1×10^5 metric tons of lead, 1×10^4 metric tons of arsenic, 1×10^3 metric tons of cadmium, and 8×10^2 metric tons of mercury were brought to the surface.

Eruptions also represent a significant human hazard (Figure 5-9).

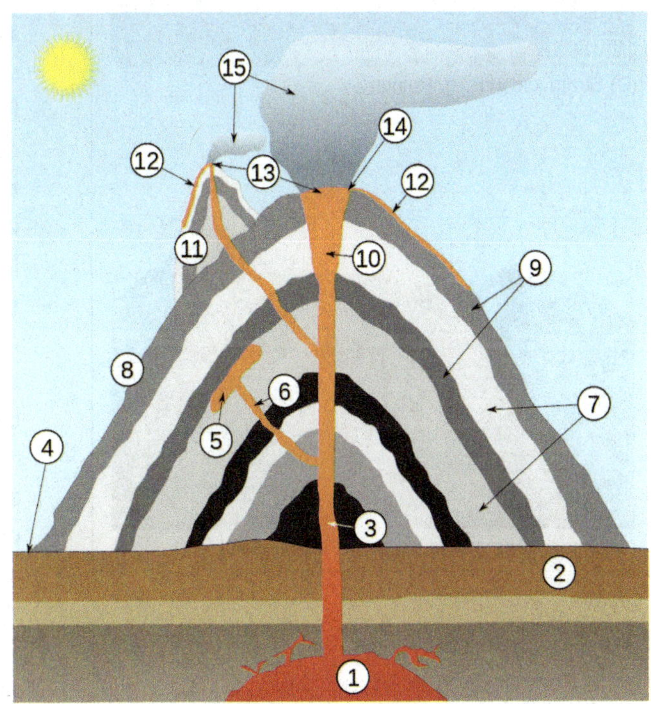

Figure 5-10 Components of an idealized volcanic cone and associated structures: (1) magma chamber; (2) surrounding bedrock; (3) pipe, or conduit; (4) base; (5) sill; (6) dike; (7) ash layers; (8) flank of cone; (9) lava layers; (10) throat; (11) parasitic cone; (12) modern lava flow; (13) vent; (14) crater; (15) ash cloud.

(A) Shield volcano: Mauna Loa, Hawaii.

(B) Stratovolcano: Mt. Rainier.

(C) Cinder cone: Krafla Volcanic Field, Iceland.

Figure 5-11 Types of volcanoes.

Types of Volcanoes

There are three types of volcanoes that are identified on the basis of differences in the volcano's cone (Figure 5-10). Which type of volcano forms depends on the nature of the material erupted, which in turn is controlled by the properties of the magma. Some geologists recognize a fourth type of volcano called a *supervolcano*; as its name implies, it is characterized by devastatingly large eruptions that alter the entire Earth.

- *Shield volcanoes* are characterized by their broad profile with gently sloping sides. This topography reflects the very runny (i.e., low viscosity) basaltic lava erupted from this type of volcano. The Hawaiian Island volcanoes (Figure 5-11A) are typically of this type. Often associated with shield volcanoes are *fissure eruptions*, lava that seeps out of faults and fractures ("fissures") adjacent to the volcanic cone. Lava from kilometer-long fissure eruptions flow out over the Earth's surface, forming sheets of extrusive volcanic rocks. Much of Iceland is built on this type of material.
- *Stratovolcanoes* (also called *composite volcanoes*) have steep sides (Figure 5-11B), a topography that reflects the growth of the cone from alternating lava flows and pyroclastic deposits. *Pyroclastic material* consists of ash (pulverized bits of small [<2 millimeters] pieces of blown-apart rocks and minerals) and ash mixed with water (called "lahar"). While shield volcanoes often "leak" lava frequently, stratovolcano eruptions are often spectacular and very hazardous, because of the more viscous magma and the pyroclastic material it produces. Mt. Fuji, Mt. Vesuvius, and Mt. Rainier (Figure 5-11B) are all examples of stratovolcanoes.
- *Cinder cones* are composed entirely of pyroclastic material. As gas dissolved in the

magma rises toward the surface, it expands in volume. The magma conduits inside the volcano can't contain the outward pressure, and the surrounding rock explodes, pulverized into the small, jagged ash particles. Cinder cones (Figure 5-11C) are smaller than either shield or stratovolcanoes are. They often occur with both of these larger volcanoes, occurring as small vents on their sides, although they may also "stand alone." The Craters of the Moon National Monument (Idaho) and the San Francisco volcanic field (Arizona) contain numerous cinder cones.

What Controls Which Type of Volcano Forms?

The type of volcano reflects the material which erupts. Shield volcanoes are composed of low viscosity lava flows, while cinder cones are entirely ash; stratovolcanoes contain both. So, the real question is what determines whether lava flows or ash forms? The difference between these materials arises from the viscosity of the magma that forms them.

Recall that viscosity is a measurement of the ease with which a substance flows. Maple syrup flows slowly when cold, and is very runny when warm. Temperature must therefore influence viscosity; warm liquids are less viscous than cold. Hotter magmas have fewer early-formed crystals and thus they will flow more easily; cooler magma start to crystallize quickly and the many crystals slow the flow. The other variable influencing the "runniness" of lava is how much silica it has in it. The more Si you have, the more likely you are to start building the complex silicate tetrahedron, and this molecule tends to "stick together" with other tetrahedrons. The more volatiles that are dissolved in the magma, the less viscous it is. The "bubbles" of water (steam) and gas keep the molecules apart from one another and flowing more easily.

Figure 5-12 Crater Lake caldera. The caldera at the top of Mt. Mazuma formed approximately six thousand years ago, when an explosive eruption of this stratovolcano blew away approximately 1.5 kilometers of the top of the cone.

What influences the temperature, volatile, and silica contents? The temperature of the magma is controlled by how fast the magma rises toward the surface.

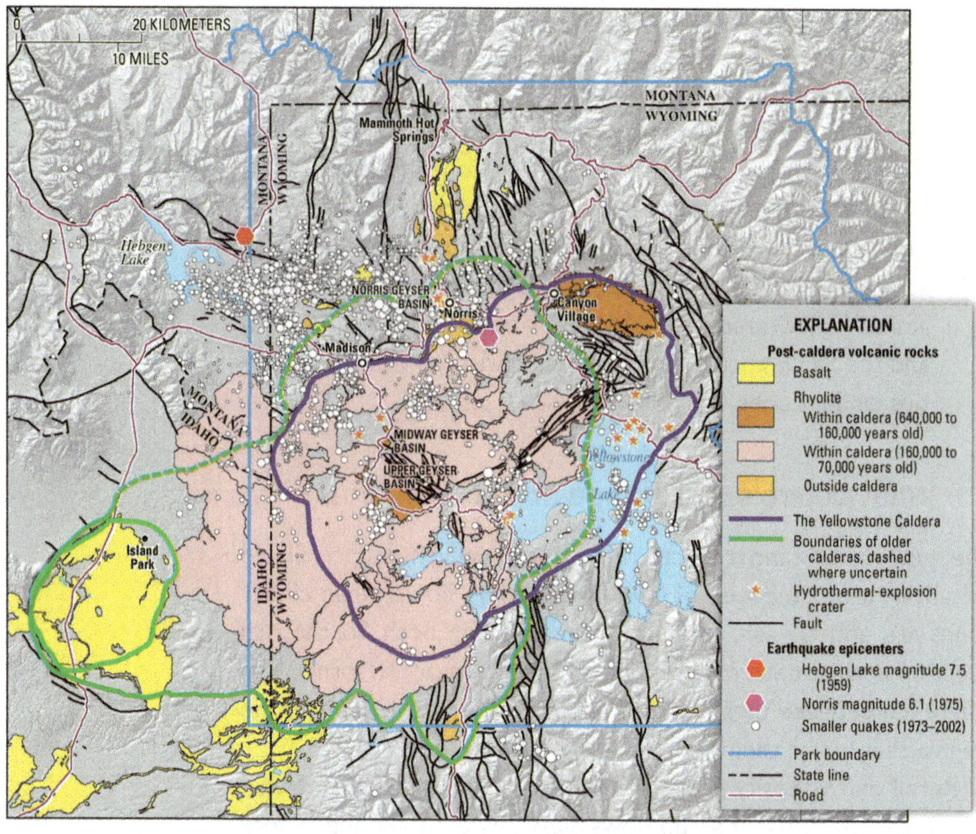

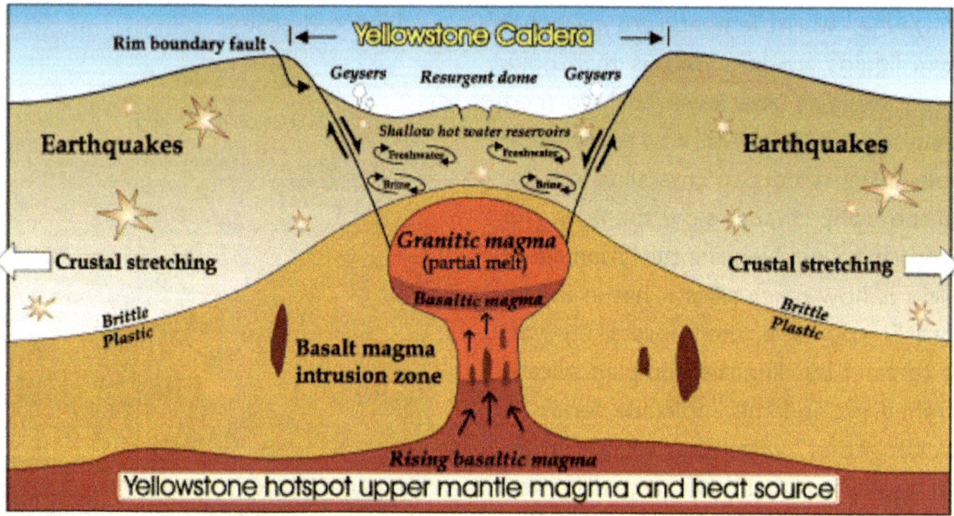

Figure 5-13 (A) The US Geological Survey geologic map of northwestern Wyoming and southwestern Montana, showing the geology around Yellowstone National Park. The park outline is shown by the light blue line. The generally circular purple line outlines the caldera formed from the last volcanic eruption, approximately 640,000 years ago. The green line shows the position of an older caldera from an eruption approximately two million years ago.

(B) Cross-section of the Earth's surface through the Yellowstone caldera. The heat source that generates the magma beneath Yellowstone is discussed in Chapter 8.

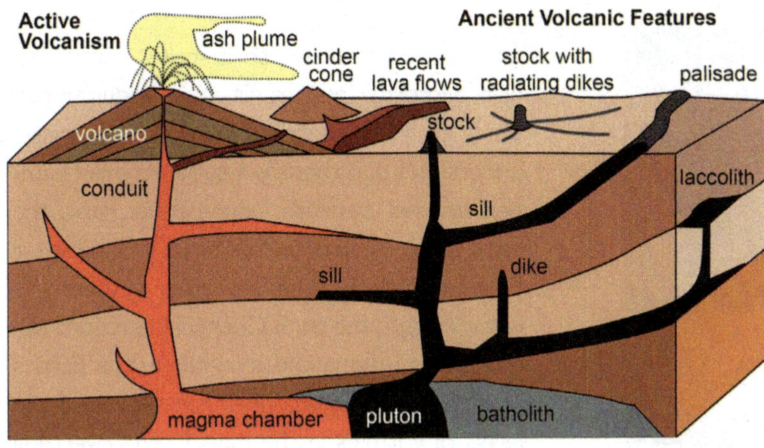

Figure 5-14 Igneous rock bodies.

USGS / Copyright in the Public Domain.

If it rises fast, it erupts before crystallization begins. If a magma sits in a magma chamber, it will begin to cool and crystallize minerals. A more slowly cooling magma may also undergo fractional crystallization, so that the magma becomes progressively more felsic prior to eruption. The material that melted to produce the magma influences its volatile and silica content. If you melt a mafic igneous rock, it will have low amounts of these components. If, on the other hand, you melt one of the other two types of rocks (sedimentary and metamorphic), it is likely that your magma will be rich in silicate minerals, such as quartz and potassium feldspar.

Supervolcanoes

Only in the past thirty years have geologists recognized that periodically there are spectacularly large eruptions of highly explosive stratovolcanoes. Following an eruption, all that is left on the Earth's surface is a very large caldera; some of these are such large features that they themselves are hard to recognize! Traditionally, a caldera is the depression at the top of a volcanic cone that forms from the collapse and sinking of magma that didn't erupt. Crater Lake in Oregon is an example of a volcanic caldera (Figure 5-12). A supervolcano's caldera is the remaining wall of the magma chamber, after the top has blown off and the magma has subsided. Yellowstone National Park lies within the Yellowstone supervolcano caldera, which exploded most recently around 640,000 years ago and twice before that.

Clearly, most stratovolcanoes don't obliterate themselves in huge explosive eruptions, so what forms a supervolcano? We discussed the role that viscosity plays in causing the explosive eruptions that

(A) Palisades Sill, Englewood, New Jersey.

(B) Schoodic Point (Maine) igneous dikes. Photo by Laura Stevens.

(C) Yosemite National Park (California): Half Dome, a pluton.

(D) Spanish Peaks ring dikes.

Figure 5-15 Igneous rock bodies.

form stratovolcanoes, so it makes sense that high viscosity magma is one requirement. Using analysis of earthquake waves, geologists understand that there is a large magma chamber located beneath Yellowstone Park, the same magma chamber that heats groundwater to create Old Faithful and other geysers and hot springs in the park. Chemical reactions between the magma and minerals in the Earth's crust around the chamber have elevated the levels of volatiles in the magma. The overlying crust has "kept a lid" on the magma, which keeps the gas in solution in the magma. If the overlying crust develops fractures or faults, some of this gas can escape, and this in turn causes the magma to expand rapidly (you are familiar with this phenomenon: when you open a bottle of seltzer water quickly, it will fizz dramatically and spill out under the cap). The explosive release of gas out of the magma causes the spectacular eruption. What erupts from a supervolcano is not lava, but ash and gas, because the explosion pulverizes the rock surrounding the magma chamber. This is why supervolcano eruptions are so catastrophic. Calculations of the amount of pyroclastic material spewed into the atmosphere the last time Yellowstone erupted are as large as one million cubic kilometers! In addition to ash particles thrown into the atmosphere, blocking sunlight and triggering a "nuclear winter" scenario, toxic gases such as $S0_4$ would create months of acid precipitation. Because this region of the United States is tectonically active, the rocks are faulted and earthquakes are frequent; thus, the geologic conditions needed to trigger a supervolcanic explosion are possible. To a geologist, it is inevitable.

Igneous Rock Bodies

Because igneous rocks form in both intrusive and extrusive settings, not only is there great variability in their compositions; they also occur in "preserved magma chambers" of varying size and shape. A volcano is one obvious example of the external shape of a large pile of igneous rock material. The magma that never makes it to the surface cools and forms igneous rocks that record the "plumbing system" deep within the Earth. Geologists have a series of terms to describe these ancient magma chambers (Figure 5-14). In Chapter 10 we will talk more about the processes that uplift and expose these rock bodies.

The landscape around us contains many of the features shown on this diagram. The New Jersey cliffs along the Hudson River by the George Washington Bridge are a superb example of a volcanic sill (Figure 5-15A). Igneous dikes can be seen in western Vermont or Acadia National Park in Maine (Figure 5-15B). The spectacular Yosemite National Park represents a series of large plutons (Figure 5-15C), and the Barre granite quarries in Vermont extract rocks from a small pluton.

Sedimentary Rocks

Sediment is the detritus produced by the physical and chemical breakdown of preexisting rocks and minerals or from the accumulation of biologically-generated fragments, such as sea shells. When the former fragments are bound together (cemented) during burial, they form sedimentary rocks. Both sediment and sedimentary rocks are reservoirs in the Earth system; sediment temporarily stores Earth materials before they are buried, cemented into rock, and "recycled" to the interior.

Sidebar 5-5 Common Sedimentary Rocks

Note: Penny shown for scale.

(A) Conglomerate.

(B) Sandstone.

(C) Siltstone.

(D) Limestone.

(A) The Rio Sierpe Delta in western Costa Rica. Sediment carried by the river is temporarily deposited on the floodplain and channel bars. The delta forms from deposition of the brown-colored sediment plume seen at the bottom of the photo. Because of the climate and bedrock of Costa Rica, most of the delta sediment is fine-grained sand, silt, and clay. Photo by Ginny Schwartz.

(B) Cape Cod coastline south of Chatham, Massachusetts. The Monomoy islands (off the south end of the "elbow" of Cape Cod) are cut by numerous tidal channels moving water between the Atlantic to the right and Nantucket Sound to the left.

Figure 5-16

Sedimentary rocks, like intrusive igneous rocks, have been uplifted to or near the Earth's surface after burial and *lithification* (cementation). Just as igneous rocks contain raw materials for human use, such as ores, sedimentary rocks contain resources such as petroleum and natural gas. Sedimentary rocks also contain the fossil record of life on Earth. Much of geological study is focused on understanding where and how certain sedimentary rocks formed and whether or not they contain sufficient fossil fuel supplies to warrant their recovery.

Sedimentary rocks can be subdivided into four groups based on the nature of the materials that comprise them:

- *Clastic:* composed of small sand and clay grains; the grains themselves may be made of tiny fragments of minerals, such as quartz, or even tiny fragments of rocks. These grains record what rocks weathered apart to form the sediment. The sedimentary rocks sandstone and shale are examples of this group.
- *Biochemical:* composed of large to tiny shell fragments of organisms. The sedimentary rock type limestone is an example of this group.
- *Chemical precipitate:* composed of minerals precipitated out of solution. If you evaporate seawater, for example, you could form a variety of rock types, such as limestone, dolostone, or gypsum.
- *Biogenic:* not to be confused with the "biochemical" group, biogenic sedimentary rocks are composed entirely of organic material. Coal is the best example of this type of rock.

Formation of Sedimentary Rocks

There are several stages involved in the formation of a sedimentary rock. First, you need to produce the sediment by breaking down preexisting rocks.

Second, the sediment must be transported, or moved, from where it forms to a place on the Earth's surface where it can be deposited, or laid down. The process of moving sediment is termed erosion. Rivers transport a great deal of sediment, as do waves and wind. The sites where large quantities of sediment are deposited are called *basins*. A delta (Figure 5-15A) is an example of a basin; the modern Mississippi Delta is a basin containing a thickness of approximately 5 kilometers of sediment! Other basins include lakes and shorelines (Figure 5-15B). The common denominator to most of these places on the Earth's surface is the presence of water. Sediment left exposed on the land surface will likely be washed or blown away before significant amounts of it can accumulate, although there are exceptions to this (Zion National Park contains spectacular ancient lithified sand dunes, Figure 5-15A). Sediment dropped in water will more likely be covered with more sediment, allowing thicknesses to accumulate. The weight of the overlying sediment in the burial process helps compact and then cement the material on the bottom of the pile.

Sedimentary rocks often contain a record of the processes that transported and deposited the sediment. Features called *sedimentary structures* include preserved large and small ripples (Figure 5-17A and B), mud cracks (Figure 5-17C), and burrows or tracks (Figure 5-17D) of organisms. Because these features often form in only a few environments in a narrow range of conditions, geologists can reconstruct the environments and processes when the sediment was deposited.

Following deposition, sediment will be lithified, or cemented, which produces a hard rock from what was originally loose detritus. What material cements sediment? Most commonly, water droplets trapped between the grains precipitate a mineral, such as calcite, iron oxide, or silica.

The study of sedimentary rocks is critical to our understanding how Earth environments have changed over time. Using sedimentary rocks, geologists can determine how sea level has changed over Earth history, documenting that

(A) Cross-bedding.

(B) Wave ripples.

(C) Mud cracks, which form from the drying, shrinking, and cracking of clay.

(D) Dinosaur footprint (with pen shown for scale).

Fig. 5-17A: USGS / Copyright in the Public Domain.

Figure 5-17 Sedimentary structures.

Sidebar 5-6 Sedimentary Rock Classification

Texture	Grain Size of Fragments	Rock Name
Rock fragments or minerals	Gravel (> 2 mm)	Conglomerate
As above	Mostly sand	Sandstone
As above	Mostly silt	Siltstone
As above	Mostly mud	Shale
Crystalline, often with fossils: highly reactive with HCl	Variable	Limestone
Crystalline, but need to powder to make reactive with HCl	Finely crystalline	Dolostone

fluctuations in the order of 100 meters has been common. The minerals in sedimentary rocks can be analyzed to help reconstruct ancient climates. Sedimentary rocks often contain fossils, the remains of ancient life forms and by synthesizing environmental information with fossil types and abundance we can reconstruct ecosystems over the past half billion years.

Figure 5-18 Photograph of a gneiss, a high-grade metamorphic rock. Note that the dark and light minerals are segregated into bands. The large crystal size, another characteristic of metamorphic rocks, is visible in this photograph. Penny shown for scale.

Identifying and Classifying Sedimentary Rocks

Sedimentary rocks are distinguished from igneous rocks by their texture. Igneous rocks are composed of interlocked crystals, while sedimentary rocks are composed of "broken bits" that, with the exception of the chemical precipitate group, are not crystalline. Fossils, the preserved remains of ancient life, may be present, as well as some of the sedimentary structures illustrated in Figure 5-16. Sedimentary rocks are classified by the size and composition of the detritus; the most common sedimentary rocks are included in Sidebar 5-6.

Sedimentary rocks are important reservoirs of many Earth materials. For example, limestone, composed of $CaCO_3$ (the mineral calcite), is one of the largest reservoirs for carbon on Earth! The mining of limestone to make cement releases 5% of the global emissions of CO_2 per year. As a result of the economic boom in many previously agricultural societies (ex, China), the manufacture of concrete for bridges, roads, airport runways, and buildings has been increasing on average of 2.5% per year, with resulting

Sidebar 5-7 Metamorphic Grade

Unfoliated: quartzite (A) and marble (B)

(A)

(B)

Low-grade, poorly developed foliation: slate (C) and phyllite (D)

(C)

(D)

High-grade, well-developed foliation: schist (E) and gneiss (F)

(E)

(F)

increase in CO_2 emissions. In addition to the CO_2 emissions, the manufacture of concrete consumes huge amounts of energy. The production of one ton of concrete requires the energy produced by nearly 400 pounds of coal and it generates nearly one ton of CO_2! In Chapter 13 we will discuss the role of CO_2 in the atmosphere in controlling Earth's climate.

Sand and gravel- unconsolidated sediment not yet lithified into rock (sandstone and conglomerate) - are also important building materials. Collectively called "aggregate," this natural resource is so important to the economy that the U.S. Geological Survey monitors its supply. Aggregate is used in the cement industry, as

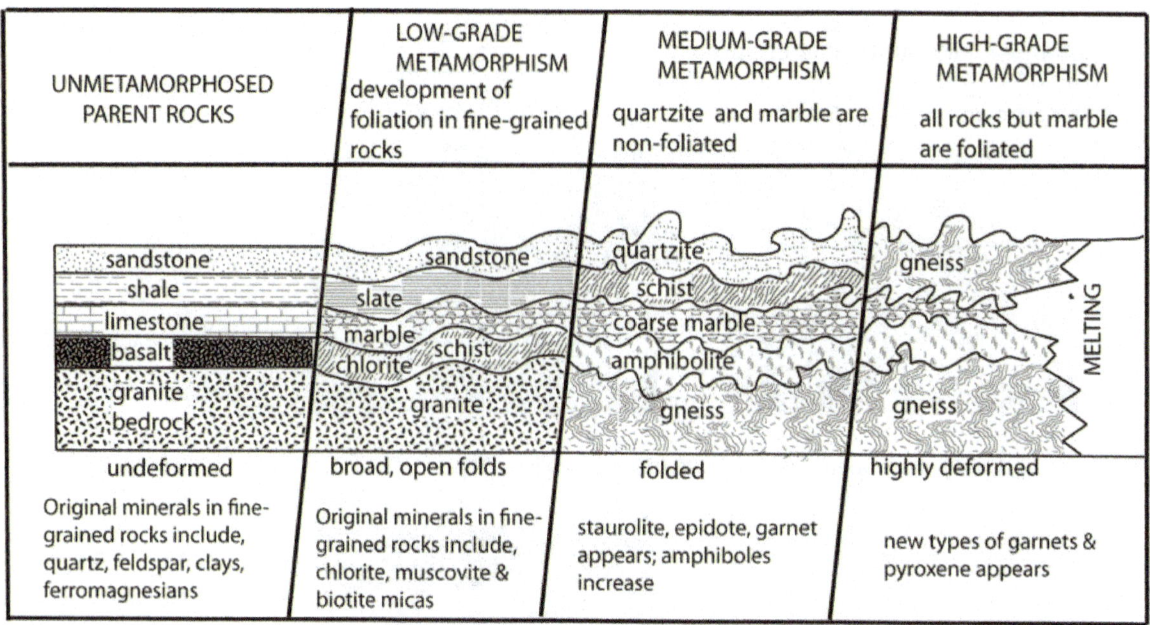

Figure 5-19 Generalized cartoon illustrating the differences in mineralogy and deformation with the increasing heat and pressure of metamorphism.

road and construction fill and additive to many other building materials. Sand and gravel mining was a $7 billion business in 2014. Sedimentary rocks such as limestone, sandstone and dolostone are quarried for use as building stone and monuments.

Petroleum and other fossil fuels are hosted in sedimentary rocks. This ancient organic material accumulates in specific environments on the Earth's surface, and much geological research focuses on the study of sedimentary rocks to deduce ancient environments that might contain these deposits.

The Difference between Sediment and Soil

What is soil? *Soil* is a mixture of sediment and organic material, with variable amounts of moisture and gas present. We can convert sediment to soil with the addition of organic material, and with the chemical interactions that will result. The production of soil is one of the best examples of the interaction between the geosphere and the biosphere. As a result of these interactions, material such as iron and calcium is transferred from the geosphere (minerals) to organic material. The biosphere, in turn, fluxes elements such as oxygen into the atmosphere and removes elements such as nitrogen and carbon from the atmosphere.

Metamorphic Rocks

As the etymology of the word "metamorphic" reveals, "meta" means "changed" and "morph" refers to shape or form. So, a *metamorphic rock* is a rock that has changed, and the primary agents of this change are heat and pressure, which act to alter the composition and texture of the original or parent rock. Metamorphic rocks are among the easiest to recognize because the heat and pressure that act on a preexisting rock cause most rocks to become coarser-grained and well-layered. The layering in a metamorphic rock is termed *foliation*, and it results from the alignment of minerals (Sidebar 5-7).

How can we tell that a rock is foliated? The alignment of minerals often causes them to line up in parallel bands or layers. As we discuss below, the degree of foliation in a metamorphic rock is related to its metamorphic grade.

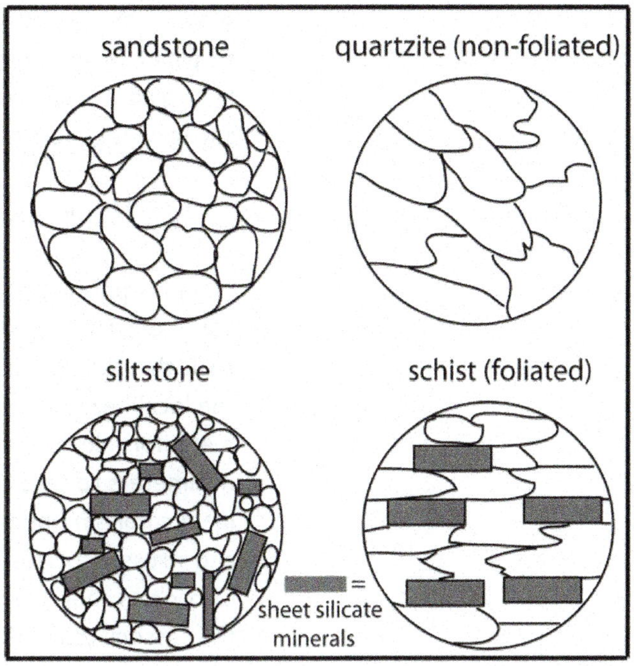

Figure 5-20 Recrystallization during metamorphism alters the size of grains and crystals and their fabric, or arrangement. In non-foliated rocks, such as marble and quartzite, we see grain boundaries suture, or weld together. In foliated rocks, such as schist, clay grains and micas align in parallel sheets. In both cases the size of crystals in metamorphic rocks increases relative to the protolith.

The Concept of Metamorphic Grade

As Figure 5-19 illustrates, as you move away from the axis or core of a mountain range, the amount of deformation (folding and faulting) and metamorphism decreases until you can see unaltered parent rock types. Conversely, as you move toward the core of a mountain range, you move into zones of progressively deformed rocks caused by increasing amounts of heat and pressure from deformation. We call these different zones of alteration *metamorphic grades*. Thus, a high-grade metamorphic rock is one that is very altered in original mineralogy and texture, and it will exhibit well-developed foliation. It was buried very deeply in the Earth from the forces of plate collision. A low-grade metamorphic rock is one that has changed little, with foliation barely developed. These rocks were on the flanks, or periphery, of a plate collision or near a large magma chamber within the Earth. There is a continuum between high-grade metamorphic rocks and complete melting, in which case a magma is generated and the resulting rock is considered igneous.

Regional versus Contact Metamorphism

Metamorphism acts on two very different scales. *Contact metamorphism* occurs immediately adjacent to a hot igneous intrusion. The rock through which the hot magma intrudes becomes altered by the local heat and hydrothermal (hot groundwater) fluids. Contact metamorphic zones tend to be very small, anywhere from a few millimeters to tens of meters away from the igneous body. *Regional metamorphism* operates on the opposite end of the scale. The heat and stress associated with plate collisions will alter thick sequences of rock as they are folded and faulted. Regional metamorphism is thus associated with the formation of mountain ranges. The uplift of metamorphic rocks is another way in which material from the Earth's interior is brought to the surface. Hydrothermal fluids often serve as a catalyst for chemical reactions between the fluids and the hot rock. These form veins of minerals, some of which are ores, such as gold and silver.

Fabric	Metamorphic Rock Name	Parent Rock	Color	Mineralogy	Distinctive Features
Non-foliated	Marble	Limestone	Light	Calcite	Reacts with dilute HCl
	Quartzite	Sandstone	Light	Quartz, some feldpar	Hard, glassy appearance
Foliated	Slate	Shale	gray, black, red, green	Clay minerals	Dense, splits into thin slabs
	Phyllite	Shale, siltstone, or slate	Gray or gray-green	Chlorite and micas	Shiny, satiny surface
	Schist	Shale, phyllite, slate, volcanic rocks	Variable	Muscovite, biotite, quartz, feldspar,	Well-developed foliation
	Amphibolite	Mafic igneous rock	Dark	Hornblende, feldspar	Banded layers
	Gneiss	Granite, slate, schist	Dark and light bands	Quartz, feldspar, mafic minerals	Coarsely crystalline, banded

Figure 5-21 Metamorphic rock classification.

We know that minerals are stable under very specific temperature conditions; thus, we can predict that metamorphism will change the mineral composition of a rock. However, pressure alters the temperature at which a mineral would be stable. For example, the mineral olivine might be stable in the mantle at temperatures of thousands degrees Celsius; however, the great pressures that exist deep within the Earth are optimal for another iron and magnesium–rich mineral called spinel. Olivine cannot exist at high pressures, even if the correct elements are present. Spinel, whose crystal structure is stable at high pressures, forms instead. This example is cited to remind you that mineral stability is a function of both temperature and pressure, and minerals will recrystallize into other minerals that are stable at the existing conditions. *Recrystallization* is a fundamental process of metamorphism. Recrystallization involves one or both of the following processes: (1) a mineral completely dissolves and new minerals are precipitated, and/or (2) the old minerals don't dissolve and new ones crystallize, but rather the old mineral crystal structure reconfigures into a new crystal structure, hence a new mineral. This process is termed *solid-state* change. Regardless of how recrystallization occurs in metamorphism, it is necessary for us to look through this process to determine the parent, or original, rock.

Metamorphic rocks are important to our Earth system because the high temperature fluids that are generated during metamorphism create many minerals that are economically important. These are termed ore minerals, or ores. Gold, silver, copper, tin and many other ores form in veins: thin seams that represent minerals precipitated out of hot fluids moving through a host rock. These fluids are generated during metamorphism. Chapter 10 will discuss how metamorphic rocks, formed deep within the Earth, are brought to the surface, where humans can mine and retrieve these minerals.

Metamorphic Rock Classification

Figure 5-21 presents a simplified metamorphic rock classification. Note that the first observation you must make is to determine whether the rock is *foliated* or not. If it *is* foliated, the degree of foliation determines the metamorphic rock name. If it is *not* foliated, the metamorphic rock type is determined by composition.

Summary

Minerals, the reservoirs of elements on Earth, are combined in a variety of forms to compose rocks. There are three types of rocks, distinguished by the ways in which they (and the minerals that comprise them) form. Igneous rocks form from the cooling and solidification of magma, or melted rock. Sedimentary rocks form from the physical and chemical breakdown of other rocks to form fragments termed sediment. Metamorphic rocks form from the alteration of pre-existing rocks by heat and pressure within the Earth. The three types of rocks, along with magma and sediment, are reservoirs for Earth materials in a closed system termed the rock cycle. The fluxes that move material from one reservoir to another in the rock cycle are geologic processes such as weathering, burial, and volcanic eruptions.

Experimental studies have shown that minerals crystallize, or solidify, from magma in a predictable sequence that reflects both the elements available in the magma, the rate of cooling and the ease of forming different crystalline structures. One well described sequence of mineral formation from the cooling of melted mantle-derived magma is termed Bowen's Reaction Series, after the geologist who described this phenomenon. The first minerals to form in this sequence are those with the simple isolated silicate tetrahedral structure (olivine) and progressively more complex silicate structures develop sequentially. As different crystals form and these elements are removed from the hot liquid the remaining magma composition evolves in composition in a process termed fractional crystallization.

With the exception of the liquid outer core, the Earth's interior is solid so in order to melt rock and make magma there must be regions where temperature and pressure conditions cross the threshold condition where the melting point is reached. Partial melting describes the sequential melting points of different minerals in a rock; the different melting points of minerals reflects their chemical bonding.

Just as bond strength controls the temperature and pressure conditions where a mineral will melt, so does bond strength affect the stability of minerals at the Earth's surface. The differential chemical breakdown of minerals, through reaction with air and water, as well as physical abrasion, causes different rocks to weather apart at different rates. The breakdown of rocks produces sediment, which is then carried by wind, water and gravity to different environments on the Earth's surface. These sediments are ultimately compressed and cemented together to form sedimentary rocks whose characteristics often record the environment in which they formed. The composition of the grains records the rock(s) that weathered apart to make the sediment.

If rocks at the Earth's surface are buried deeply within the Earth heat and pressure will alter the way in which minerals are packed together, producing a fabric termed "foliation." Heat and pressure also cause recrystallization of minerals. Recrystallization does not involve the melting of material but it restructures crystal lattices into new forms. Because each mineral has a unique crystalline structure, recrystallization forms new minerals within metamorphic rocks.

The rock cycle represents the movement of Earth material from the planet's interior to the surface and back down again. The processes involved in this cycling are visible all around us, for example in the rivers that carry sediment to the sea or the volcanoes that erupt on the Hawaiian Islands.

Chapter 5 Review Questions

- Using boxes and arrows, draw and label the rock cycle
- Cover up the captions for Figure 5-2, Sidebar 5-1, Sidebar 5-5 and Sidebar 5-7 and describe the features you see that would enable you to determine if a rock sample was igneous, sedimentary or metamorphic. If you determine if a rock is igneous, apply the appropriate textural terms. If a rock is metamorphic, determine if it is foliated or not.
- What are at least 3 of the reservoirs in the rock cycle?
- What is at least one flux in the rock cycle?
- What is the relationship between the size of the crystals in an igneous rock and the rate of cooling of the magma?
- What is the melting point of a rock? Explain the role that water plays in
- Define what is meant by the geotherm. Use Figures 5-3 and 5-4 to practice interpreting the temperature at various depths within the Earth and calculate the geothermal gradient (in C/km).
- What is the difference between fractional crystallization and partial melting and how can both explain how diverse magma compositions are created?

- What controls the viscosity of a magma?
- How is the viscosity of a magma related to the type of volcano that would form?
- Cover up the caption to Figure 5-11 and determine what type of volcano is imaged in A,B and C.
- If you recognize a stratovolcano, what are the two primary deposits that have constructed it?
- What physical property of the magma controls the type of material erupted from a volcano?
- Satellite images reveal the presence of volcanoes on other planets such as Mars. What can we infer about these planets as a result of noting that volcanoes are present?
- What is "sediment" and what has to occur to produce a sedimentary rock?
- What is meant by "foliation" and how is it produced? Why are many metamorphic rocks foliated?

Chapter 6

Telling Time

Time is nature's way of keeping everything from happening all at once.

—Albert Einstein

At the end of this chapter you should be able to

- Define the term *half-life*
- Interpret a half-life curve and use it to determine the age of a mineral
- Define Steno's laws and be able to apply them to determining the relative age of geologic events in a cross-section of the crust
- Define and identify (in a photograph or sketch) the different types of unconformities
- Synthesize relative and numerical dating techniques to bracket the ages of a geologic event
- Characterize the ranges of rates of geologic processes, compare these to human activities, and explain how this might impact societal decision making

Introduction

In order to understand how the Earth system works, we need to identify its reservoirs and fluxes, but we also need to be able to determine, along with their magnitude, the rates at which the fluxes operate. *Rate* is a ratio of a change in a variable, such as distance or temperature, to an interval of time. The content of Chapter 13 suggests that although climate change has happened many times in the Earth's history, the rate at which it is currently changing exceeds that previously seen. In order to be able to support such statements, we need to be able to tell time.

This chapter will cover two approaches that geologists use to tell time. The first of these is termed *relative time* and enables us to say that one event happened before or after another. The second approach, *numerical time*, enables us to attach a date to an event. Finally, this chapter will summarize some of the rates for a variety of geologic processes. Many of these are the rates of fluxes by which material is moving through the Earth system.

Relative Geologic Time

Some of the oldest principles in geology involve determining which events recorded in the rock record are older and younger than others. The "father of stratigraphy" was Nicolas Steno, a seventeenth-century naturalist employed as a tutor to the wealthy Medici family (*stratigraphy* is the study of rock layers). While walking through the Italian countryside, Steno made observations of the sedimentary and volcanic rocks around him. He formulated his observations into a series of principles:

- *Law of superposition:* in a sequence of rock layers, the oldest rock layer sits at the bottom of the sequence.
- *Law of original horizontality:* sedimentary rock layers are deposited in flat-lying layers.
- *Law of cross-cutting:* if a rock layer is cut by another rock, or a fault, the rock has to be older than the feature that cuts it.
- *Law of lateral continuity:* sedimentary rocks will extend laterally until they grade into the adjacent environment.

These principles are illustrated in Figure 6-1, a series of three-dimensional block diagrams.

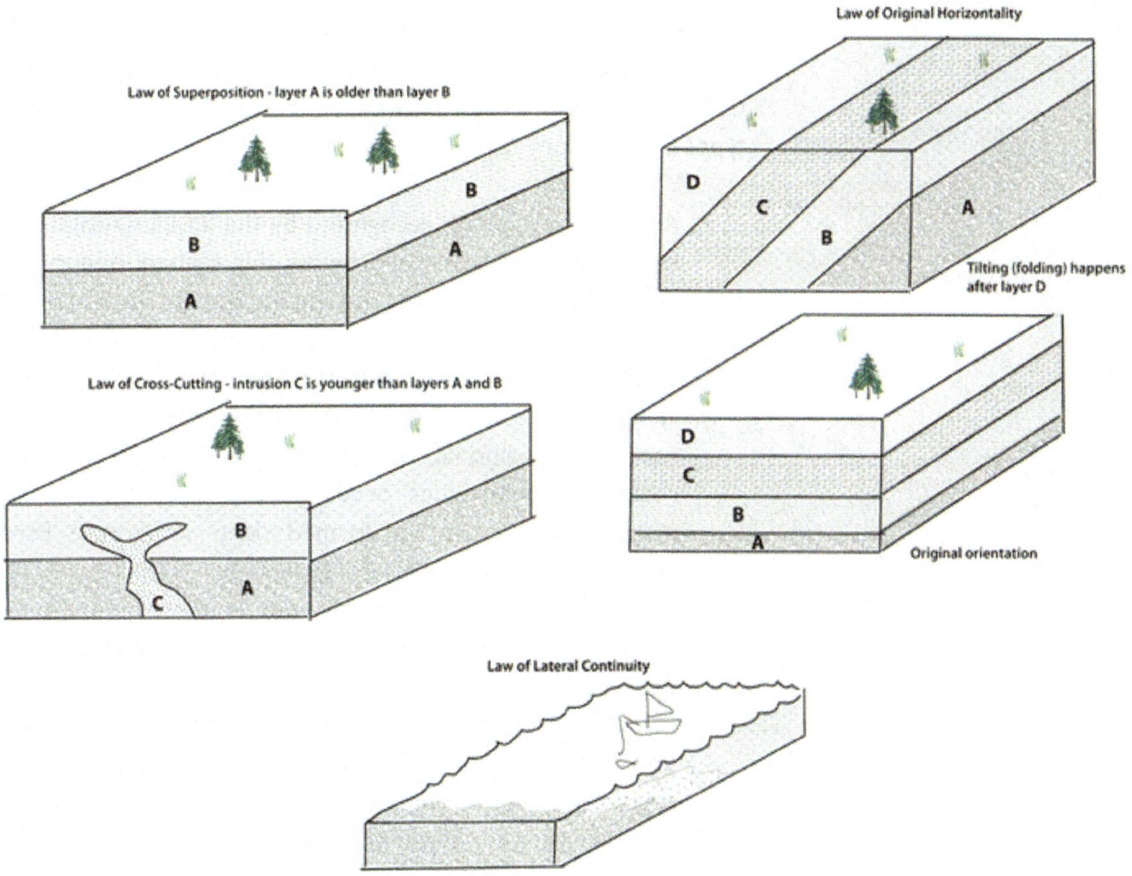

Law of Superposition - layer A is older than layer B
B
A
B
A
Law of Original Horizontality
D
C
B
A
Tilting (folding) happens after layer D
D
C
B
A
Original orientation
Law of Cross-Cutting - intrusion C is younger than layers A and B
B
A
C
Law of Lateral Continuity

Figure 6-1 Steno's laws.

We use Steno's laws to unravel complex sequences of geologic events such as those illustrated in Figure 6-2, a "cutaway view" of the near surface. You might see layering like this in a roadcut or a cliff face.

E
J
A
F
B
C
I
H
G
D

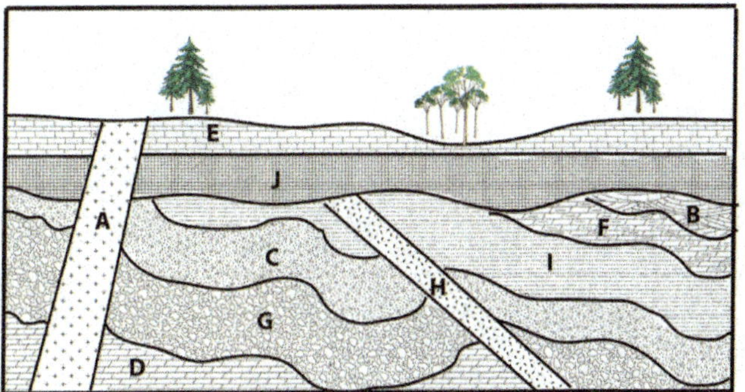

Figure 6-2 The relative timing of events shown in this figure, from oldest (first) to youngest (last), is as follows: deposition of layers D, G, C, I, F, and B; bending of all these; intrusion of H, erosion; deposition of layers J and E; and, finally, intrusion of rock A. The present land surface has eroded rock A, so A is older than the present day land surface.

Unconformities

Two events in the sequence shown in Figure 6-2 represent periods of erosion. The first occurred after the folding of layers D through B, and the second is the present-day surface. In the relative timing of geologic events, we need to note these times. *Unconformities* are periods of time when no rock was deposited, or it was deposited and removed, or eroded. We recognize three types of unconformities:

- *Angular unconformity:* recognized by the angular relationship between rocks above and below this surface. Angular unconformities record the time when the rocks were tilted, or folded. The unconformity between layers D through B and J through E is an angular unconformity.
- *Nonconformity:* recognized by the occurrence of sedimentary rocks in depositional contact with igneous or metamorphic rock. Nonconformities occur because both igneous and metamorphic rocks are formed deep within the Earth, far from the surface where sedimentary rocks form. The "missing time" represents the uplift of these rocks to the surface, where they could be covered by sediment. Sediment deposited on top of layer A would produce a nonconformity
- *Disconformity:* the most difficult unconformity to recognize, unless you have independent age information, a disconformity represents a gap in time between two layers of sedimentary rock. In Figure 6-2, if you had age information that rock layer J was twenty million years old and rock layer E was one million years old, you would infer that there was a disconformity between the two layers.

Figure 6-3 The nonconformity between the rocks of the inner Grand Canyon (Vishnu Schist), above the Colorado River, and the rims rocks of the plateau. Tapeats Sandstone is in the center of the image. The arrows point to the nonconformity. Photo by Paul Strother.

Geologic Time Scale

Since Nicolas Steno established the principles of relative timing of rock layers in the seventeenth century, geologists have been assembling a chronology of the Earth's history. Initial subdivisions of rocks identified only four groupings, from oldest to youngest, the primordial (or primary), secondary, tertiary, and quaternary; however, through the eighteenth and nineteenth centuries, much more detail on the subdivisions of geologic time was added. Figure 6-4 shows the present-day geologic time scale. The names for the various subdivisions owe their origin to locations where geologists recognized distinct assemblages of rocks and fossils. For example, the Devonian Period is named after Devon, England, where rocks of this age (younger than Silurian rocks and older than rocks of the Mississippian Period) are well exposed. By the late nineteenth century, advances in radiometric dating meant that numerical ages could be added to the geologic time scale. As a result, it is possible to describe Quaternary rocks as those that are younger than 1.6 million years before the present day. These radiometric ages are constantly being refined with improvements in instrumentation, as will be discussed below.

Numerical Dating

Numerical dating refers to the ability to determine the date or age of formation. The primary technique used to do this involves the decay of unstable isotopes, so numerical dating is achieved through radiometric decay in a type of element. In Chapter 3 we learned that *isotopes* are elements with variable atomic mass but the same atomic number. For example, the element uranium (U) has 92 protons in its nucleus (its atomic number) and 146 neutrons, for an atomic weight of 238 ($^{238}U_{92}$). Another isotope of uranium also has 92 protons and 143 neutrons, for an atomic weight of 235 ($^{235}U_{92}$). These isotopes of uranium are *unstable isotopes*, which means that over time they transform, with the release of energy, into other elements as a function of time. Uranium 238 (^{238}U) decays to lead 206 (^{206}Pb), potassium 40 (^{40}K) decays to argon 40 (^{40}Ar), and rubidium 87 (^{87}Rb) decays to strontium 87 (^{87}Sr). The starting isotope in a decay sequence is termed the "parent" and the product the "daughter." Energy in the

EON	ERA	PERIOD	MILLIONS OF YEARS AGO
Phanerozoic	Cenozoic	Quaternary	1.6
		Tertiary	66
	Mesozoic	Cretaceous	138
		Jurassic	205
		Triassic	240
	Paleozoic	Permian	290
		Pennsylvanian	330
		Mississippian	360
		Devonian	410
		Silurian	435
		Ordovician	500
		Cambrian	570
Proterozoic	Late Proterozoic Middle Proterozoic Early Proterozoic		2500
Archean	Late Archean Middle Archean Early Archean		3800?
Pre-Archean			

Geologic time scale showing both relative and numeric ages.

Ages in millions of years are approximate

Figure 6-4 Geologic time scale.

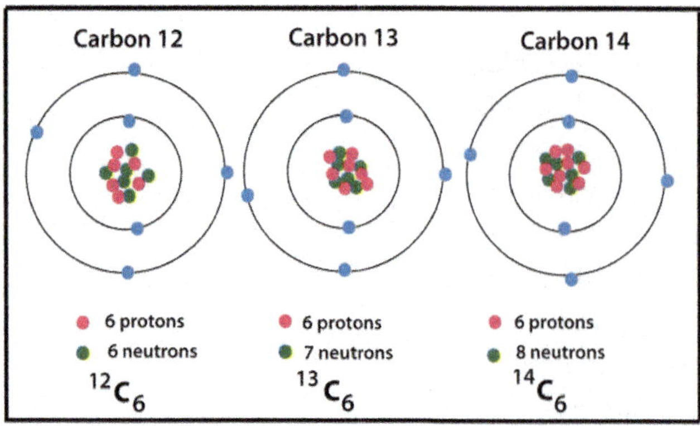

Figure 6-5 The differences between carbon isotopes.

form of particles and forms of energy, such as gamma rays, which we call "radioactivity," are released during these transitions. Carbon is an interesting element in that it has both stable ($^{12}C_6$ and $^{13}C_6$) and unstable isotopes (Figure 6-4); the unstable isotope $^{14}C_6$ decays to $^{14}N_7$. There are three isotopes of carbon: $^{12}C_6$, $^{13}C_6$, and $^{14}C_6$.

The disintegration of parent to daughter isotopes is independent of the physical environment (temperature or pressure) in which it occurs; the decay is proportional to the amount of atoms you have. The more parent atoms you have at the start, the more daughter atoms you have at the end (see Sidebar 6-1).

Using Radioactive Decay to Tell Time

How long do unstable isotopes take to decay, and how do geologists use this to date the age of a mineral? At the time of crystallization, an isotope is locked into crystal lattices, which represents the starting point for these "atomic clocks." Geologists apply the concept of a "half-life." The *half-life* of an isotope is the amount of time it takes for one-half of the parent isotope to decay to the daughter isotope, so we need to understand this concept before we apply it to dating rocks.

The half-life of each isotope is different. The half-lives of several geologically important isotopes is shown in Table 6-1. The interval of time is fixed and does not change over time. If we know an isotope's half-life and can count the number of parent and daughter atoms, we can determine the length of time, because the atoms were locked into the crystal; in other words, we can determine how old a mineral is.

Let's use an analogy of melting ice cubes to explain this. You place several ice cubes in a funnel on top of a graduated cylinder. The ice cubes will begin to melt. Someone comes into the room and notes the amount of water in the graduated cylinder and the time. That person walks away and comes back several minutes later. He or

Sidebar 6-1 Relationship between Half-Life and Decay Constant

An examination of the graph in Figure 6-6 shows a nonlinear relationship between the two variables of "number of atoms" and "time." This relationship is exponential.

Unlike melting ice, the decay of unstable isotopes is exponential, as shown in Figures 6-5 and 6-6. Growth and decay relationships are exponential in nature; the change in the quantity of parents or daughters over time is a function of the rate of change. This rate is termed the *decay constant*. If this constant is positive, you have a growth situation. For example, if you have two rabbits, the rate at which you produce offspring (an increase in the number of bunnies) will be less than if you have four rabbits. If the decay constant is negative, you have a decay constant; if you start with four rabbits and two are eaten by foxes, you will produce fewer rabbits, with two rabbits rather than four.

The mathematical relationship that describes exponential decay is

$$dN/dt = -\lambda N,$$

which translates to the change in the number of starting atoms (N) divided by the change in time (t), which equals the decay constant (λ) times the number of starting atoms. Because we are talking about decay, the change is negative. As with any equation that is a ratio of change in two variables, we can integrate it through a series of steps to get

$$N = N_0 e^{-\lambda t}$$

Which says that the number of parent atoms (N) present at any time t is a function of the number of original parent atoms (N_0) that were present when $t = 0$ times the decay constant λ.

If you know an isotope's half-life, you can calculate the decay constant as follows:

$$N = N_0 e^{-\lambda t}$$
$$½N_0 = N_0 e^{\lambda t}$$
$$½ = e^{\lambda t}$$
$$\ln ½ = \ln (e^{\lambda t})$$
$$\ln 2 = \lambda t$$
$$\mathbf{-\ln 2/\lambda = t}$$

For example, if ^{14}C has a half-life of 5,730 years,

$$T½ = 5{,}730 \text{ yr}$$
$$\lambda = -\ln 2 / T^{1/2} = -\ln 2 / 5730$$
$$\lambda = 0.000121 \text{ atoms per year} = \text{the decay constant}$$

Table 6-1 Half-lives of Isotopes Commonly Used in Geochronology

Parent Isotope	Daughter Isotope	Half-Life
Uranium 238 (^{238}U)	Lead 206 (^{206}Pb)	4.4×10^9 years
Uranium 235 (^{235}U)	Lead 207 (^{207}Pb)	7.04×10^8 years
Potassium 40 (^{40}K)	Argon 40 (^{40}Ar)	1.25×10^{10} years
Thorium 232 (^{232}Th)	Lead 208 (^{208}Pb)	1.401×10^{10} years
Rubidium 87 (^{87}Rb)	Strontium 87 (^{87}Sr)	4.88×10^{10} years
Beryllium 10 (^{10}Be)	Boron 10 (^{10}B)	1.5×10^6 years
Carbon 14 (^{14}C)	Nitrogen 14 (^{14}N)	5.73×10^3 years

Figure 6-6 Changes in the ratio of parent to daughter atoms. At the time of crystal formation, 100% of all atoms are present; in this case, twenty-four atoms are all parents. After one half-life, twelve atoms are parents and twelve atoms are daughters (this is the definition of *half-life*). After two half-lives, one-half of the remaining parent atoms have decayed to daughters, so there are six parent atoms left and eighteen daughters. After three half-lives, the six parent atoms decay to produce three parents and twenty-one daughter atoms. On the graph, the line plotting the number of parent atoms crosses the line plotting the number of daughter atoms at one half-life. Below the graph, the changing ratios of parent atoms (white) and daughter atoms (red) are shown.

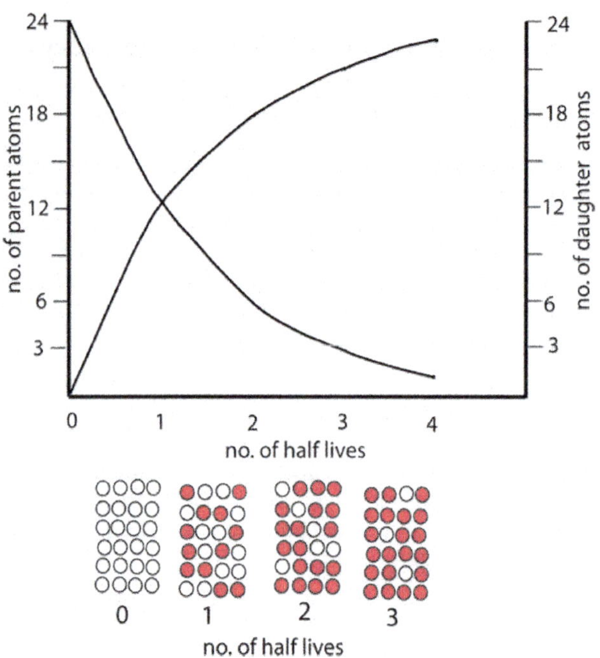

Figure 6-7 A generalized half-life curve. If you were able to determine that 25% of the atoms present in a mineral were daughter ^{206}Pb atoms and 75% were parent ^{238}U atoms, knowing that the half-life of uranium ^{238}U is 4.4×10^9 years, you could determine that your sample was 2.2×10^9 years old (one-half of one half-life).

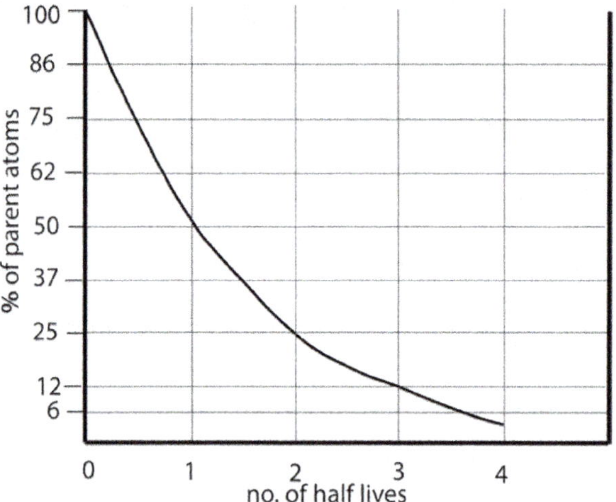

she again notes the amount of water in the cylinder and the time. The person leaves and returns a third time. You can make a graph of the amount of water in the cylinder on one axis and the time on the clock on the other. Draw a line through the data points. If you extend the line back to the origin of the graph, you would be able to interpret when the melting of the ice cubes began. Note that you did not need to know the number of ice cubes that you started with. All you know is that 100% of all the water started out as ice. What you measured was the amount of water produced as an interval of time. From these two facts you could extrapolate when melting began.

Applying this analogy to radiometric dating, we do not need to know the number of atoms present in our mineral to start with; we know that 100% of the atoms were parent atoms. If we measure the rate at which daughter atoms are produced (the decay rate; see Sidebar 6-1) and the number of daughter atoms we have right now, we can calculate how long the decay has been occurring.

Unlike melting ice, the decay of unstable isotopes is *exponential*, as shown in Figures 6-6 and 6-7. Given that decay relationships are exponential in nature, the change in the quantity of parent or daughter atoms over time is a function of the rate of change. This rate is termed the *decay constant*. Sidebar 6-1 briefly summarizes the mathematical relationship between a radiometric isotope's half-life and its rate of decay.

If we can (1) measure the numbers of current parent and daughter atoms of an unstable isotope in a mineral at this time in Earth history, (2) measure the decay rate, and (3) know that half-lives are constant, we can work backward to determine the length of time that decay has been occurring. If you examine the half-lives of the common isotopes used in geochronology, it's obvious that because of the long lengths of time involved, we don't measure half-lives in the lab. What we *can* measure, however, is their decay rate. Essentially, that's what a Geiger counter does: "clicks" represent escaping particles of energy with each parent-to-daughter transition. The instrument we use to count atoms is termed a *mass-spectrometer*. This instrument is capable of distinguishing atoms of different elements based on pathways through a magnetic field. Based on their different masses, they behave differently. A collector plate can count the number of heavy (parent) and light (daughter) atoms.

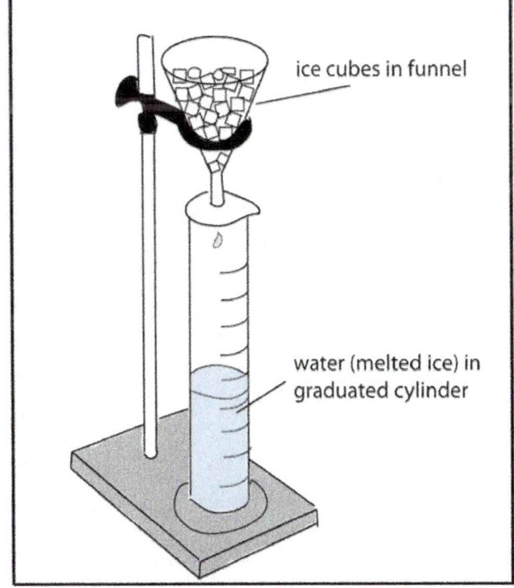

Figure 6-8 Melting ice analogy for radiometric dating. Ice cubes represent parent isotope and melted ice (water) represents daughter isotopes.

Assumptions in the Radiometric Dating Process

Because they are based on laws of atomic physics, some parts of the radiometric dating process are immutable. These include the immunity of the decay process to physical variables such as temperature

Sidebar 6-2 Sample Problem: Determining the Age of a Mineral

We collect a rock sample of granite and discover that it contains minute amounts of zircon. As is often the case, uranium is present in trace amounts inside the zircon minerals. We break up the rock and isolate the zircons. Placing our sample in a mass spectrometer, we determine the parent to daughter ratio and find it to be 75:25. In other words, if we had 100 atoms, 75 of them (75%) would be parent and 25 of them (25%) would be daughter. How old are the zircons and the rock that contains them?

We could estimate the age from the half-life curve of Figure 6-6: 75% parent atoms is one-half of one half-life. If the half-life of ^{238}U is 4.45×10^9 years, this means that the zircons should be half of this length of time (2.225×10^9 years old). Interpreting off a graph is OK but subject to error. Let's see what we can determine mathematically:

Step 1: Determining decay rate:

$$K = -\ln(2) / t_{1/2}$$

$$K = -\ln(2) / 4.47 \times 10^9 = -1.55 \times 10^{10} \text{ years}$$

Step 2: Determining age:

$$K = -\ln(2) / 4.45 \times 10^9 = -1.55 \times 10^{10}$$

$$N_o = 0.75 = N_o e^{-\lambda t}$$

$$0.75 = e^{-(-1.55 \times 1010t)}$$

$$\ln 0.75 = 1.55 \times 10^{10} t$$

$$t = 1.86 \times 10^9 = 1.86 \text{ billion years old}$$

and pressure and the constancy of the half-life of an isotope. There are some assumptions that we make that we need to acknowledge. First of these is that the decay system is closed: the sum of parent and daughter atoms is 100%, and all atoms of the parent are locked into a crystal lattice at the time of mineral crystallization. This is called the *closure temperature*. If a rock is not heated above this temperature, we make the assumption that parent + daughter = 100% of all the atoms. The system would not be closed if parent atoms were added or daughter atoms lost. Under what conditions could this happen? If temperature of crystallization is part of our discussion, you should recall that this means we are staying at temperatures below the melting point of a mineral. If a rock is heated above this temperature, atoms will break the bonds holding the crystal lattice together, and elements can enter and leave the solid-liquid solution.

As this discussion implies, igneous rocks with a simple cooling history are the most reliable to date. Not all igneous rocks can be dated; enough minerals containing unstable isotopes must be present. You might also predict that metamorphic rocks might be difficult to date, as elevated temperatures are often part of

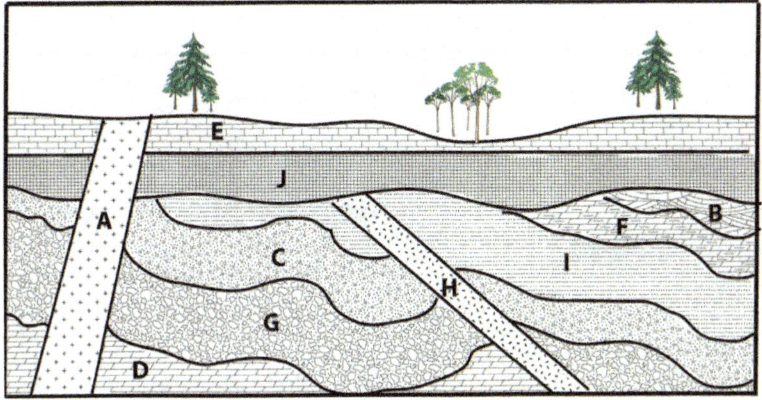

Figure 6-9 Synthesize relative and numerical dating techniques to establish a sequence of geologic events. Biotite minerals collected from igneous rock H contain potassium 40 and argon 40 atoms in the following proportions: 45% daughter isotopes and 55% parent. Minerals collected from igneous rock A contain biotite minerals of 80% parent and 20% daughter abundance. Given the half-life of potassium 40 of 1.25×10^9 years, use the half-life curve in Figure 6-7 to bracket the age when folding of the rocks could have occurred.

metamorphism. Sedimentary rocks are slightly different. Because they are composed of lithified sediment grains, what you are dating when you recover a zircon grain in a sandstone, for example, would be the age of the zircon in the rock that existed before it broke down to produce the sediment. In Sidebar 6-2 we had an example of a zircon grain in granite that was 1.86 billion years old. If this granite weathers apart to produce grains of sand, these grains will include zircon grains. If we date the zircon grains in a sandstone, we know this sedimentary rock has to be younger than the zircon grains contained in it. This is still a useful age to know, but it only gives us a maximum age for the sedimentary rock.

Synthesizing Relative and Numerical Ages

As the example above suggests, not all rocks are suitable for radiometric dating. However, we can obtain much useful information if a sequence of rocks contains even one datable rock. Figure 6-9 shows how we can use two radiometric ages to constrain a sequence of geologic events.

Using Steno's laws of superposition, original horizontality, and cross-cutting, the following sequence of events can be determined

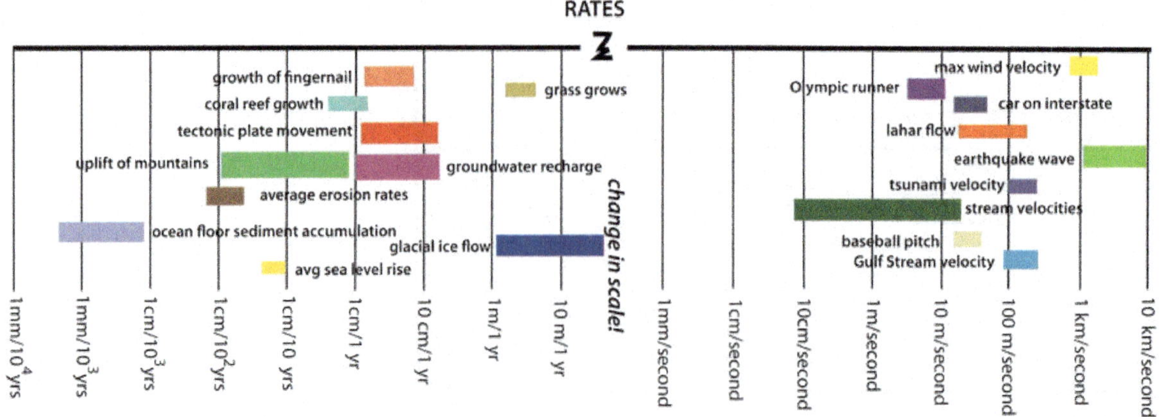

Figure 6-10 Rates of processes. Note the change in scale in the middle of the graph.

(from oldest to youngest): deposition of layers D, G, C, I, F, and B; folding of these layers; intrusion of igneous rock H; folding of all these; erosion and production of an angular unconformity; deposition of layers J and E; intrusion of igneous rock A; and, finally, erosion of rock layers A and E to produce an unconformity. Igneous rock H is approximately 1.0×10^9 years old, and igneous rock A is approximately 6.0×10^8 years old. This means that the tilting of rock layers D through B occurred between six hundred million and one billion years ago.

Rates of Geologic Processes

Now that we have established how geologists can tell time, we can use this information to determine the rates at which various geologic processes occur. One of the challenges to being a crew member on *Spaceship Earth* is dealing with the wide range of rates at which various fluxes in the Earth system occur. Some phenomena are very slow (rates of plate movement, rates of the uplift of mountains, erosion) while others are spectacularly fast (earthquake wave movement, volcanic eruptions). Some of these are so slow that they are easy for humans to ignore. Others are so fast they are hard to respond to.

Another aspect of our rate discussion is that of the frequency with which these processes occur. Some, like those in the left half of Figure 6-10, are inexorable, occurring every minute of every day. Thus, they can be easy to ignore. Others, in the right half of

Figure 6-10, occur once in a human lifespan, or perhaps only once in several human generations. When events happen so infrequently or occur at such slow, constant rates, it is difficult to remember that they *are* happening or *are* about to happen and prepare appropriately.

Summary

In order to be able to determine the rate or frequency of geologic processes geologists need to be able to tell time. There are two approaches to being able to do this. The first, determining the relative age of events, distinguishes which events happened before or after other events. The procedure for determining relative timing is based on Steno's Laws: superposition, original horizontality, cross-cutting, and lateral continuity. The geologic time scale is based on the application of Steno's Laws. The second approach "telling time" is numerical dating, which uses the decay of unstable isotopes present in certain minerals, to determine an age of formation of the mineral. Recall that an isotope is an element with variable atomic mass. Unstable isotopes emit matter and energy over time as the elements seek a stable state. The subatomic particles and energy emitted are known as radioactive decay. The half life of an isotope is the amount of time it takes for one-half of the starting, or parent, isotope to decay to the product, or daughter, isotope. If we can measure the number of parent and daughter isotopes present now, and we can measure the rate of decay of parent to daughter, we can work backwards to determine how long the decay process has been occurring. This is the age of crystallization of the mineral. Unstable isotopes with half lives ranging between seconds and billions of years means that a many different geologic processes can be dated. With age information, geologists are able to determine the rates at which many different events occur, for example, the rate at which tectonic plates move (avg. 2–5cm/yr). Interestingly, geologic processes occur at slow rates (mm to cm per year) but do so over many millions of years or they occur almost instantaneously (km/sec) in extremely short snippets of time. An earthquake is an example of this.

Chapter 6 Review Questions

- What are the four Steno's Laws?
- What is an unconformity?
- Examine a photograph of the Grand Canyon. Apply Steno's Laws to demonstrate that the Colorado River is younger than the rocks of the walls of the canyon.
- What is the difference between relative and numerical geologic time?
- What is an "unstable isotope"?
- What is the definition of the "half life" of an isotope?
- Apply Figure 6-7 to determine the age of a mineral where 75% of an isotope is "parent" and 25% is "daughter and the half life of the isotope is 4.4 × 109 years.
- Solve the question posed in Figure 6-9

Solve the following rate problem: we know that South America and Africa were joined together 200 million years ago, however they are now 3,000 km apart. What is the rate of spreading of the South Atlantic Ocean (in cm/yr)?

Chapter 7

Earthquakes, Seismology, and the Layered Earth

At the end of this chapter you should be able to

- Describe why seismic waves are like and unlike water waves
- Draw the layered Earth to scale
- Describe the properties of the layers of the Earth
- Describe why earthquakes happen
- Describe the characteristics of body and surface waves
- Describe how seismic waves are used to determine the Earth's internal structure
- Correctly position the P and S wave shadow zones on a cross-section of the Earth.

- Describe how the elastic rebound theory is related to earthquakes
- Explain how seismograms would appear if the Earth were not layered
- Explain how the epicenter of an earthquake can be determined
- Explain how we describe the size of an earthquake
- Explain the origin of the layered Earth
- Discuss what can we say about the interior of the moon based on our observations of "moonquakes"?

Introduction

As discussed in Chapter 5, we have directly explored only the uppermost 4 kilometers of the Earth's interior. How do we learn about the remaining 6,000-plus kilometers? One important technique for doing this involves using the passage of waves through the Earth, in other words, the sutyd of earthquake (seismic) waves.

When rocks break, energy is released in the form of a seismic wave. The entire Earth vibrates in the same way in which a bell vibrates when it is hit. The movement of the Earth in response to the passage of seismic waves is termed an *earthquake*. Where the earthquake originates within the Earth, the origin of the seismic waves is termed the *focus* (or *hypocenter*). The *epicenter* is the area on the Earth's surface directly above the focus (Figure 7-1).

What Causes an Earthquake?

When rocks are stressed (pushed or pulled), they store the stress, producing potential energy. When the rocks fail, or break, the potential is converted to kinetic energy: the rocks move, and energy is released. This energy is transmitted as a series of different wave forms. You might ask, "Which processes stress rocks?" We will discuss in detail the large-scale geologic processes involved in plate tectonics later in the text (Chapter 9), but other earthquake-generating stressors might be familiar to you. Some regions of our globe experienced a major glaciation geologically recently—millions of years ago. By one million years before the present day, New England, for example, was buried under more than 1 kilometer of ice. The weight of this ice pushed down on the underlying crust, causing it to sink (imagine stepping on a trampoline and noticing how the surface under your foot bows down). When global climates warmed and the ice retreated northward, the removal of that weight of ice cause the crust to "rebound," but unlike a trampoline, it did not do so immediately or geographically uniformly. The vertical up-and-down movement of the crust occurred at different times along "planes of weakness" in the rock, such as major changes in rock type. Nearly ten thousand years after the ice has gone, New England and southern Canada still experience earthquakes from slips on the rocks "rebounding" to their elevation before the weight of ice.

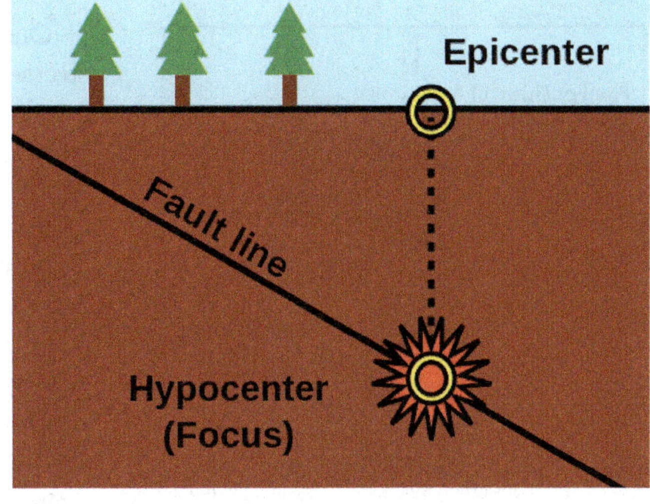

Figure 7-1 The region within the Earth where an earthquake originates is termed the focus (or the hypocenter). The point on the Earth's surface directly above the focus is termed the epicenter of the earthquake.

Our second example of earthquake-producing stressors is more controversial. You might have heard about clusters of small earthquakes associated with the technique of fossil fuel extraction termed "hydrofracking." This process involves drilling into rocks and injecting fluids under pressure, which causes the adjacent rock to crack, or fracture. The fluids contain a chemical concoction that can be pumped out, along with the fossil fuel material. There are reports of small earthquakes resulting in areas where "fracking" has been widespread. Why might this be? There is good evidence that the new cracks created by fracking, along with remnant fluids left behind, have weakened the rock. If the geologic history of the region in question might have generated stresses (such as from ice loading), this weakening of the rock could result in its "failure" and the generation of a small earthquake. In general, the vertical movement of hot fluids, such as magma, will trigger earthquakes in overlying rocks.

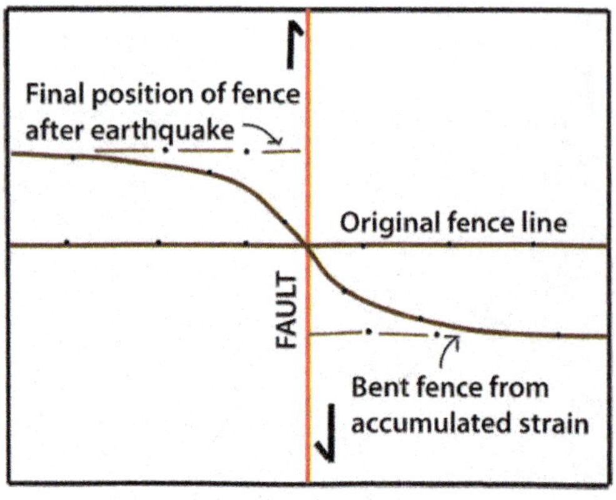

Figure 7-2 A deformed fence line illustrates the elastic rebound theory.

Our final example of earthquake-producing stresses is related to the motions of the tectonic plates that comprise the Earth's surface. As these large slabs of rock move in response to heat flow within the Earth's interior, stresses build up. Much of this stress is localized at the margins or plate boundaries; however, rocks can transmit stresses horizontally, so some earthquakes may occur away from plate boundaries. Rocks are also capable of storing stress (potential energy) for millions of years before releasing it in the form of an earthquake. Because humans may not be aware of the geologic history of the region in which they live, these earthquakes may take people by surprise.

Figure 7-3 A broken fence line in California illustrates the slow displacement across the San Andreas fault following the 1906 earthquake. Before the fault moved, the fence was bending as in Figure 7-2.

Why Do Earthquakes Occur?

If an earthquake represents the stored (potential) energy released when a rock breaks, we need to understand the conditions under which this happens. A theory describing how rocks respond to being stressed is termed the *elastic rebound theory* (Figures 7-2 and 7-3). The analogy for this process would be a rubber band that you pull. For a period of time the rubber band accommodates being pulled apart, but at some point it snaps in two. In the laboratory we can place cylinders of rock in a "vise grip," apply pressure, and see how the rock responds. The elastic rebound theory was developed by geologist Henry Reid, who documented the slow, progressive warping of the Earth's surface prior to an earthquake, which then physically moved the rocks on either side of the broken (faulted) rock (a fault is a fracture, or crack, in a rock where there has been movement; see Sidebar 7-1). The larger the movement of the fault, the greater the energy that is released.

Stress is a force applied per unit area. *Compressional stress* squeezes a rock (like our rock in the vise grip). *Extensional stress* pulls the rock apart (like our rubber band analogy). *Strain* describes how the rock responds—the change in shape of the rock—to being stressed. How a rock responds to stress depends on

- *Temperature:* at low temperatures, a stressed rock will break, while at high temperatures, it will bend.
- *Confining pressure:* at high confining pressures, the bonds within minerals will stay intact, resisting fracturing. At low confining pressures, the bonds will break more easily. and the rock will fracture.
- *Composition:* some minerals are more prone to bending because of the shape and bonds in their crystals. Calcite, for example, bends when stressed. Other minerals, such as quartz and feldspar, will break when stressed. The presence of volatiles in a rock will also impact its response to stress, making a rock more prone to bending.
- *Rate:* if a rock is stressed slowly over a long period of time, it will tend to bend compared to a rock that is instantaneously stressed, which will fracture.

Sidebar 7-1 Earthquakes Occur along Faults

A *fault* is a fracture in rocks where there has been movement. There are three types of faults:

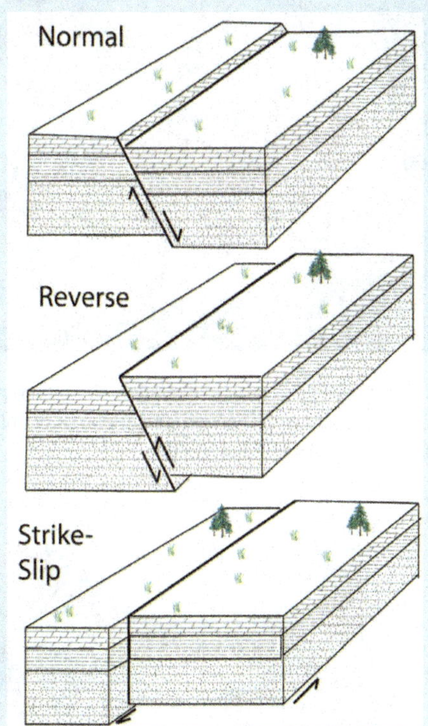

Geologists are pursuing research to try to predict when stressed rocks might fail and trigger an earthquake. Satellite technologies developed at NASA are promising. Interferometric synthetic aperture radar (InSAR) is able to detect minute amounts (2–3 millimeters) of land displacement, which are thought to occur immediately prior to the rupture of rocks. Shown in Figure 7-4 are two interferograms showing land surface elevation prior to (A) and after (B) a 1999 earthquake in California.

The "after" image in Figure 7-4B shows the impact of the earthquake on the surrounding land surface. The new satellite technology is part of a proposed network of satellites, the Global Earth Satellite System (GESS), that would employ InSAR technology to monitor faults around the world.

How Do Geologists Study Waves in the Interior of the Earth?

Because geologists can't travel to the interior of the Earth for data collection, our understanding of how waves move through

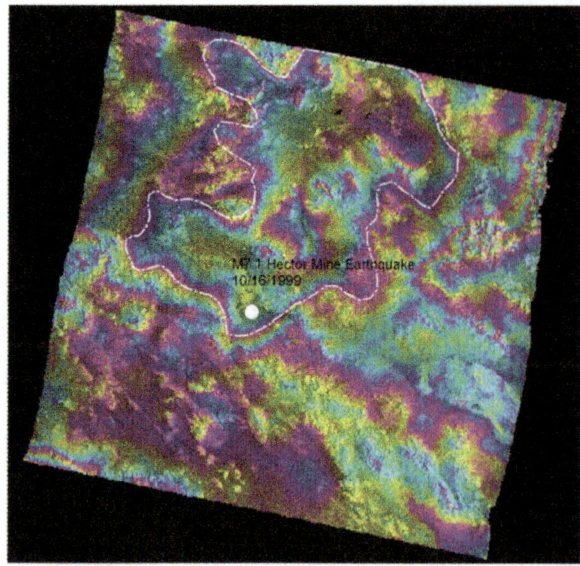

(A) InSAR image of ground elevation one month prior to the earthquake. The faint white line denotes the area of uplift. The epicenter of the earthquake is the white dot.

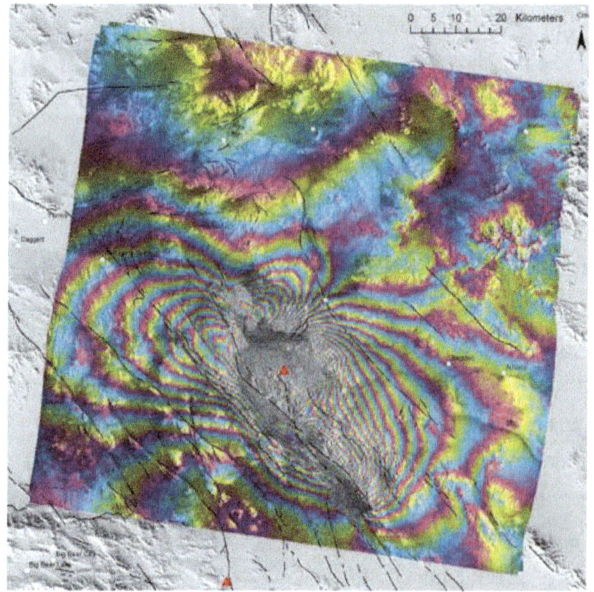

(B) InSAR image of ground elevation after the earthquake; the red triangle is the epicenter.

Fig. 7-4A: USGS / Copyright in the Public Domain.
Fig. 7-4B: USGS / Copyright in the Public Domain.

Figure 7-4

the interior of the Earth is obtained indirectly. Politicians and the military unknowingly designed scientific experiments for geologists when they authorized atomic testing in the 1950s. Explosions of atomic weapons generated "shockwaves" through the Earth in exactly the same fashion as the waves released by snapping rock in an earthquake. Geologists had the benefit, however, of knowing the time of the explosion and exactly where it happened. They also had a general idea of the types of rocks around the explosion site. By analyzing when the shockwaves arrived at instruments, termed *seismometers*, around the world, they were able to work backward to solve the problem: if the explosion happened at *xx* time in *yy* location, and it took *tt* seconds to read our seismometer zz kilometers away, it must have traveled at *pp* kilometers/second. If you are thinking carefully, you'll realize that there are many assumptions here, foremost among them the possibility that the nature of the material through which the waves traveled varied. Could this influence their speed? Here is when our second data set is critical: we are able to construct experiments in the lab to see how wave velocities vary as a function of changing both rock type and pressure. As we saw in Chapter 5, geologists have a starting point with which to model the possible rock types in the Earth's interior, so the list of options to use in the wave velocity experiments is constrained.

Seismic Waves: Shockwaves through the Earth

You are familiar with waves in water. Possibly you've make waves by throwing a pebble in a pond, or you may have seen waves on the surface of the ocean. Waves are the result of displacement of molecules in the material as the energy is passing through. How are seismic waves like ripples in water?

Fill a glass with water and place it in the sink, under the faucet. Perhaps the faucet leaks, resulting in the formation of a water droplet. The water droplet, suspended above the glass, is storing potential energy equal to the mass of the water droplet times the acceleration caused by gravity times the height of the droplet above the glass. As the droplet grows, so does the amount of energy it is storing. As some point the mass of the droplet will exceed the cohesion of water molecule's ability to maintain the droplet, and it will fall into the glass. Ripples will radiate out away from the droplet; these ripples represent potential energy converted to the kinetic energy of moving water molecules in the glass (Figure 7-5).

Figure 7-5 Water ripples formed from a pebble striking the water surface.

While you see the ripples on the surface of the water, there is also disturbance of the water below the surface; ripples are three-dimensional features.

In rocks being stressed, potential energy originates as a result of the rock's incompressibility. The rocks are being squeezed, but the molecules can't rearrange themselves in the solid state; the potential energy is stored in the crystal lattices of minerals. Eventually the stress accumulates to the point that the rock fractures, at which time the stored energy is released in the form of kinetic energy. We feel this kinetic energy in the form of earthquake waves. Like our water ripples, the seismic energy travels out in all directions in a series of wave fronts.

The accumulation of potential energy in a dripping faucet occurs over a matter of seconds to perhaps minutes, compared to years to millions of years in rocks. Water ripples also originate from gravity acting over a vertical distance, such that the water droplet falls, creating ripple falls from above. Seismic waves mostly originate from horizontal stresses within the Earth, not from above its surface. The deepest earthquake waves originate from 300 to 700 kilometers in depth within the Earth. The biggest difference between water ripples and seismic waves, however, is the nature of the material that they pass through. Water is a homogenous surface, so the wave ripples travel in uniform concentric paths outward

Sidebar 7-2 Refraction of Waves

All images from http://www.iris.edu/hq/programs/education_and_outreach/animations#BB.

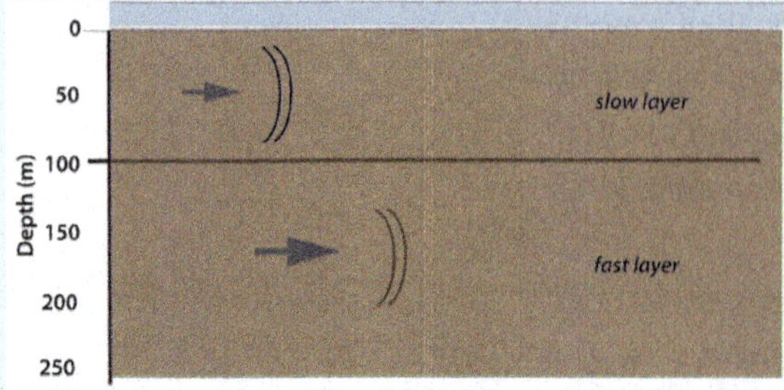

(A) Waves in parallel layers with different densities will move at different velocities.

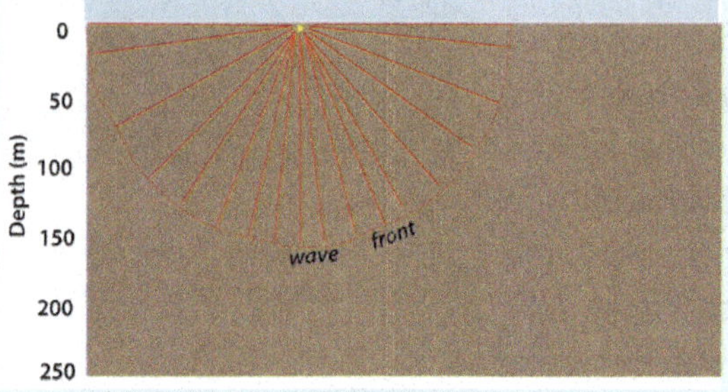

(B) Earthquake waves moving outward from the focus of the interior of the Earth were uniform.

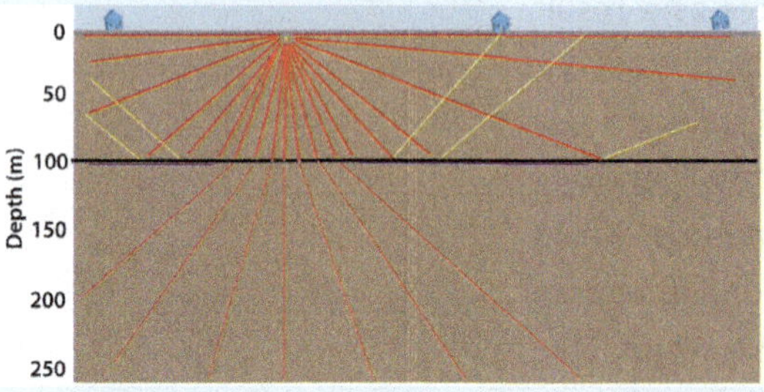

(C) The change in the pathway of a wave when it encounters a more dense layer; the wave path is bent (refracted) or reflected.

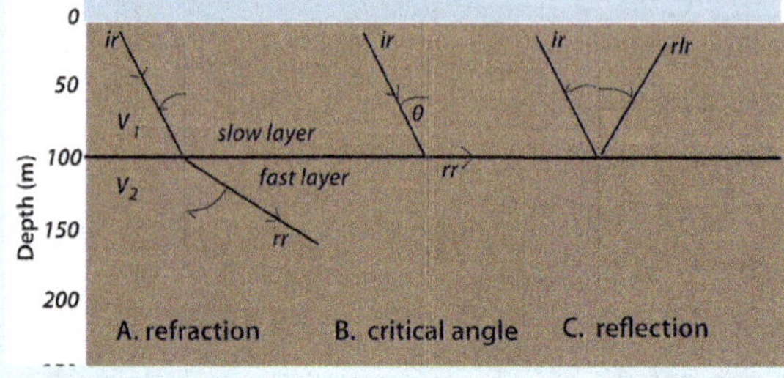

(D) Simplified illustration of wave fronts encountering a more dense layer at depth and the variety of reflected and refracted waves that can result.

from the impact point. The interior of the Earth is not homogenous; thus, the paths only approximate concentric pathways. In fact, it is the deviation from the concentric pathways that we use to model depths within the Earth where the nature of the material changes.

Because of increasing pressure (from the weight of the overlying material) from the surface inward, the density of the rocks increases. As a result of higher densities, the crystal lattices of minerals are composed of shorter bonds that are more efficient at transmitting kinetic energy. Thus, seismic waves speed up with depth (Sidebar 7-2A). As waves traveling down through the Earth encounter layers of increasing density, they speed up. However, they also change direction as they encounter layer boundaries, and refraction, or bending, results (Sidebar 7-2B and C). Other layers encounted at depth reflect changes in composition, and some layers are less efficient at transmitting kinetic energy; these layers slow down seismic waves, and once again, the wave pathway is refracted, or bent, as a result. While water waves move at a wide range of speeds, they are generally in the order of magnitude of meters per second. Seismic wave velocities, however, travel at speeds of 165 to 167 kilometers/hr. Because 1 meter/second = 3.6 kilometers/hour, seismic wave velocities are approximately one hundred times faster! Seismic velocities are usually expressed in kilometers/second and range within the Earth between 2 and 13 kilometers/second.

Types of Seismic Waves

Geologists who study earthquakes and seismic waves (seismologists) have identified many wave forms; however, three are important to understand: P, or *primary waves*; S, or *secondary waves*; and *surface waves*. Each of these differs in how it transmits energy as it travels through the Earth (Figure 7-6). P waves are "push-pull" waves: the rock material deforms in directions parallel to the direction of wave travel. S waves are shear waves: the rock material moves vertically perpendicular to the direction of wave travel. Some surface waves (Love waves) exhibit material movement sideways (transverse) to the direction of wave travel, while others (Rayleigh waves) "roll" in elliptical paths.

How do seismologists distinguish P, S, and surface waves from one another? As Figure 7-6 shows, how the deforming material moves as the seismic wave passes is different in each case. Analysis of a *seismogram*, the record of arrival of seismic waves at a recording station's *seismometer* (the instrument itself), allows for identification of this Earth motion and seismic wave identification.

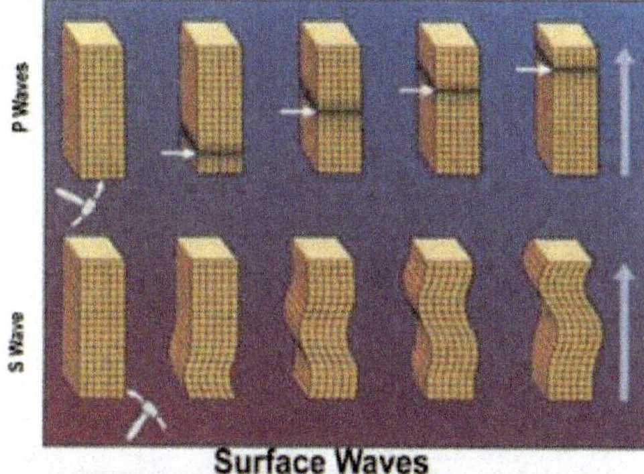

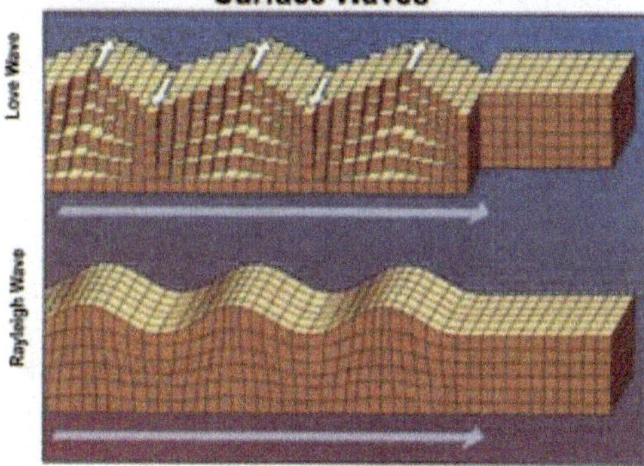

Figure 7-6 Types of seismic waves. In this photo, a hammer strikes the bottom of a column of rock. The arrow on the right indicates the direction of propagation of the wave. P waves compress and expand in the same direction in which the energy is moving. S waves move side to side, at right angles to the direction of movement. In the two types of surface waves shown, the rock moves in a series of sideways dislocations (Love waves) and rolling wave motions (Rayleigh waves).

Image from http://earthquake.usgs.gov/learn/glossary/images/PSWAVES.JPG.

Seismic Wave Velocities

The different seismic waves move at different speeds. P waves travel fastest (6 to 14 kilometers/second), S waves travel between 3 to 6 kilometers/second, and surface waves between 1 to 5 kilometers/second. With this range of values, you'd probably predict that P waves are the most damaging, but other factors discussed below are more important than velocity in determining damage. Sidebar 7-3 illustrates what differences in P and S wave velocities are used to determine the distance between an earthquake's epicenter and a seismometer.

The velocity values given above have broad ranges in kilometers/second. Why is this the case? The velocity of seismic waves is

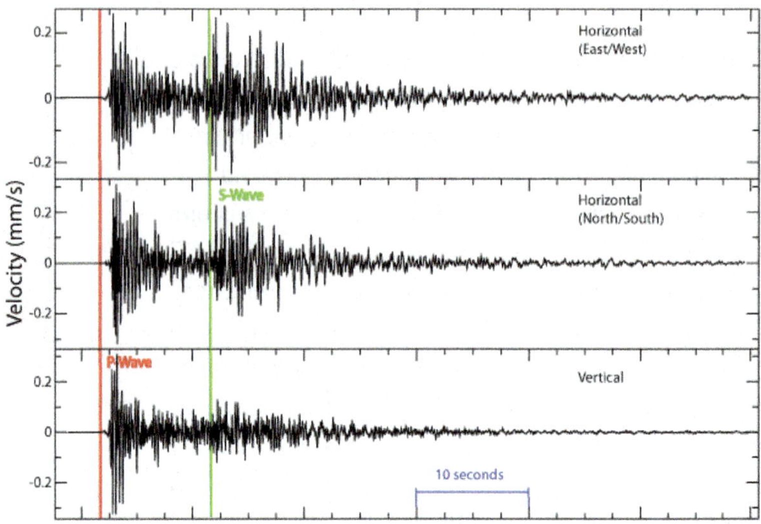

Figure 7-7 The record of the Earth's motion as a function of intervals of time is termed a *seismogram*. The seismogram records motion in three dimensions: horizontal east-west plane, horizontal north-south plane, and vertical. The initial "spike" in the seismogram records the arrival of the first wave (P wave), and the smaller second "spike" records S wave arrival. The gap in time between arrival of P and S waves at a network of stations can be used to determine the location of the earthquake's epicenter.

a function of several variables, notably density, and this in turn is controlled by rock type and pressure.

Pressure is a function of depth within the Earth. This is an easy concept to understand if you've ever swum underwater. The deeper you dive down, the more pressure you've probably felt in your ears. This pressure is the result of the weight of the overlying water. The same is true for the interior of the Earth; the deeper you go, the greater the pressure from the weight of overlying rock, so pressure increases toward the Earth's interior. Why does pressure affect seismic velocities? The more densely packed the crystals within a rock, and the shorter the bonds, the more effectively the rock can transmit the energy of the seismic wave.

At the Earth's surface, P waves travel more slowly than they do at hundreds of kilometers of depth. The same is true for S waves. However, the rocks found near the Earth's surface are also different. Could the slower velocity be the result of this difference? Yes, rock composition is very important, but this indirectly reflects differences in a rock's density. Different minerals have different

Sidebar 7-3 Using Travel Time Curves to Determine the Distance to an Epicenter

Because P, S, and surface waves travel at different velocities, the further they travel from the epicenter of an earthquake, the further apart their arrival times are at seismometers. By analogy, if two cars get on the interstate at the same time and one travels at 70 miles per hour while the other only travels at 55 miles per hour, the faster car will get to the destination first. The further the destination is from the entrance ramp, the longer the gap in time between when the faster and slower cars arrive. We know the average speed of P, S, and surface waves and are able to generate a graph that shows the relationship between arrival times of these waves at a seismic station and the distance that they traveled to get there.

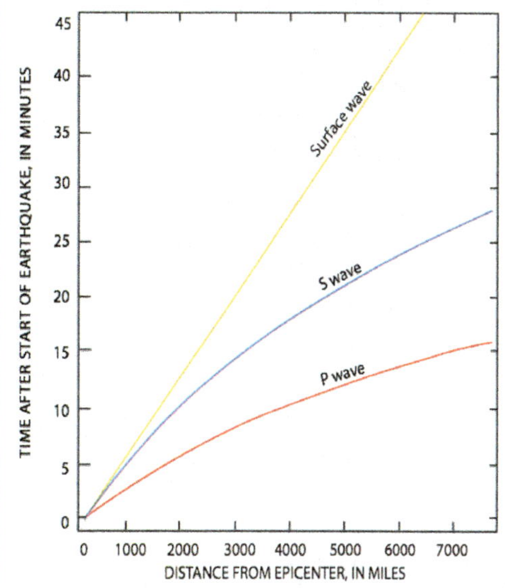

The arrival times for any seismic wave can be read off a seismogram, so the difference in arrival times between P and S waves can easily be determined. Using the graph, you can convert the difference in arrival times to distance to the epicenter. However, because seismic waves travel out in all directions, the earthquake epicenter could lie in a 360° arc away from you. A minimum of three seismic stations are needed to triangulate the solution. The epicenter will be located where the distances intersect, as shown in the image at right.

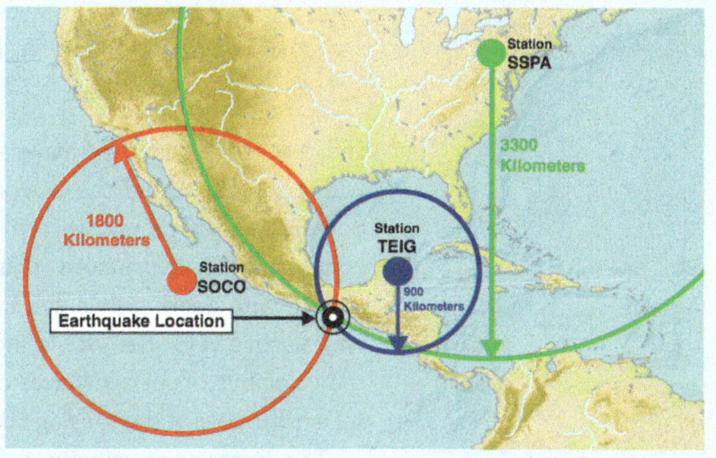

SB 7-3A: Source: Incorporated Research Institutions for Sesimology (IRIS).
SB 7-3B: Source: Incorporated Research Institutions for Sesimology (IRIS).

densities, depending on the chemical elements that form them. Minerals composed of iron, magnesium, and nickel, for example, have a higher specific gravity than do minerals composed of aluminum, potassium, or calcium. As we discussed earlier in this section of the text, if the Earth is compositionally layered, and less dense minerals are more concentrated at the Earth's surface, we would expect the rocks composed of these minerals to be less effective at transmitting the energy of seismic waves. Higher-density minerals

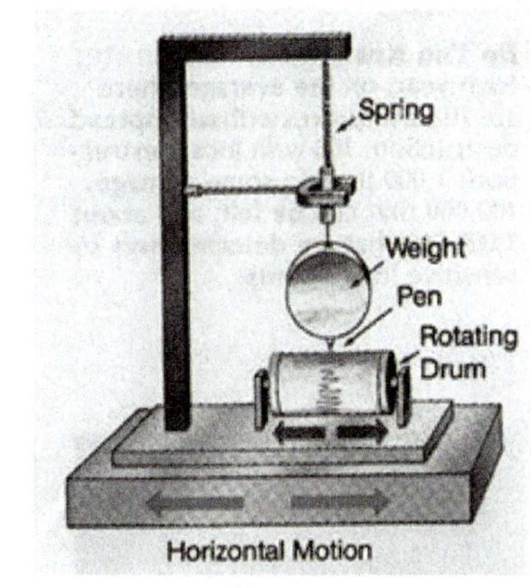

(A)

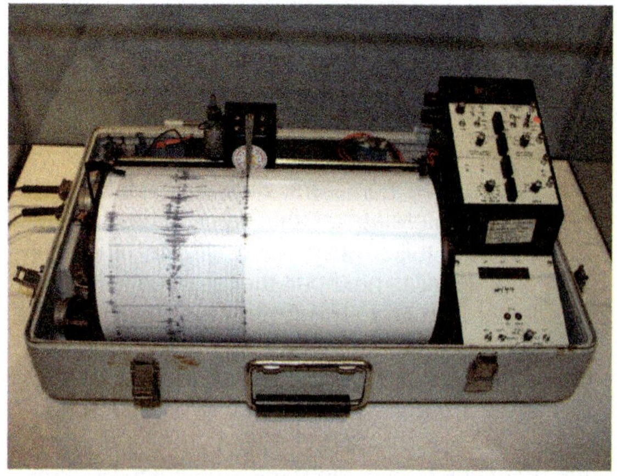

(B)

Figure 7-8 (A) The basic workings of any seismometer. A pen hangs from a spring suspended from a mounting bracket secured to a rock at the Earth's surface. As the rock moves below the pen during the passage of an earthquake wave, a record is made on paper (the seismogram).

(B) A small portable seismometer used by the US Geological Survey. The original mechanical seismometer has now been replaced by electronic instruments.

and the rocks that they make would be found at greater depths within the Earth, and these would be more effective at transmitting seismic waves. So, we are left with the possibility that both composition and pressure influence seismic velocities, and both change with depth within the Earth. Thus, there are good reasons why seismic velocities would increase with depth. The variation in the velocities of surface waves can't be because of depth but must reflect the role of the varying composition of the rocks themselves.

The increase of velocity with depth is not uniform for P and S waves (Figure 7-9). At several depths within the Earth, seismic wave velocities "jump." The first of these is between 5 and 70 kilometers (averaging 30 kilometers), an abrupt increase interpreted to represent a major compositional change in rock types between the crust and mantle. First recognized by Croatian scientist Andrija Mohorovicic, this boundary is termed the *Moho* in his honor. A second jump in seismic wave velocities is between 80 and 300 kilometers in depth, in the upper mantle. Here, P and S wave velocities drop; hence, this region within the Earth is termed the *low-velocity zone* (LVZ). This is interpreted to represent a change in the physical properties of upper-mantle rocks. Recall that higher-pressure dense rocks are most effective at transmitting seismic waves, so you could predict that the LVZ reflects a region of less dense rocks.

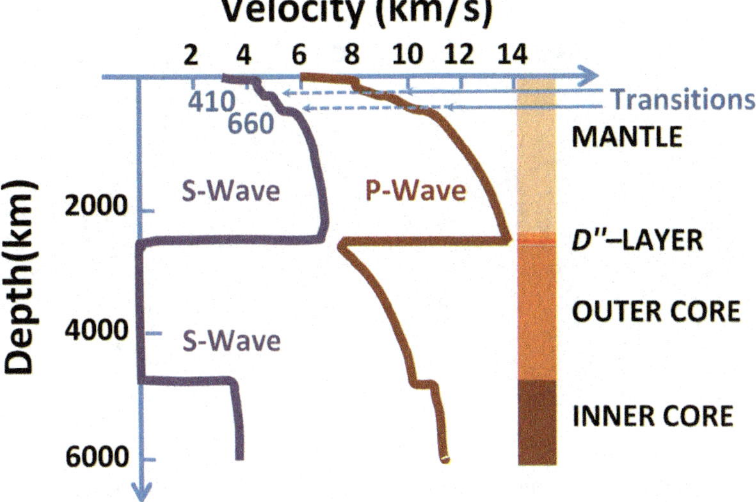

Figure 7-9 P and S wave velocity with depth. There are several horizons within the Earth where seismic wave velocities abruptly change. One of these is at a depth of 2,900 kilometers, and marks an important transition in layers between the mantle and core. Another is the transition from the outer to inner core at 5,150 kilometers depth. The transition from the crust to mantle occurs at an average depth of 35 kilometers, a depth so shallow that it appears as the sharp velocity change near the X axis of the graph.

The lower density cannot be because of a "low-pressure" zone at this depth; thus, there must be another cause. Laboratory models suggest that the density decrease is not because of differences in the amount of high-density minerals but because of their slight (approximately 1–2%) partial melting. The low-velocity zone will turn out to be very important to explaining how plate tectonics works. The rocks above the low-velocity zone comprise a layer of the Earth termed the *lithosphere*. The low-velocity zone itself marks the top of another layer termed the *asthenosphere*. Below the asthenosphere, there are step-like increases in seismic velocities at 450- and 600-kilometer depths. These are interpreted to represent changes in the crystalline lattice structure, from increasing pressure, in iron-silicate minerals comprising the mantle. These changes in mineralogy may influence heat flow in the mantle because they represent density changes that may absorb or release latent heat. Below these transition zones, P wave velocities increase gradually until an abrupt increase at a depth of around 5,150 kilometers, which marks the boundary between the outer and inner core.

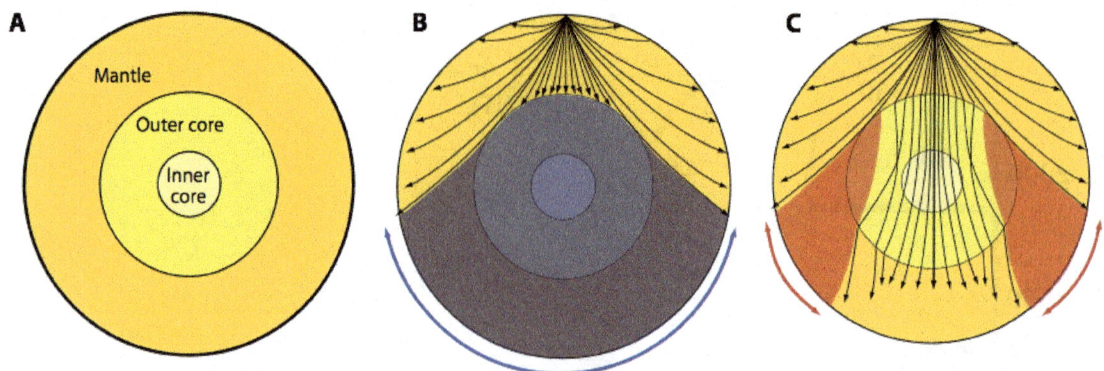

Figure 7-10 Figure A shows the layers of the Earth and their relative thicknesses; the crust is represented by the slightly thicker black line of the outer sphere. B. All illustration of the S wave shaddow zone for an earthquake epicenter near the top of the sphere. No S waves are received by seismometer stations anywhere in the gray shaded area 104 degrees from the epicenter, which lies on the opposite side of the liquid outer core. C. The P wave shadow zones are shown in orange; between 103 to 142 degrees from the epicenter. These result from reflection and refraction of P waves traveling through the mantle and encountering the mantle/core boundary at the critical angles.

Perhaps the most significant change in seismic wave velocities with depth is seen at a depth of 2,900 kilometers, the Gutenberg discontinuity. Secondary, or shear, waves cannot propagate through liquids (which have no strength). Geologists noticed that seismometers located on the opposite side of the Earth from an earthquake focus did not pick up direct S waves. Because the type of dislocation associated with the passage of secondary seismic waves cannot occur in liquids, this "shadow zone" is the primary evidence that Earth possesses a liquid outer core (see Figure 7-10). P wave velocities also slow in the liquid outer core, because while the "push-pull" motion characteristic of P waves can pass through liquids, this material is a less efficient transmitter.

Describing the Size of an Earthquake

The size of an earthquake is determined in two ways: (1) by the amount of energy released, a concept termed *magnitude*, and (2) the damage caused, termed *intensity*. Magnitude is calculated by measuring the amount of P wave oscillation on a seismogram at a station after correcting for its distance from the epicenter, a technique developed by Charles Richter. The Richter scale is logarithmic; the difference in shaking between a Richter magnitude-2

Table 7-1 The Mercalli Scale

Modified	Mercalli Intensity	Scale
Mercalli	Observations	Richter Scale
Intensity		Equivalent (magnitude)
I	felt by very few people; barely noticeable	1.0–2.0
II	felt by a few people, especially on upper floors	2.0–3.0
III	noticeable indoors, especially on upper floors	3.0–4.0
IV	felt by many indoors, few outdoors; mistaken for heavy truck	4.0
V	felt by almost everyone, sleepers awakened. Shaking	4.0–5.0
VI	felt by everyone. Difficult to stand, furniture may move; chimneys slightly damaged	
VII	Slight to moderate damage in well built structures, considerable damage to poorly built structures	6.0
VIII	considerable damage to ordinary structures, some walls collapse	6.0–7.0
IX	wholesale destruction, landslides	7.0
X	most structures destroyed, landslides,	7.0–8.0
XI	total damage, few structures remain, large cracks in earth	8.0
XII	total damage, objects thrown in air	8.0–>8.0

and magnitude-3 earthquake is ten times, which also represents a thirty-two-times increase in energy released. Huge, damaging earthquakes are usually greater than 6.5 in magnitude. The Hiroshima nuclear bomb released the same amount of energy as a magnitude-6 earthquake would have. Richter's original calculations of amplitude have been refined, and now earthquakes are usually reported as "moment magnitude" or "modified Richter magnitude." These terms represent measurements of energy released based on the strength of the rocks that ruptured as well as the length of the fault surface that slipped. If we go back and examine some of the larger earthquakes in history, such as the 1964 Alaskan Good Friday earthquake, the reported Richter magnitude of 8.6 would be a modified Richter 9.2.

Intensity describes the nature and amount of damage done by an earthquake and is reflected by the *Mercalli scale* (Table 7-1). Qualitative in nature, the Mercalli scale is very useful for representing the geographic variation in an earthquake-affected region that results from the variables responsible for the destruction. The extent of Earth shaking is obviously a function of how close the area is to the epicenter: the closer, the more intense the shaking, because the energy has not dissipated through the Earth. Less obvious is the role of the type of substrate. When shaken, rock remains coherent. It can break apart, but is much less prone to doing so than is unconsolidated (loose) soil, gravel, and sand. When this material is shaken, grains slide and roll past one another, expelling fluids and compacting. Significant ground sinking results, and on slopes, landslides are common. Some of the most destructive earthquakes ever recorded resulted from high-density urban development on unconsolidated substrates. The 1995 Mexico City earthquake is the textbook example of this. Although the epicenter was 350 kilometers away, on the west coast of the country, the city was built on unconsolidated sediments of the former Lake Texcoco. Most of the ten thousand fatalities resulted from buildings collapsing from sinking into the liquefied sand produced from 2.5 minutes of shaking. As foundations sank, buildings toppled over.

What Earthquakes Tell Us about the Internal Structure of the Earth

Analysis of seismic waves is the primary source for information about the Earth's interior; however, geologists also examine fragments of meteorites that reach Earth to model the composition of the interior. Meteorites are thought to represent debris left over after the formation of the asteroid belt where the Earth is located. Meteorites have a range of compositions; some appear similar to the peridotite rocks of the mantle, but others are very enriched in iron-nickel alloys. Thus, this material could also be present somewhere within the Earth. These elements are not abundant in most of the Earth's surface rock layers, so they must be concentrated within the interior. Support for this idea comes from astronomical calculations of the Earth's average density (5.52 g/cm^3), which greatly exceeds the average density of crustal rocks. This suggests that the interior must be composed of very high-density minerals. Combining these different data sets, geologists can characterize the depths, compositions, and physical properties of the Earth's internal layers (Figure 7-10A).

Crust: The outermost layer of the Earth ranges between 5 and 10 kilometers in thickness in the oceans and 30 and 70 kilometers in the continents. The thinner ocean crust is composed of mafic igneous rock, such as basalt and gabbro; average densities of these rocks are approximately 2.8 g/cm^3 to 3.0 g/cm^3. The thicker, less dense (2.7 g/cm^3) continental crust is composed of felsic rocks, such as granite and gneiss. These rocks contain minerals with the largest concentrations of unstable isotopes, such as uranium, in the Earth. The crust is separated from the mantle beneath by the Moho, the boundary where seismic velocities speed up.

Mantle: The mantle extends from the base of the crust to 2,900 kilometers; however, it can be subdivided into several different regions. Shallower than 450 kilometers, the mantle is composed of mafic rock rich in olivine, pyroxene, and garnet. Between 450 and 600 kilometers, the composition is essentially the same; however, the crystalline structure of these minerals is altered to a more densely packed structure. These mineralogical transitions and their effect on seismic wave velocities has been confirmed through laboratory studies. Despite elevated temperatures with increasing depth (Figure 5-4), increasing pressure with depth maintains the mantle in a solid state; the partial melting that occurs at approximately 450 kilometers (the low-velocity zone) has already been noted. The bulk of the mantle, although solid, exists as a "plastic" (which means that when it is deformed, it alters its shape). Plastic substances appear solid; however, they respond to stress by "flowing." The density of the mantle increases with depth from approximately 3.3 g/cm^3 near the Moho to approximately 6 g/cm^3 near the core-mantle boundary. The mantle comprises approximately 85% of the Earth by volume (about 15% is core, and the crust is barely a trace).

Outer core: The core-mantle boundary occurs at a depth of 2,900 kilometers. The inability of S waves to pass through the outer core indicates that this layer is liquid. Despite the pressure, the elevated temperatures at this depth within the Earth (in excess of 5,000°C) cross the melting points for outer-core materials, thought to be a combination of primarily iron and nickel minerals and "impurities" (lower-mantle iron-silicate minerals). The temperature difference between the mantle and core is thought to be large enough to be responsible for significant heat flow from the core to mantle, a process that we will discuss more in Chapter 9. The density of core material exceeds 12 g/cm^3. In Chapter 8 we will discuss the origin and significance of the Earth's magnetic field, which originates from circulation in the liquid outer core.

Inner core: The inner core extends from its boundary with the outer core at approximately 5,150 kilometers to the center at 6,300 kilometers. There is not a huge temperature change between the outer and inner core, so the transition from liquid back to solid is thought to reflect a compositional difference: the "impure" iron-nickel composition of the outer core versus an inner core composed of more pure iron-nickel minerals with higher melting points. The elevated temperatures within the core are thought to reflect (1) heat generated from the gravitational collapse and melting in the earliest stages of planetary formation, (2) heat generated from radioactive decay, and (3) heat generated from chemical reactions within the core. The overlying mantle has served as an effective thermal insulator to this primordial heat. It is also possible that uranium is present within the core and some "modern" heat is being generated through its decay.

The Lithosphere and Asthenosphere

The division of the Earth into crust-mantle-core layers is based on a combination of compositional and physical differences. Two other layers can be identified based solely on their response to stress. The *lithosphere* is the outermost 100 to 150 kilometers of the Earth; thus, it is composed of the crust and uppermost mantle. The lithosphere is brittle, meaning that when it is stressed, it breaks. The *asthenosphere* exists below the lithosphere, between 150 and 450 kilometers, although the base of the asthenosphere is poorly defined and at variable depths within the mantle. The asthenosphere-lithosphere boundary is identified by the slowing of seismic velocities (low-velocity zone). These slower velocities are interpreted to be the result of this layer's slightly melted physical state. As a result, the asthenosphere is a mostly plastic solid. When stressed, these rocks flow. The lithosphere comprises the "plate" in "plate tectonics," and the behavior of the lithosphere will be the subject of Chapters 9 and 10. The asthenosphere represents the region in the underlying mantle material that the plates move across. In terms of geologic processes that operate in the geosphere part of the Earth system, the lithosphere and its behavior are critically important. They will be discussed in more detail in Chapter 9.

Origin of the Layered Earth: Differentiation

Why does the Earth have its layered structure? Are other planets in our solar system also layered? These are important questions when trying to understand how our planet has evolved over time.

In Chapter 3 we discussed the origin of planetary material in the moments following the big bang, nearly fifteen billion years ago. The process of nucleosynthesis inside stars produced many chemical elements, and the remainder were produced from successive star formation and subsequent explosion into supernovae. The cosmic material produced from these stellar explosions condensed into nebula—a disc-shaped swirling cloud of gases, particles, and energy. The interior regions of nebulae form stars, while the outer regions of the swirling cloud condense into regions that form protoplanets. Condensation happens as a result of collisions between the elemental particles; some are called "volatiles," to describe their highly reactive nature, while other elemental particles are "refractory" materials, which are heavier. Initially, the protoplanetary disk contained a homogenous mixture of volatile and refractory elements throughout, but as the central sun grew and emitted heat, the volatile and refractory elements segregated: gases moved outward in the disk. During these initial stages, elemental particles also began colliding with one another and adhering, forming over time large enough chunks of solid matter to form "planetesimals," which in turn would collide with one another to ultimately form "protoplanets." Protoplanets are the size of planets in our solar system; they differ in their internal structure (not developed) and temperature (cold). Computer models of this process suggest that it happened over a very short time span, geologically speaking (one hundred thousand years). The final stage, forming a planet, was a slower process (hundreds of millions of years), as it represents the stages of gravitational collapse, heating, cooling, and solidifying. It is this final stage to which we owe our internal Earth structure.

With gravitational collapse and melting, differentiation of Planet Earth began. The more dense refractory materials (silica, aluminum, iron, potassium, magnesium, etc.) sank into the interior, while the volatile elements (hydrogen, oxygen, nitrogen) remained on the surface. Were it not for gravity, these gases would have escaped from Earth entirely. Trapped, they form our planet's atmosphere.

Within the hot Earth interior, the most dense materials sank inward to form the core; the remaining less dense materials formed the mantle. The oldest felsic continental crust represents the lowest-density residue from cooling of the "magma ocean" that was the bulk of the Earth's surface. The Earth has been cooling ever since this time, offset by heat generated from radioactive decay. New evidence suggests that the oldest rocks (cooled magma) on Earth are approximately 4.4 billion years old, an age that is in close agreement with dated meteorites that indicate that the planets formed approximately 4.6 billion years ago, or roughly 10 billion years after the big bang.

Where Earthquakes Occur

So far our discussion has focused on the role that seismic waves play in our understanding of the structure of the Earth's interior. In Chapter 9 we will examine in great detail where on Earth earthquakes are more likely to occur and why this is the case. For the present time, however, you should be able to apply "first principles" to answering the question "Why does an earthquake occur?"

Summary

When rocks are stressed the store this stress as potential energy. When the rocks break, this energy is converted to kinetic energy released in the form of a seismic, or earthquake, wave. The location where an earthquake originates within the Earth is termed the focus or hypocenter. The location on the Earth's surface directly above the focus is termed the earthquake epicenter. Seismic waves move through the Earth like vibration moves through a rung bell. Instruments that record the vibrations of the internal Earth are termed seismometers and the visual representation of the earthquake waves recorded by the seismometers are termed seismograms.

There are several different types of seismic waves, distinguished by how the waves move through the Earth and their speed. The major types of seismic waves are body waves, which move through the Earth's interior, and surface waves, which as their name implies, move along the Earth's surface. Two types of body waves are primary (P) and secondary, or shear (S) waves. P waves which travel between 6 and 14 km/sec and move in a push-pull motion parallel to the direction in which the wave is traveling. S waves vibrate at right angles to the direction of propagation and travel 3 to 6 km/sec. Two types of surface waves are Rayleigh and Love waves which travel between 1-5 km/sec. Love waves move in a series of multiple dislocations while Rayleigh waves exhibit a rolling motion. Because of the proximity of these waves to structures such as roads and buildings and the types of Earth motions involved, surface waves create much damage on the Earth's surface.

Analysis of seismic waves provides much of our information about the interior of the Earth. Earthquake waves traveling through the Earth's interior speed up with increasing depth due to increasing pressure and density in the Earth's interior. They also change their path of motion as they travel, due to refraction and reflection as they encounter layers within the Earth's interior. Analysis of seismograms on the side of the Earth opposite the focus reveals an absence of S waves in a region termed the S wave shadow zone. Because shear waves can't travel through liquid, the S wave shadow zone tells us that a layer of the Earth, termed the outer core, must be liquid. Two P wave shadow zones also occur and these represent regions where wave refraction off the outer core boundary redirects primary waves. Important layers of the Earth include the crust, mantle, outer and inner core. The boundary between the crust and mantle is termed the Moho. The "plate" of "plate tectonics" is termed the lithosphere, a layer that consists of the crust and uppermost mantle. Beneath the lithosphere is the asthenosphere, a layer in the mantle where seismic velocities decrease, suggesting that this layer, while solid, may be "soft." The layering of the Earth reflects the compositional and density segregation that occurred following the accretion and melting of early Planet Earth.

Chapter 7 Review Questions

- Define the elastic rebound theory
- What are the different types of faults and how are they distinguished?

- What is the difference between the epicenter and focus of an earthquake?
- Describe the differences between the types of motion characteristic of the four main types of seismic waves
- How fast do P and S seismic waves travel? Is this speed constant as the wave travels through the earth? Explain your answer.
- What is the difference between wave refraction and wave reflection?
- What is the difference between a seismometer and a seismogram?
- What is the Moho and how is it recognized?
- How is the lithosphere/asthenosphere boundary identified?
- What property of the outer core is identified by the behavior of seismic waves?
- What are seismic "shadow zones"?
- Draw a cross section of the Earth, with a scale, and layers labeled
- On a cross section of the Earth, place an earthquake focus near the outer edge (surface) of the Earth at the position of "2 o'clock." Draw the position of the outer core in the Earth's interior and then draw the P and S wave shadow zones.
- How is a travel-time curve used to determine the location of an earthquake epicenter?
- How do we describe the size of an earthquake?
- What variables control the amount of damage done by an earthquake?
- Does the speed of a P wave increase, decrease, or stay the same when it passes from the crust to mantle?
- Does the speed of a P wave increase, decrease, or stay the same when it passes from the lithosphere to asthenosphere?
- Does the speed of a P wave increase, decrease, or stay the same when it passes from the mantle to the outer core?
- Complete the following sentences:
  a. "The _________ of an earthquake describes the damage levels done at a particular location"
  b. "The difference in energy released between a Richter magnitude 2 and 3 earthquake is _________________"

Chapter 8

Other Characteristics of the Earth's Interior

At the end of this chapter you should be able to

- Explain the origin of the Earth's magnetic field
- Describe patterns of in the flux of heat and interpret what this indicates about the Earth's interior
- Describe how geologists use seismic wave velocities to determine the Earth's internal heat distribution
- Compare and contrast convection cells and mantle plumes
- Define the attributes of a magnetic field
- Describe how the Earth's magnetic field changes during a polarity reversal
- Define what the Curie point of a rock is and why it's important

- Predict which types of rocks might be most likely to preserve the Earth's magnetic field
- Define what is meant by a magnetic anomaly and describe how it is used to determine geologic processes at mid-ocean ridges

Introduction

In earlier chapters we discussed the origin of the internal heat of the Earth (radioactive decay) and the trend in temperature as a function of depth within the Earth (the geotherm). We also know that analysis of the speed of earthquake waves as they move through the interior of the Earth is the primary data source that geologists use to infer the Earth's internal structure.

All cycles need an energy source to operate, and the energy source for the Earth is heat. Heat energy is derived from external solar radiation as well as internal heat generated within the Earth. We will discuss the role of solar radiation later in the text. This chapter presents more information on the distribution of heat within the Earth and the nature of energy flow. In terms of the Earth system, the movement of heat energy within the Earth drives the movement of the tectonic plates and thus fuels many geologic processes that we see around us. Finally, this chapter examines the Earth's magnetic field and its change over various time scales.

Heat Flow within the Earth

In Chapter 5 we discussed the change of temperature with depth and the geothermal gradient that results (refer back to Figures 5-3 and 5-4). We noted that there is a shallow or surface geotherm in the crust of approximately 30°C per kilometer, however, if this rate were maintained over the entire depth of the mantle, extreme temperatures would quickly be achieved that exceed the temperature at which any rock would melt, even at the elevated pressures of the Earth's interior. We know from seismic data (Chapter 7) that the Earth's interior is mostly solid, so clearly the geotherm through the mantle is much lower. Temperatures at various depths in the Earth's interior represent the maximum that can be achieved at the calculated pressure at a particular depth, such that the mantle remains solid rock. A geotherm of approximately 0.5°C/kilometer though the mantle would achieve this (Figure 5-4).

Source of the Earth's Internal Heat

Chapter 6 summarized the possible sources of heat for the internal Earth. While radioactive isotopes are most concentrated in the felsic rocks of the granite crust, their lower abundance in a volumetrically larger mantle means that this layer generates much of the Earth's internal heat. The remaining heat source is concentrated in the core from initial gravitational collapse and melting. How much internal heat is generated, and how is it transferred?

The Earth's internal heat budget is estimated to be 47 terawatts of energy (a terawatt is one trillion, or 10^{12}, watts; a watt is a unit of energy = 1 joule/second). This seems like a huge number, but as we will see in Chapter 12, solar radiation provides Planet Earth with 173,000 terawatts of energy! The Earth's internal heat, however, is significant enough to "run" the geologic processes in its interior, including plate tectonics. Figure 8-1 illustrates how heat

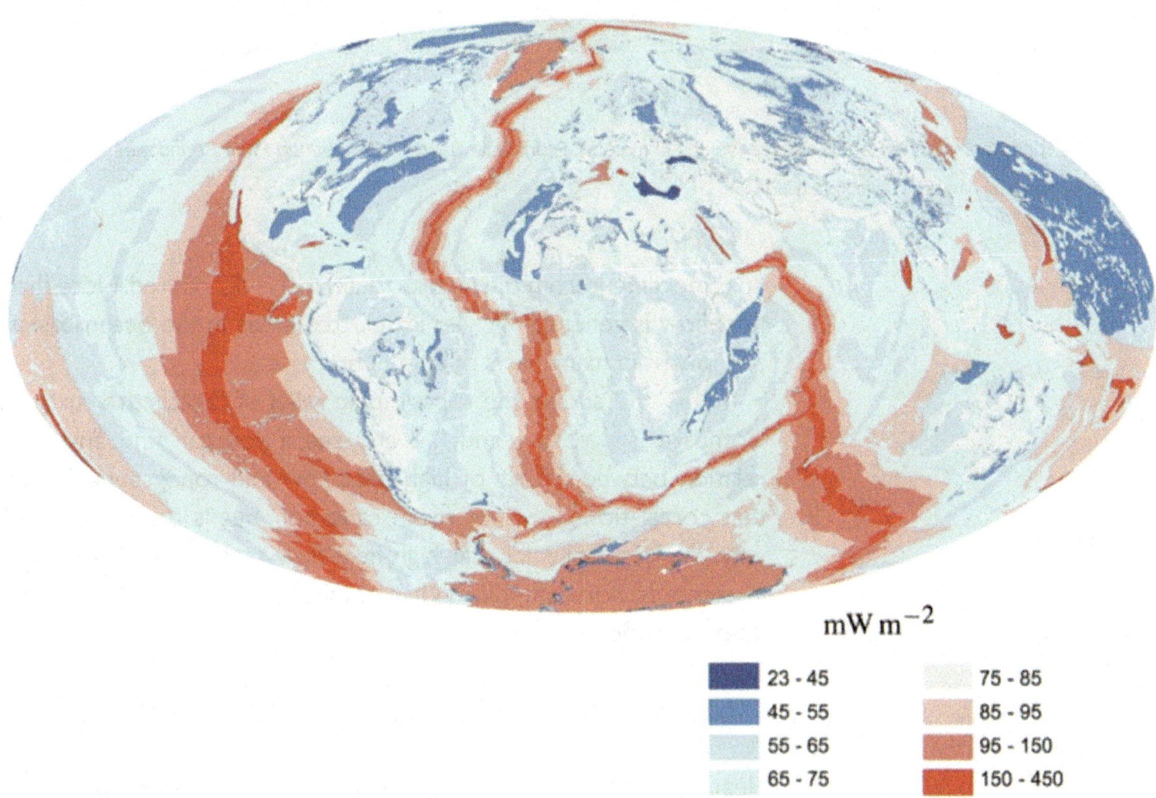

Figure 8-1 Heat flux from the Earth's interior in milliwatts per square meter (from Davies and Davies, 2010).

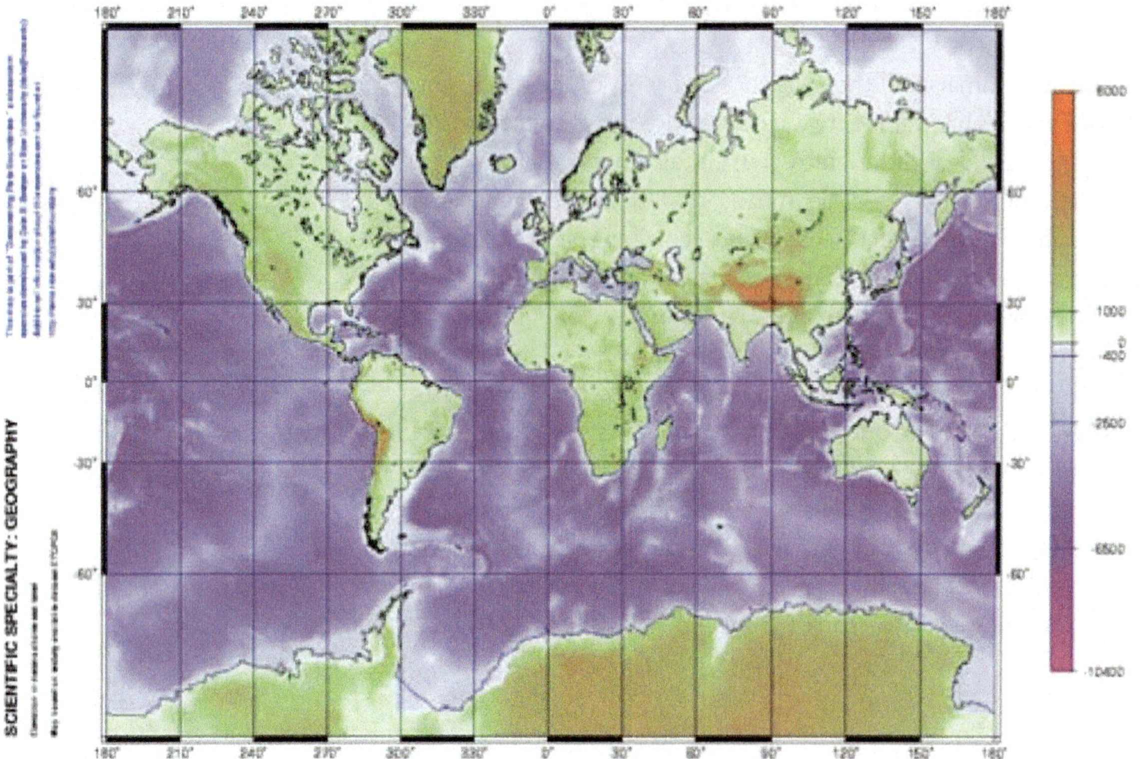

Figure 8-2 Topography of the Earth's surface (values in meters). From Dale Sawyer, *Discovering Plate Boundaries*.

Source: http://plateboundary.rice.edu/topo.72.gif.

flux (movement from the interior outward) is distributed. Note that heat flow is generally highest in selected areas in the ocean basins and lowest in many parts of continents.

Figure 8-2 shows the topographic relief of the Earth's surface. Compare Figures 8-1 and 8-2 and see if you can find any correlation between areas of high heat flow and topography. Recall that correlation does not mean causation, but is there a reason why you might predict that high heat-flow regions would be high in elevation, for example? What physical property or properties does heat affect?

How Does Heat Energy Move within the Earth?

Heat is a form of energy and follows the laws of thermodynamics. Energy cannot be created nor destroyed, but is transferred. Heat energy is transferred in one of three processes: conduction, convection, and radiation. We will discuss radiation transfer in

Chapter 12. When considering the internal Earth, we will focus on the processes of conduction and convection.

Recall that temperature is a measure of the vibration of atoms. Elevated temperatures describe a condition in which atoms are rapidly vibrating, while lower temperatures induce less motion in atoms. Realizing that temperature describes motion may clarify why heat is a form of energy; we are really discussing the energy of atomic motion within a substance.

Conduction describes the transfer of heat energy from one warmer substance in contact with a cooler material. The energetically moving atoms in one substance induce increasing motion in the adjacent atoms. You utilize conduction when you occurs in a pan of water placed on a lit stove burner. In the Earth, the core conducts heat to the base of the mantle. Magma generated within the mantle (from decompression melting) can melt adjacent rock from conduction melting (Figure 5-7). Conduction, however, is not efficient for the bulk transfer of heat within the Earth. Convection, on the other hand, is.

Convection describes the transfer of heat energy through the physical movement of warmer material. Because temperature influences density, warmer substances will be less dense than will colder substances. Just as heating the air within a balloon will cause it to rise, heating rock will also trigger its upward movement. Conversely, cooling a substance will increase its density, enhancing its ability to descend, or sink. You are familiar with convection, as it occurs in a pan of water placed on a stove to heat it up. The water at the base of the pan is heated by conduction from the hot pan bottom. This hot water rises upward to the water-air surface, where it spreads laterally. At the top of the pan the water is cooler, so it sinks. The "closed loop" produced is called a "convection cell." Convection operates in our atmosphere, but it also operates within the mantle.

It is difficult to conceive of hot rock moving vertically within the Earth, but because the outermost Earth is fragmented into the tectonic plates, and these plates move around, this allows for "upwelling" of material from deeper in the mantle. We see this upwelled material in the form of active volcanoes of the shield and cinder cone varieties, as well as fissure eruptions. The magma erupting represents mantle that melted from decompression melting accompanying the upwelling of a mantle convection cell. If we re-examine Figure 8-1, we can now hypothesize that the regions of the Earth where heat flux is greatest might represent the surface location of this upwelling of hot mantle. The higher heat flow would lower the density

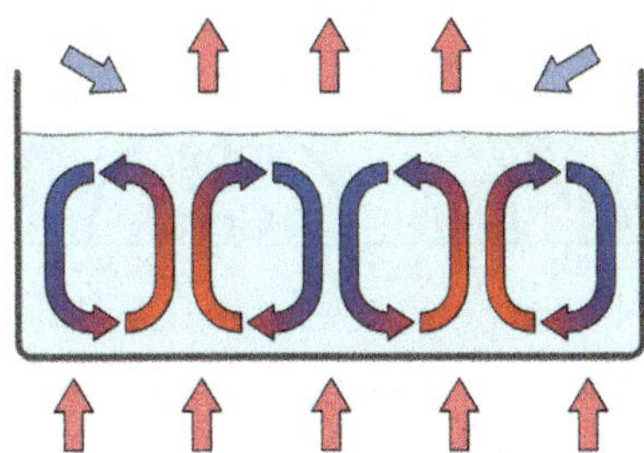

Figure 8-3 Simple convection cells.

of the lithosphere in these regions, and they would be more topographically elevated as a result. These areas should also exhibit active volcanism.

Following our convection-cell boiling-water model, if hot, lower-density material is rising in one location then cools, more dense material must sink elsewhere. If we look at Figure 8-1, we can hypothesize that the downward sinking of cooler material might happen at regions where heat flux values are much lower. There is a very large region of the western Pacific where ocean floor rocks have very low heat flux values, so we might predict that this region might be experiencing "downwelling." Are these areas of higher density lithosphere characterized by low topographic relief (Figure 8-2)?

Figure 8-4 is a schematic illustration of the idea that the mantle is divided into a series of convection cells that represent the flux of heat from deep within the Earth toward the surface and the return of cooler material to the interior. As this illustration shows, the upwelling and downwelling limbs of the convection cells are in alignment with plate boundaries. In other words, the edges or boundaries of the tectonic plates appear to be strongly influenced, if not controlled, by temperature differences. We need to test this idea further.

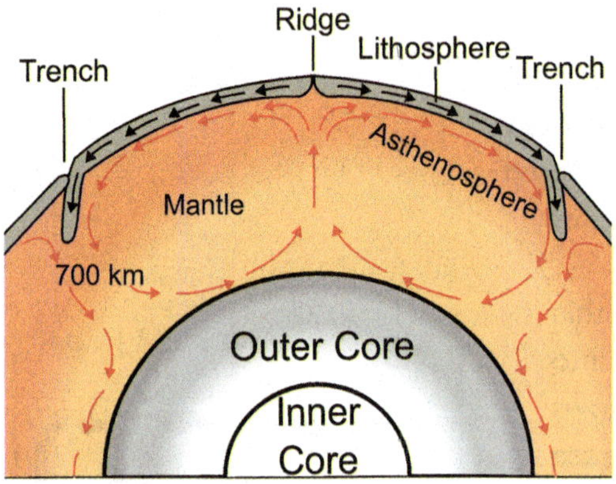

The convection-cell model for moving plates is attractive because we can model convection in the laboratory and confirm that it could occur at geologically reasonable temperatures, pressures, and compositions of the mantle. If you compare Figure 2-3, a map of the present-day tectonic plates, to Figure 8-1, you will see a very high degree of correlation. Recall that correlation does not mean causation, so we need to look at the places where we don't see this relationship. The scale of the map of Figure 8-1 doesn't show this detail, but the Hawaiian Island chain is one of the anomalies, with heat flow of greater than 60 mW/m^2, or approximately 20% greater than the surrounding region, yet the islands are sitting in the middle of a tectonic plate. Can we reconcile this with convection cells, or is something else happening?

Figure 8-4 Cutaway view of the Earth showing the schematic presence of convection cells within the mantle.

Mantle Plumes

We can utilize seismic wave analysis to help us examine the nature of heat flow within the shallow (hundreds of kilometers in depth) mantle. Seismic wave velocity is influenced by temperature (cold = dense = fast; warm = less dense = slow), and if we can eliminate compositional change as a variable, variation in the speed of P and S waves will reflect their thermal properties. Using a network of seismic stations to reconstruct three-dimensional images, a technique termed *seismic tomography*, we have been able to identify isolated thermal *plumes*, or hot spots in the mantle.

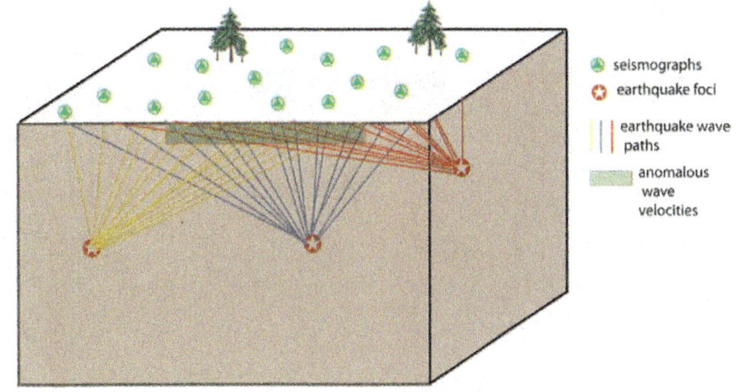

Figure 8-5 The cartoon shows 15 seismometers capturing P wave velocities from 4 earthquakes. The area outlined in red represents a region where the P wave velocities were anomalously slow, suggesting the presence of elevated temperatures.

Seismic tomography produces three-dimensional images from computer synthesis of seismic wave velocities from a global network of seismometers. The noninvasive medical procedure termed a CAT scan (computerized axial tomography) employs a similar type of analysis: a three-dimensional image is produced by the merging of X-ray images. Seismic tomography is analogous to a CAT scan of Planet Earth (Figure 8-5).

On the Earth's surface above these hot spots are active volcanoes, and seismic tomography helps us visualize the size and location of the magma chambers beneath them. Hot spots can

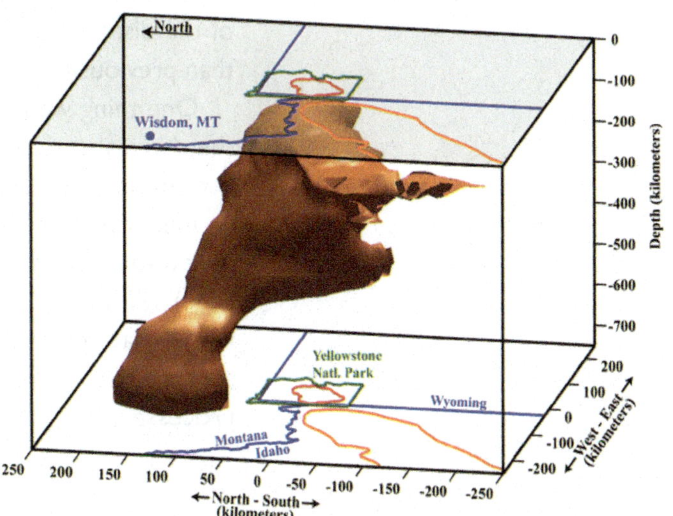

Figure 8-6 The Yellowstone plume as modeled by seismic tomography. The top and bottom of the image show geographic references points of the Wyoming/Idaho region and the outline of Yellowstone National Park. The "brown blob" represents the region beneath the surface where P wave velocities are anomalously slow, indicating that the rock is very hot. The region indicated on this cross-section is interpreted to be a crustal magma chamber, which was responsible for past eruptions of the Yellowstone "supervolcano" (see Chapter 5). The magma body at the top of the plume is thought to be 90 kilometers long and 5 to 17 kilometers deep; note that the plume itself is identifiable to more than 500 kilometers of depth in the mantle.

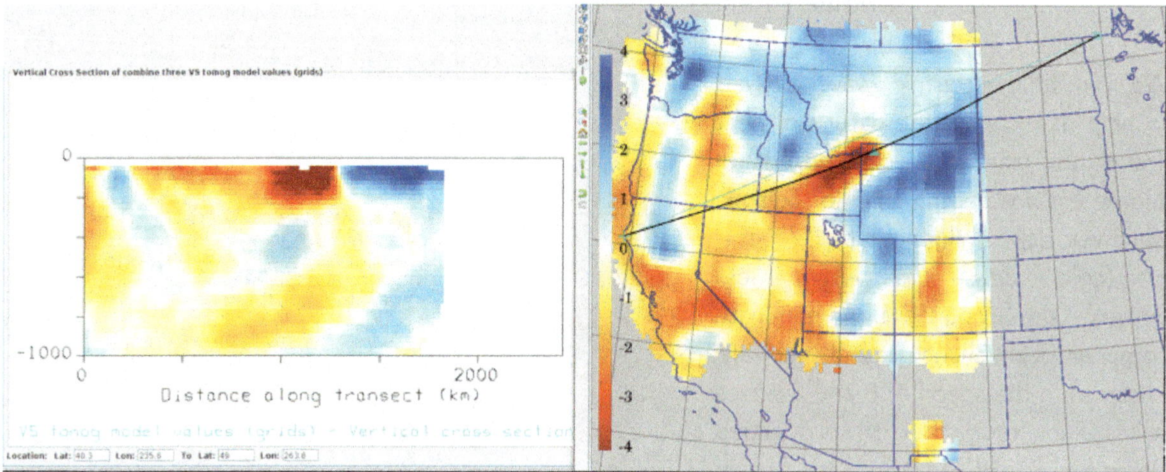

Figure 8-7 Map view of the western United States with seismic wave velocity anomalies. The outlines of states are shown in light blue lines. Bright red colors indicate slower than expected P wave velocities (i.e., hot rocks); blue colors indicate faster than expected velocities (i.e., cold rocks). The black arcuate line represents the orientation of the cross-section profile seen on the left. Note the highest red anomalies beneath the Yellowstone region, at a depth of approxiamtely 200 kilometers, the top of the mantle plume; the yellow region at a depth below this (to approximately 1,000 kilometers) represents its base.

Source: http://westernexplorers.us/WesternUSGeophysics-tomog.html#overview.

be identified under the Hawaiian Islands (Figure 8-8), Iceland, Yellowstone (Figure 8-7), and more than a dozen other locations on Earth. The mantle plumes are stationary features, forming at or near the core-mantle boundary, nearly 3,000 kilometers in depth. The tectonic plates move across the Earth's surface, so the interaction of the fixed heat source and the moving slab of lithosphere is responsible for the production of linear tracks of volcanoes (Figure 8-8B), such as the Hawaiian Island chain.

The development of high-speed computers and an extensive global computer network have revolutionized our understanding of the distribution of heat within the mantle. It is more complex than previously thought (Figure 8-9).

One thing we can say for certain is that our understanding of the thermal state of the Earth's interior is still evolving. This is a very important subject to explore, however, because it is really an exploration of how Planet Earth is cooling. Some geologists believe that our simple three-layered Earth model (crust, mantle, core) will be revised as a result of seismic tomography! We do know that this research is providing details about the energy source that "drives" plate tectonics and many other geologic processes.

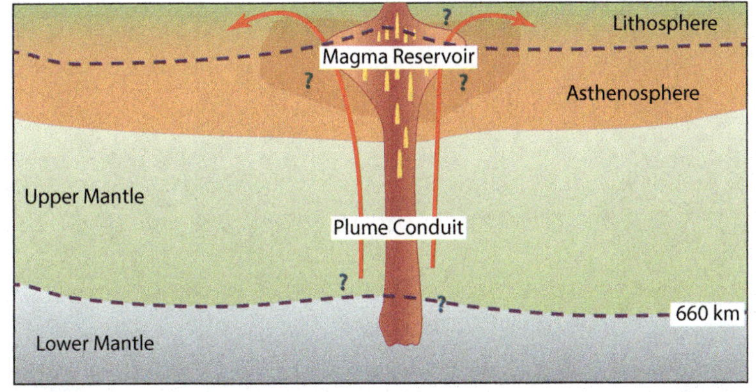

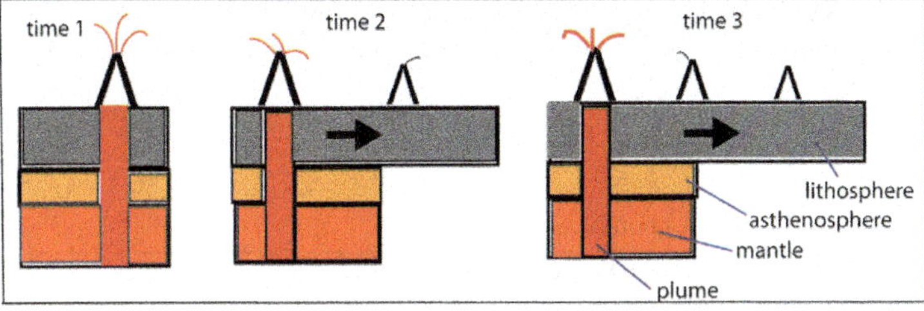

Figure 8-8 (A) Schematic of Hawaiian Island volcanic chain and the location of a mantle plume, or hot spot. The Pacific plate is moving 10 to 18 centimeters/year to the northwest. (B) How a hot spot track is produced. The youngest volcano is located immediately over the mantle plume. The volcano furthest away from the hot spot is the oldest.

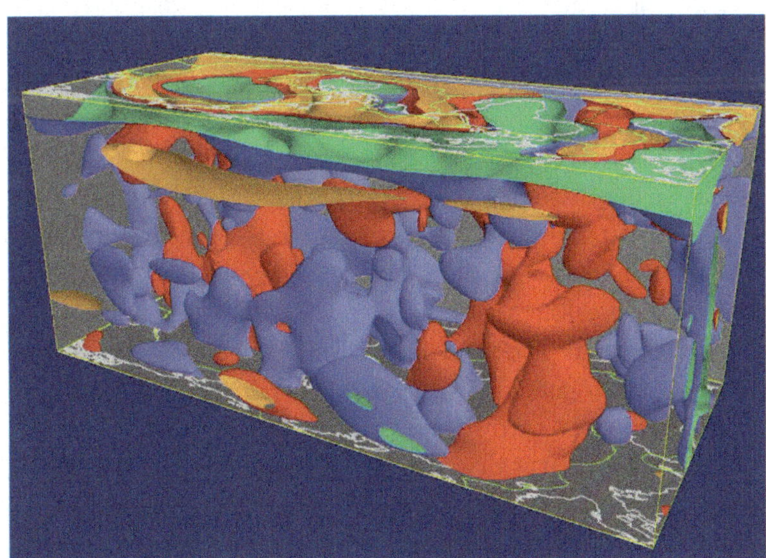

Figure 8-9 Schematic rendition of seismic tomography data of the mantle. On the top of the image, shown in white, are geographic reference points. The perspective is from the north. On the extreme right of the block are North and South America; Eurasia lies in the middle, and the Pacific Ocean and westernmost North America lie on the left side of the block. The large red blob representing the upwelling limb of a convection cell lies beneath the mid-Atlantic region. Another well-developed cell is thought to be located under the East Pacific Rise.

Figure 8-10 The aurora borealis.

Magnetism

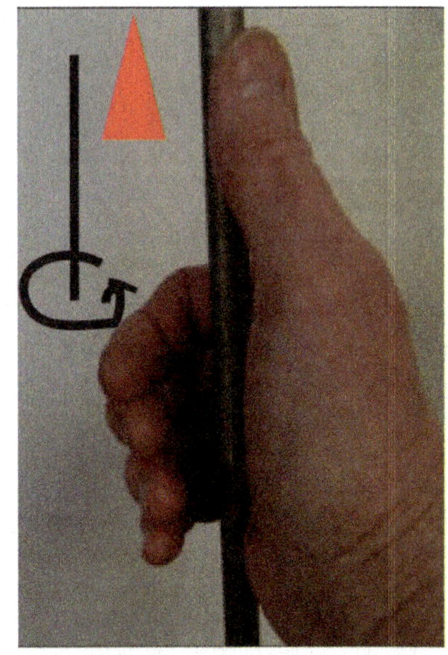

Figure 8-11 The right hand rule: a electric current (black line) induces a magnetic field oriented at right angles to it. In the illustration the current is flowing parallel to the direction of the thumb on a right hand, and the resulting magnetic field is oriented perpendicular to the extended thumb, as represented by the curled fingers of the observer's hand.

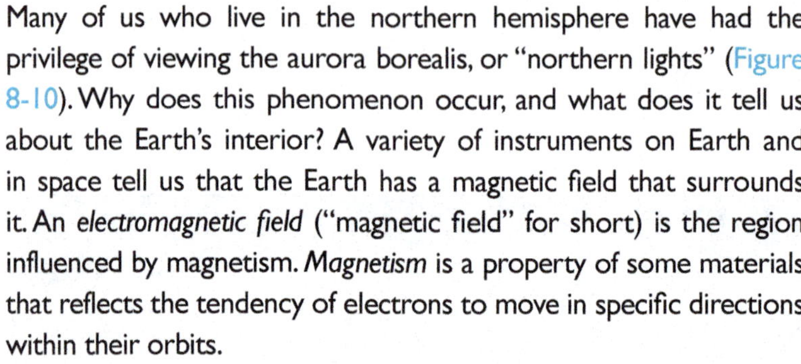

Many of us who live in the northern hemisphere have had the privilege of viewing the aurora borealis, or "northern lights" (Figure 8-10). Why does this phenomenon occur, and what does it tell us about the Earth's interior? A variety of instruments on Earth and in space tell us that the Earth has a magnetic field that surrounds it. An *electromagnetic field* ("magnetic field" for short) is the region influenced by magnetism. *Magnetism* is a property of some materials that reflects the tendency of electrons to move in specific directions within their orbits.

The Earth's magnetic field protects it from highly charged solar particles that travel through space. You see the light energy emitted from the collision of the magnetic field with solar particles as the aurora borealis.

Laws of physics describe the relationship between electric currents and magnetic fields (Figure 8-11). If an electric current is flowing in the direction shown by the red arrow, a magnetic field is created that is oriented at right angles to this (the black arrow). You can imagine this relationship with your hand (the extended thumb = the direction of flow of the electrical current, and the curled fingers = the orientation of the induced magnetic field). In Chapter 12 we will explore *electromagnetism*, the self-propagating interaction between electric currents and magnetic fields, as it relates to the Earth's external energy budget and climate.

Because seismic waves tell us that the Earth has a liquid outer core, and a variety of evidence suggests that it is composed of iron and nickel, it has the proper material to create electrical currents. Its rotation would induce a magnetic field around the Earth.

It's an oversimplification, but the Earth's magnetic field behaves like a simple bar magnet, with two poles (north and south); it is, in other words, a *dipole*. Figure 8-12 shows a bar magnet with poles labeled, as well as the orientation of iron shavings sprinkled around the magnet. You can see that the iron grains align themselves relative to the dipole. Their orientation describes the relationship of the magnetic field to the magnet. By convention, physicists describe the north end of the dipole as the region where the magnetic field lines "start" or emanate, and the south end of the dipole as where the magnetic field lines "come to closure."

Just as we now know that the simple convection cell model for heat flow in the mantle is oversimplified, the same can be said for treating the Earth's magnetic field as a dipole (Figure 8-13), but it does generally describe many observations.

The following figure illustrates a computer-generated model for what the magnetic field encompassing the Earth looks like (Figure 8-14). Recall the convention for identifying the ends of the dipole: the north end = emanating field lines, and the south end = closing field lines. Examine Figure 8-14 and note the position of the blue (closing) and orange (leaving) lines. Yes, they are opposite the *geographic* North and South poles! For the present-day Earth, a dipole within the outer core is oriented such that the south end of the dipole would be positioned toward the north geographic pole. The north end of the dipole would be oriented toward the geographic south pole.

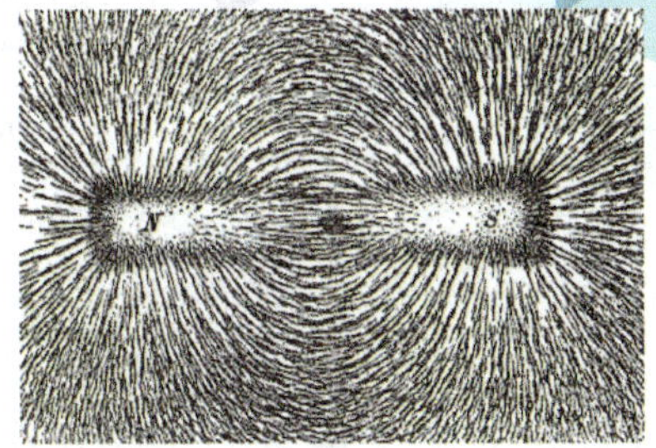

Figure 8-12 Orientation of the field lines around a dipole.

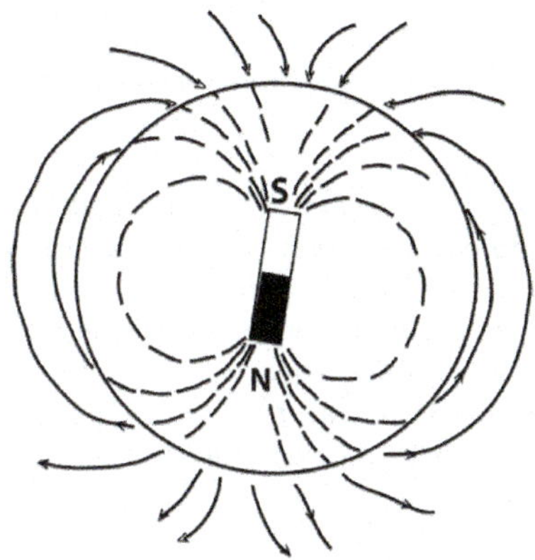

Figure 8-13 Simplified model of a dipole in the core. A dipole would explain most observations about the Earth's magnetic field.

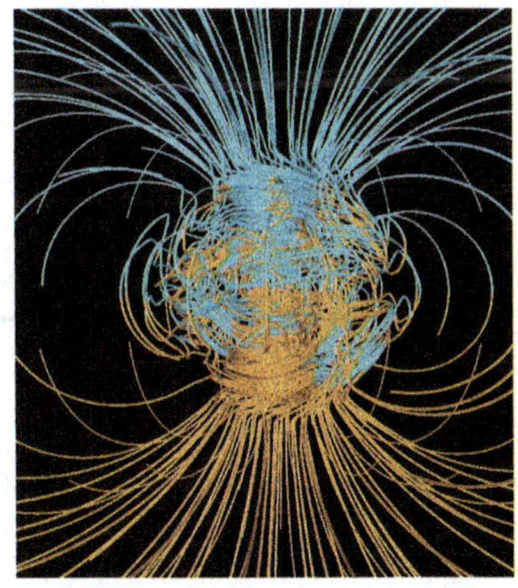

Figure 8-14 Glatzmaier-Roberts geodynamo computer model of the Earth's magnetic field. Field lines are blue where the field is directed inward and orange where directed outward.

Why Does a Compass Point Toward North?

Even when we don't view the northern lights, we have evidence of the Earth's magnetic field at our fingertips. A *compass* is a magnetized filament of iron that aligns itself parallel to the Earth's magnetic field. A compass points toward north because it is aligning itself parallel to the Earth's magnetic field lines. Because of the "opposites attract" principle of dipolar magnetism, the north end of the compass magnet (the needle) points toward the south end of the Earth's

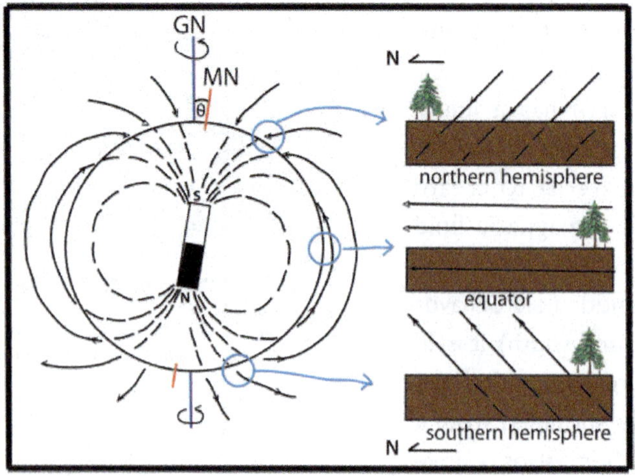

Figure 8-15 Orientation of the Earth's magnetic field. The magnetic field encircles the globe. Note that there is a relationship between the angle of orientation of the Earth's magnetic field and the Earth's surface: parallel at the equator, perpendicular at the poles, and rotating from 0° to 90° between these. MN = magnetic north pole; GN = geographic north pole (also the pole of rotation of the Earth). Θ (theta) is the angle between the geographic and magnetic poles

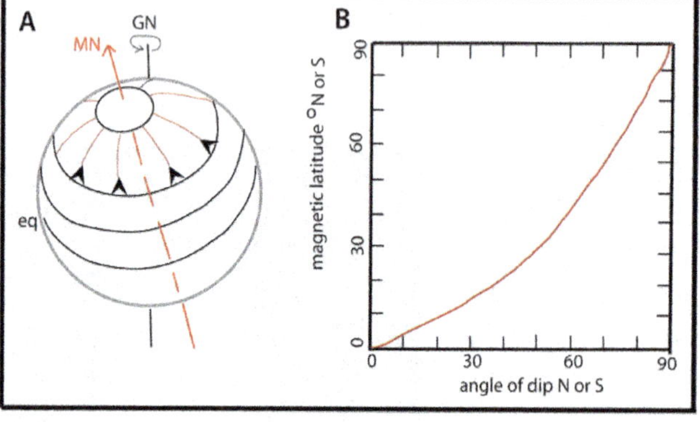

Figure 8-16 A. globe with lines of latitude shown. Everywhere along a line of latitude a compass would point towards north (direction) with the same angle of inclination. The Earth is not a perfect sphere, so a plot of latitude versus angle of inclination (or magnetic dip) produces a *nearly* straight line (B).

dipole (geographic north). (Figure 8-15) shows the orientation of the Earth's magnetic field relative to latitude of the Earth's surface. Regardless of latitude, the compass points toward geographic north (the south end of the magnetic dipole). Compasses tell us that one attribute of a magnetic field is direction. Figure 8-15 also shows us that there is a second property as well, the *angle of inclination* of the magnetic field relative to the Earth's surface.

Note that the magnetic field lines are parallel to the Earth's surface at the equator. In the northern hemisphere the field lines are tilted downward, while in the southern hemisphere they are tilted upward. If you look at the globe in Figure 8-15, you should note that between the equator and the poles the angle of inclination progressively steepens. Confirm this relationship by looking at Figure 8-12. The iron shavings are vertical at the magnet's poles and parallel at the middle of the magnet. The relationship between latitude and the angle of inclination can be plotted on a bivariate graph (Figure 8-16).

The relationship between the angle of inclination and the orientation of the magnetic field is very important. It allows geologists to view some rocks as "ancient compasses" that have preserved the Earth's magnetic field at the time of their formation. If we can measure the angle of inclination, we can determine the latitude at which the rock formed.

Another attribute of magnetism to note is the angular difference between the orientation of the rotational, or geographic, pole and that of the dipole.

This is termed the magnetic *declination.* Figure 8-17 illustrates the difference between inclination (α) and declination (Θ).

Finally, we need to note that the strength and orientation of the Earth's magnetic field changes over time, a phenomenon termed *secular variation.* Both the strength and declination vary over time scales of 10^1 to 10^3 years. At the latitude of London, England, for example, the angle of declination has varied by 35° over approximately four hundred years. There have also been variations in the strength of the Earth's magnetic field. Data by the Natural Resources Canada tell us that the strength of the field at Toronto has decreased by 14% over 150 years of measurement. The origin of secular variation is thought to be variation in the same processes that generate the field to begin with: electrical currents in the core.

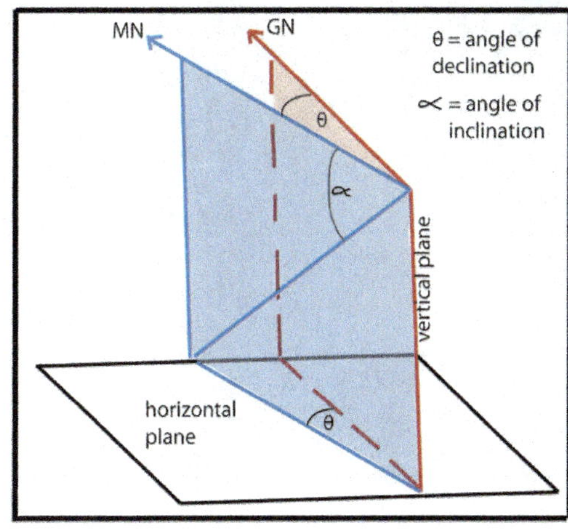

Figure 8-17 Attributes that describe a magnetic field. Theta (Θ) describes the angular difference between geographic north (red) and magnetic north (blue). Alpha (α) describes the angle of magnetic field lines relative to the Earth's surface, a horizontal plane at any point on the surface.

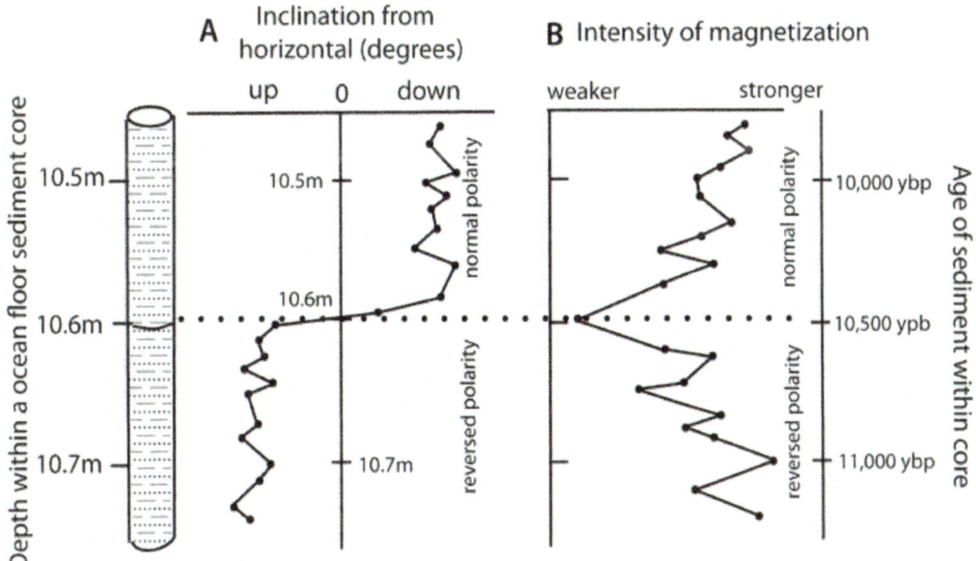

Figure 8-18 A core in ocean sediment illustrates the record of changes in inclination (α) and field strength during a polarity reversal. The actual "flip" in polarity is geologically very rapid, perhaps a matter of 10's to 100's of years. The magnetic field strength collapses and rebuilds over a longer time span of ~1,000 to 100,000 years. Ages shown are estimates that vary as a function of sedimentation rate.

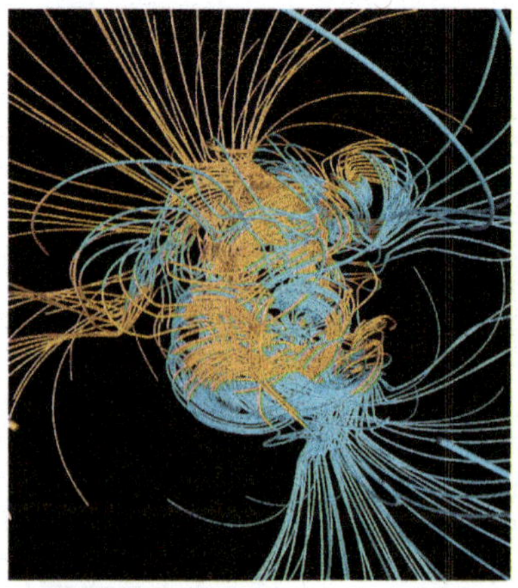

Figure 8-19 Glatzmaier-Roberts geodynamo computer model of the magnetic field orientation during a polarity reversal. Compare to Figure 8-14.

Magnetic Polarity Reversals

A far more significant change in the orientation of the Earth's magnetic field are changes in the polarity of the dipole. *Polarity* refers to the orientation of the north and south ends of the dipole. *Polarity reversals* describe the apparent 180° "flips" in polarity that occur on a roughly one–million-year frequency. When the Earth's magnetic field flips, its strength decreases before flipping, after which it re-strengthens over the same interval of time (Figure 8-18). How much time is does this reversal take? That depends on the amount of time represented by the drill core of sediment. The rate at which sediment accumulates on the seafloor can vary between 1 millimeter/1,000 years to 100 millimeter/1,000 years. Thus, the 10-centimeter interval in the core before a polarity reversal represents between one thousand and one hundred thousand years. This seems like a huge range, but both figures represent very brief intervals of time, geologically speaking.

Like secular variation, polarity reversals probably owe their origin to the behavior of electrical currents in the core. Figure 8-19 shows a computer model of the Earth's magnetic field in a

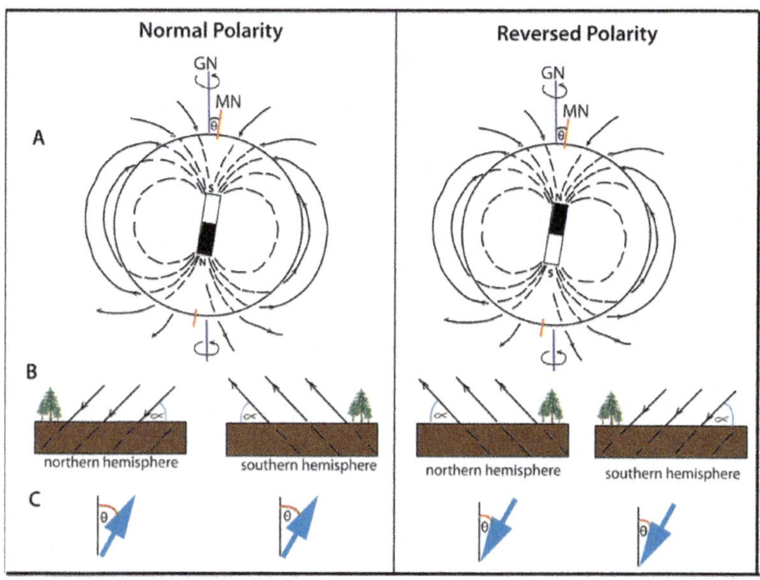

Figure 8-20 Summary of the differences in the orientation of the Earth's magnetic field during normal (left) and reversed (right) polarity events. A represents the idealized orientation of the dipole in the Earth's interior. B represents the angle of inclination of the Earth's magnetic field at similar latitudes on the Earth's surface. C represents changes in the direction towards magnetic north.

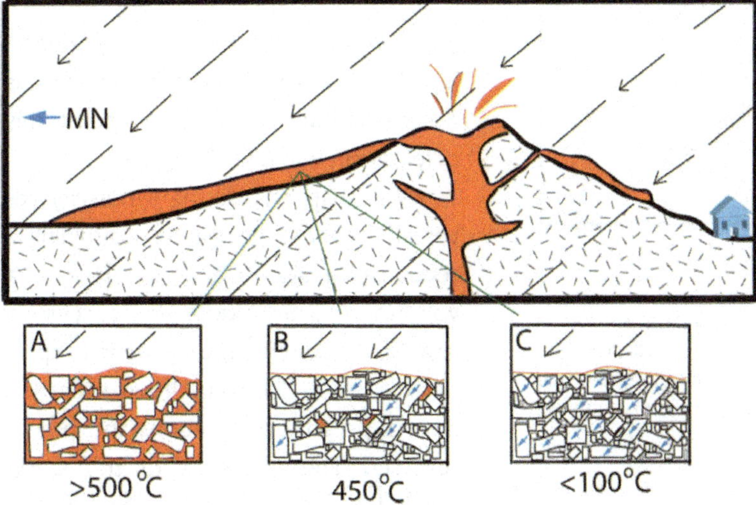

Figure 8-21 Cooling a mafic lava through the Curie Point and "locking in" the magnetism of the present day. Top sketch shows erupting volcano with lava flows. The orientation of the Earth's magnetic field lines indicates that this volcano is located in the northern hemisphere. Panels A through C illustrate how the Earth's magnetic field is fixed within cooling lava. A. crystals have started to form surrounded by droplets of lava (red). B. As the lava cools through the Curie Point the orientation of the Earth's magnetic field is fixed into crystal lattices. C. With further cooling late-forming crystals also record this orientation and the rock is now magnetized.

reversal, and Figure 8-20 illustrates the difference in direction and inclination between normal and reversed times.

How Is the Orientation of the Earth's Magnetic Field Preserved in Rocks?

Iron-rich rocks, such as mafic igneous rocks, have the potential to preserve the Earth's magnetic field. Figure 8-21 illustrates the conditions under which this happens. Shown in the figure is an erupting volcano. The dashed lines represent the present-day magnetic field lines. From this information we can see that the volcano is located in the northern hemisphere, about halfway between the equator and the pole. As the magma cools and minerals begin to crystallize (Figure 8-21A) around 500°C, the electron orbits in the iron-silicate minerals will begin to align themselves with the magnetic field at this time. As cooling continues, to approximately 450°C, most iron silicates have formed (Figure 8-21B), and the rock has recorded the magnetic field at the time of cooling (Figure 8-21C). The range of temperatures between 450 and 500°C, where the

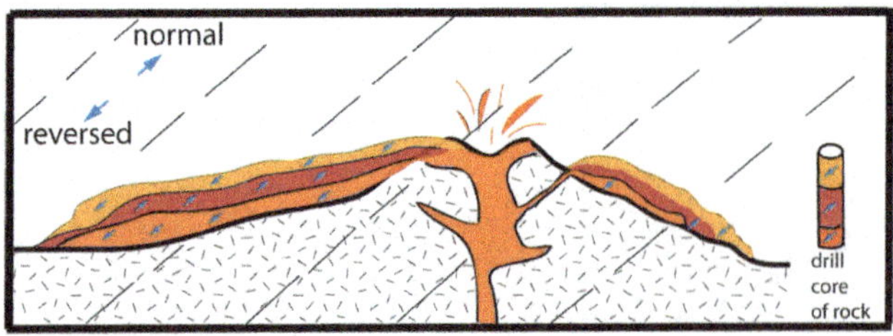

Figure 8-22 Polarity reversal during successive lava flows. During the polarity reversal the orientation of the Earth's magnetic field relative to the Earth surface (angle of inclination) stays the same but "flips" 180°. A drill core of rock through the sequence of lava flows records the reversals.

rock's remnant magnetism is produced, is called the *Curie point*. If the rock is not heated above the Curie point, this orientation will be preserved.

What happens during a polarity reversal? Japanese geologists examining a sequence of lava flows on the flanks of a volcano in Japan noticed that successive lava flows had different magnetic properties (Figure 8-22). Some flows recorded normal polarity orientations and others reversed. The existence of the Curie point for ferromagnetic minerals had been established by geophysicists for some time, so these geologists assumed that the rocks had been reheated and the original remnant magnetism had been "reset."

Study of other volcanoes in Japan with lava flows of the same age, however, indicated that the variation from flow to flow occurred elsewhere as well and thus was not likely to be the result of localized heating. These studies, in the 1960s, (see Takewuchi et al, 1970) were important because it demonstrated that polarity reversals occurred and if the rocks containing them can be dated by radiometric analysis, a time scale could be developed. The cartoon of Figure 8-22 shows that the Earth was in normal polarity (i.e., today's orientation) at some time in the past (orange layer), then reversed polarity (red layer), and finally switched to normal polarity (yellow layer). When we used radiometric decay techniquest to date these rocks the *polarity reversal time scale* was being developed.

(A) 

(B) 

Figure 8-23 (A) A. Lava flows in Iceland B. Hverfjall cinder cone, Iceland (B) Hverfjall cinder cone, Iceland.

Some sedimentary rocks will also preserve the Earth's magnetic field. Grains of iron-rich sediment fall through the water to settle on the seafloor. As they settle, they behave like tiny parachutes, orienting themselves parallel to the field lines. Unlike the crystals of igneous rocks, which are fixed in place once they form, grains of sediment can be reoriented in the compaction and lithification process. Animals can churn through the sediment, looking for food. Despite these problems, the large numbers of grains that will be generally correctly oriented means that these rocks will usually still have a reliable record of the magnetic field at this time and place, as long as they are not heated above the Curie point. This relatively low temperature for alteration of the preserved magnetism explains why metamorphic rocks are not useful for this type of analysis.

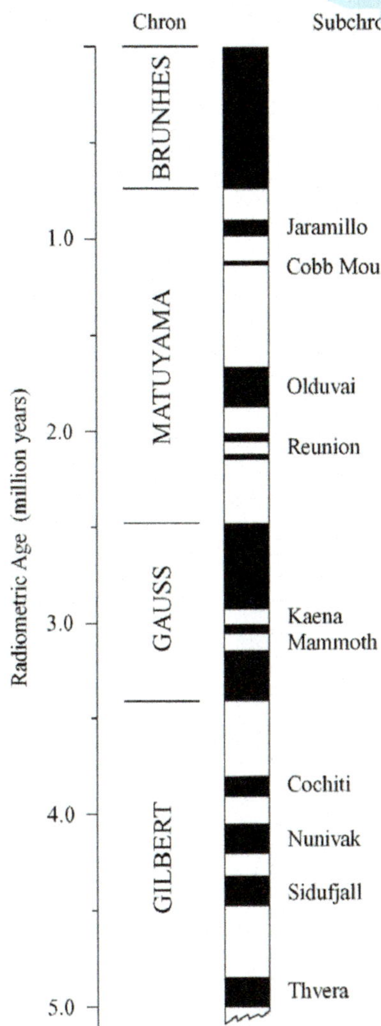

Figure 8-24 The geomagnetic polarity time scale. Numerical ages were obtained through radiometric dating. Black bands represent normal polarity periods, and white bands represent periods of time when polarity was reversed. *Chrons* are intervals of time characterized by predominantly one polarity; they are named after famous geophysicists. Shorter intervals within these are named *subchrons*; these names reflect geographic areas where rocks record these events.

Image from http://geomaps.wr.usgs.gov/gump/common/MPTS.html.

How We Sample Rocks for Their Magnetic Record

There are several assumptions that we need to make if we are going to use data on the magnetism preserved in rocks. First, the assumption that the rocks were formed horizontally is critical, because our measurement of the angle of inclination assumes that we are starting with a dip of 0°. If this is the case, the angle of inclination that we measure in a rock will be a function of the latitude at which it formed. In some cases, such as lava flows across the land (Figure 8-23A), this is generally correct, while in others, such as the flanks of volcanoes (Figure 8-23B), it is clearly incorrect.

Once you have identified an iron-rich rock that you would like to sample for analysis, you need to measure its orientation so that you can restore it to the horizontal. You then take the sample to a special laboratory containing a cryogenic magnetometer, an instrument that enables geophysicists to measure the very faint permanent magnetism of the rock inherited at the time of its formation.

Magnetic Polarity Reversals Over Time

The Japanese geologists sampling the lava flows on the flanks of volcanoes were the first to recognize that there was a preserved history of polarity reversal events. While the flanks of active modern volcanoes might go back in time a few million years, the ocean floor sediments have been recording the Earth's magnetic field for

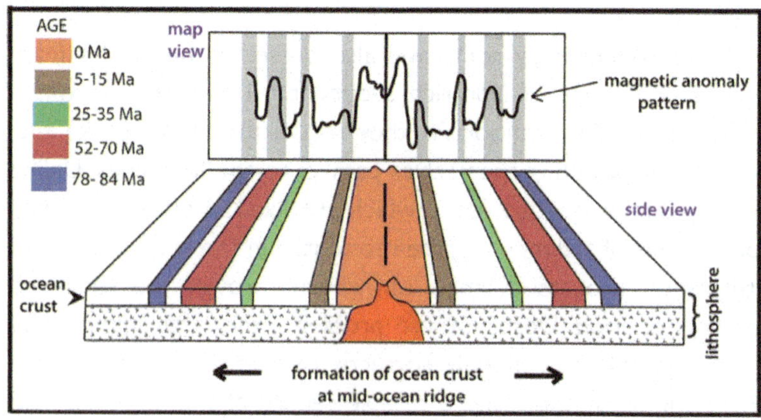

Figure 8-25 Production of magnetic anomalies at a mid-ocean ridge. Rocks of normal polarity are represented by colored bands while reversed polarity layers are white. Side view shows upwelling magma from mantle produces ocean floor basalt at the mid-ocean ridge, which spreads laterally away from ridge axis. Map view shows data viewed from oceanographic magnetometer surveys. Upward spike represent positive anomalies and downward spikes, negative anomalies.

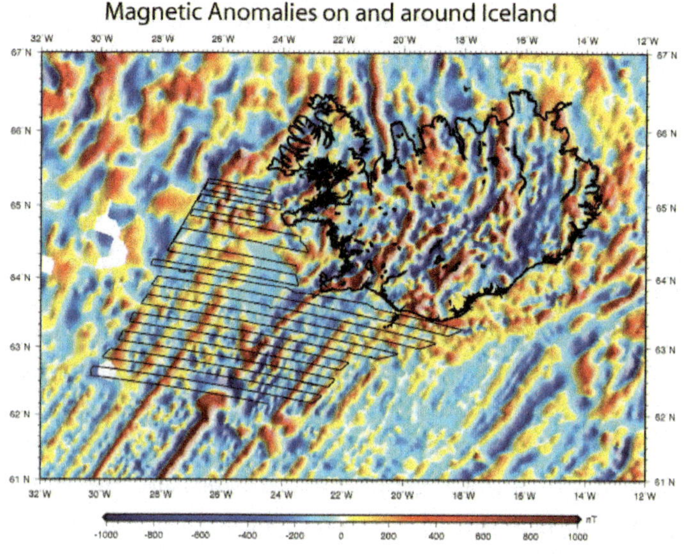

Figure 8-26 Magnetic anomalies on the Mid-Atlantic Ridge, Iceland. Brightly colored bands represent different magnetic anomaly strengths. The heavier black line describes the outline of Iceland. Lighter thin black lies in a parallel array mark the location of transects over the seafloor. The parallel "tape recorder banding" pattern seen in Figure 8-24 is harder to see on the upper left and lower right corners of the map because of the effect of ocean floor sediment (itself magnetized) overlying magnetized ocean floor basalt. Closer to the active mid-ocean ridge, the sediment cover is thinner and the anomalies more distinct.

Source: http://www.soest.hawaii.edu/HIGP/Faculty/hey/rr2007/images/RR2007-figAmag_map.jpg.

tens of millions of years through the slow, steady accumulation of mud on the seafloor. The Deep Sea Drilling Project (DSDP), active in the late 1960s through 1980s, and the more recent Ocean Drilling Program (ODP) and Integrated Ocean Drilling Programs (IDOP) have continued research in the ocean basins around the world. Among other important data, cores of sediment and rock

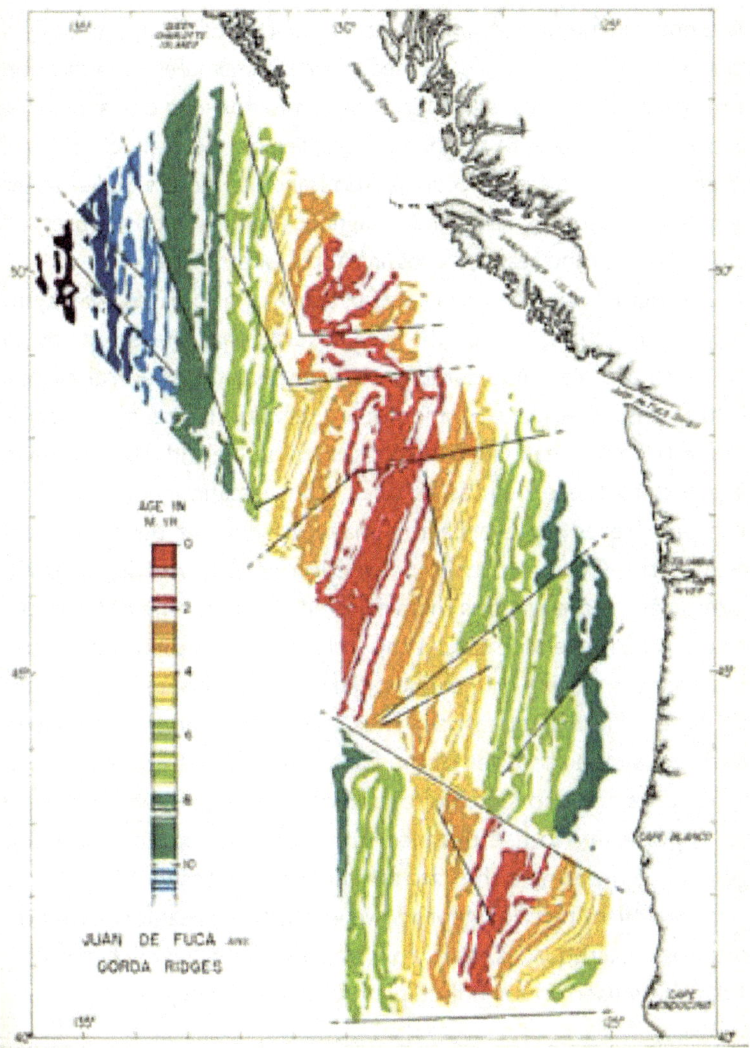

Figure 8-27 The magnetic anomalies on the seafloor off northwestern North America. The ridge axis (red) is currently erupting mafic lava. Normal polarity rocks are colored; white represents reversed polarity. Older rocks, with different anomaly values, are color-coded for their ages. Note the symmetrical pattern about the ridge axis. Offsets in the color bands denote magnetic anomalies offset by faults on the seafloor.

Image from http://walrus.wr.usgs.gov/infobank/programs/html/training/IBDemo/rawjpg/Magnetics.jpg from the ocean floor all over the world have been analyzed, which has led to the creation of a much more complete polarity reversal time scale that extends back eighty million years. Figure 8-24 shows the most recent portion of the documented polarity reversals.

Magnetic Anomalies

Differences in the strength and/or direction of the Earth's magnetic field from what is expected are termed *magnetic anomalies*. Identification of magnetic anomalies is used in prospecting for iron ore and other reserves. The pattern of magnetic anomalies on the seafloor is useful for identifying processes occurring thousands of meters under water. Figure 8-25 illustrates a simplified magnetic anomaly pattern observed on a portion of the seafloor crossing a mid-ocean ridge.

Geologists observed that the pattern of anomalies was symmetrical on either side of the ridge. Recall that these regions of the Earth are also associated with high heat flux. Several of our illustrations of volcanoes and fissure eruptions are also associated with mid-ocean ridge settings, such as Iceland. The evidence all points to these regions as ones where mafic igneous rock is erupting, preserving the Earth's magnetic field as the lava cools through the Curie point. As convection cells move the tectonic plates apart, more igneous material is erupted, and the older rocks are displaced to either side. If the polarity of the Earth's magnetic field changes during this process, successive igneous rock layers can have different polarities. Thus, the ocean floor acts like a continuous tape recorder of the Earth's magnetic history. Because these rocks are also suitable for radiometric dating, we now have the opportunity to use distance (from the mid-ocean ridge) and age data to determine the rate at which the plates move.

In the following two chapters we will synthesize information on volcanoes, earthquakes, heat flow, and magnetism, which consists of data sets used to formulate plate tectonic theory.

Summary

The internal heat of the Earth is one of the two energy sources that drives the Earth system. Most of this heat is generated by the decay of unstable isotopes. Much of these heat-generating material is concentrated in the crust, however the lesser abundance in the mantle is compensated by its greater volume. Thus, the mantle is the source of most of the Earth's internal heat. The amount of heat generated through radioactive decay produces ~0.001 that of the sun, but it is sufficient to power plate tectonics! Heat loss from the interior is focused in narrow belts on the surface which correspond to plate boundaries. Because of the relationship between temperature and density, the density of crustal rocks in "hot regions" lessens, and the rocks "float" higher on the mantle beneath, which gives them a slightly higher elevation on the Earth's surface. There are three ways to transfer heat energy: conduction, radiation and convection. Convection involves the movement of heat energy by the physical movement of hotter material. An over-simplified model of convection is a circular cell of rising and sinking limbs. This implies that within the mantle, hot rock rises towards the crust, spreads laterally, cools and sinks. Where the upwelling limb of the convection cell occurs, the overlying crust heats, rises, cracks and a divergent plate boundary forms. Where a downwelling limb of a convection cell occurs the mantle sinks, "tugging" the overlying lithospheric plate down. In addition to mantle convection cells, there are isolated "plumes" of heat within the mantle that create "hot spots" on the Earth's surface. Volcanoes in the middle of plates, such as the Hawaiian Islands, are examples of hot spots. The linear chain of islands records successively younger volcano formation as the Pacific Plate moves across the stationary plume. Computer synthesis of seismic wave data from around the world allow us to construct three dimensional perspectives of temperature anomalies within the mantle. Convection cells and plumes can be identified but they are idealized representations of a complex reality of heat distribution.

Seismic wave analysis tells us that the outermost core is liquid. Analysis of meteorite composition and calculations of the Earth's average density all point to a liquid outer core composed primarily of iron and nickel. On a rotating earth, this material will generate electric currents which in turn create a magnetic field. Observation of the Earth's magnetic field suggests that it can be modeled as a dipole (two poles: north and south). A compass works because the tiny needle aligns itself parallel with the Earth's magnetic field, pointing to what we call the "north (di)pole." The north (di)pole is slightly out of alignment with the rotational pole of the Earth, an angular difference termed the declination. The third aspect of the magnetic field is the angle of inclination: the angle the Earth's magnetic field makes with the Earth's surface. The angle of inclination varies as a function of latitude. All of these attributes, along with the strength of the magnetic field, are preserved in iron-rich rocks when they form. Particularly useful "rock compasses" are mafic igneous rocks. By studying the orientation of the Earth's magnetic field back through time in layers of volcanic flows we can track the movement of the tectonic plate where the volcano occurs. We also observe that the orientation of the Earth's magnetic field has "flipped" 180 degrees at various times in the past. These "polarity reversals" occur at fairly regular intervals and they follow a predictable sequence: slow reduction in the strength of the magnetic field over several thousands of years, "rapid" polarity reversal (100's to 1,000's of years) and then slow buildup of field strength. We don't know why polarity reversals occur. A magnetic anomaly represents a layer of rock with an unexpectedly greater or lesser magnetic strength or orientation. Because of the continuous production of new ocean crust at divergent plate boundaries, they preserve a complete record of magnetic anomalies over the history of the ocean basin. This enables us to determine the direction and rate of sea floor spreading extending back nearly 200 million years.

Chapter 8 Review Questions

- Examine Figure 8-1 and make at least two observations regarding the data exhibited on this map.
- Use the color key of Figure 8-2 to determine the elevation above sea level of the Himalayan Mountains. What is the maximum depth below sea level of the Atlantic Ocean sea floor at 30 degree North latitude, 300 degree longitude?
- Compare Figures 8-1 and 8-2 and make at least one observation about possible correlations between the two data sets.
- What is the source of the Earth's internal heat?
- What is the relationship between temperature and density of a substance?
- What is the difference between how heat is transferred in conduction, radiation and convection?
- Convection cells are associated with plate boundaries and plumes are not. What is another difference between the two?
- There are 3 types of plate boundaries, classified on the basis of their relative movement: divergent, convergent and strike-slip. What parts of convection cells are related to which type(s) of plate boundaries?
- Complete the following sentence: "A "hot spot track," or a line of volcanoes of increasing age, forms because a plate_________"
- What is "seismic tomography" and what does it enable us to say about the Earth's interior?

- What is "magnetism" and why are some materials "magnetic"? What is a magnetic field?
- Sketch a cross section of the Earth and the orientation of the dipole in the outer core. Label the dipoles and label the geographic north and south poles of the Earth. Summarize the present day relationship between these two.
- How does a compass work?
- There are 3 attributes that, along with strength, describe the orientation of a magnetic field. List and define these.
- What happens to the Earth's magnetic field during a polarity reversal?
- Are all rocks equally suitable for studying the Earth's past magnetic field? What makes a "good rock" for paleomagnetic study?
- What is the Curie Point of a rock?
- What is a magnetic anomaly?
- What observations can we make about the magnetic anomaly pattern on the sea floor that tells us that new ocean crust is being created at mid-ocean ridges?

Part III

How the Solid Earth Works

Chapter 9

Plate Tectonics

At the end of this chapter you should be able to

- Define the theory of plate tectonics and explain how it is different from continental drift
- Describe the characteristics of plate boundaries, including volcanism, seismicity, topography, and heat flow
- Draw and label cross-sections through each of the three plate boundary types
- Define what is meant by a Benioff-Wadati zone and explain its importance in describing the processes occurring at convergent plate boundaries
- Define what is meant by a "subduction zone"
- Define "ridge push" and "slab pull" and explain what these have to do with plate motion

- Predict the type of ancient plate boundary preserved in a region if you are provided with information on rock types and geologic structures
- Identify different types of volcanoes in a photograph and predict the plate boundary type they are associated with
- Describe what is meant by a "passive margin"

Introduction

Philosopher and historian of science Thomas Kuhn published an important book in 1962 in which he suggested that new discoveries in science did not occur by the slow, steady accumulation of data prior to an "aha!" moment. Instead, Kuhn proposed that the bulk of scientific research operated within the existing model or framework (which Kuhn called a *paradigm*), which confirmed existing beliefs while filling in gaps and providing details. Quantum theory in physics and evolution in the biosciences are examples of two modern scientific paradigms. Kuhn proposed that new scientific breakthroughs arise only when the daily workings of science produce results that the existing paradigm cannot explain. Initially, the proposal of a new model would be rejected by the scientific community, particularly because it fails to conform to the existing paradigm. However, after struggling to fit the new data into the old model, scientists *might* acknowledge that the old model may need to be "tweaked," and only when further irreconcilable data force them to do so will they acknowledge that a shift to a new paradigm is needed.

Kuhn's theory "On the Nature of Scientific Revolutions" played out in the geosciences in the 1960s. Prior to the development of our modern plate tectonic paradigm, the geoscience model for how the Earth worked involved a crust (the rocky, hard outer shell) that moved vertically, producing topographic differences such as ocean basins and mountains. Through the late nineteenth and early twentieth centuries, a wide range of hypotheses was suggested for the processes that would cause these vertical movements. The first scientist to suggest that the Earth's surface might also experience significant horizontal movements was a Danish geographer by the name of Alfred Wegner. While a professional glaciologist and climatologist, Wegner was an observant scientist of the world around him. He suggested in 1912 that similarities of the outlines of the coasts bordering both sides of the Atlantic Ocean and the similarity of rocks and fossils on both sides of the Atlantic suggested that these landmasses were one joined into a "supercontinent" he termed *Pangaea* (Figure 9-1). Pangaea broke apart, and the continents moved to their current positions. The theory was called *continental drift*.

Figure 9-1 Reconstruction of Pangaea.

Although Kuhn did not use this geologic revolution as an example in his book, the sequence of steps in the scientific process that followed Wegner's proposal followed Kuhn's description. Alfred Wegner's proposal was soundly rejected by the geologic community. There were some meaningful criticisms, namely Wegner's failure to suggest a viable mechanism for how and why large segments of the crust would move across the Earth's surface. However, some of the criticism was exceptionally harsh because Wegner's professional expertise lay outside of the disciplines in geology that study crustal movement (i.e., he was not an "expert"). It was not until the great mass of oceanographic data on the nature and age of the seafloor became widely known in the 1960s, along with discoveries of the record of the Earth's magnetic field preserved in rocks at roughly the same time, that a paradigm shift was needed. Thus, a new theory was proposed (the Vine-Mathews-Morley hypothesis for seafloor spreading, 1962), which, after further data synthesis, spawned our present plate tectonics paradigm.

This chapter explores plate tectonics, the current paradigm of how the Earth works. *Plate tectonics* describes processes that explain how material moves not only horizontally across the Earth's surface but vertically as well. Plate tectonic processes explain why volcanoes and earthquakes occur where they do, why the style and composition of volcanic eruptions vary, how mountains are uplifted, and how material from the Earth's interior rises to the surface and surface materials are recycled into the interior.

How Plate Tectonics and Continental Drift Differ

The theory of plate tectonics says that the outermost Earth is divided into a series of rigid slabs that move and interact in response to heat flow from the Earth's interior. *Continental drift* describes the horizontal movement of the Earth's crust, so one major difference between the two theories is "what's moving." The "rigid slabs" of the plate tectonic definition is the layer of the Earth called the *lithosphere*. The lithosphere consists of the crust plus the outermost 100 to 150 kilometers of mantle. In other words, the crust and outermost mantle are in motion. This may seem like a subtle difference to you, but it is an important one. The *crust* is the uppermost layer of the lithosphere, a passive passenger on segments of lithosphere that move and interact. As its name implies, *continental drift* places its focus on continental landmasses. The theory of plate tectonics arose in large part from studies on the nature of the ocean floor, especially its youth: the oldest ocean crust is only 180 to 200 million years old, compared to the age of the oldest continental rocks, 4.4 billion years. A comprehensive theory for how the Earth works must be able to explain observations and data from both the continents and ocean basins.

What is a Plate?

The "plate" of plate tectonics refers to the outermost 100-to-150-kilometer-thick layer of the Earth termed the *lithosphere*. This layer is defined not by composition, as the uppermost portion of the lithosphere may be continental or ocean crust overlying the mantle beneath. What distinguishes the lithosphere is the physical behavior of the rocks. This is reflected in the part of the definition referring to the lithosphere as the "rigid slab" of Earth. *Rigid* is a term that describes the physical nature (stiff, unyielding) of a substance; antonyms would include "limp" or "drooping." When stressed, rigid substances are more likely to deform brittlely: bending up to a point and then breaking. Non-rigid substances lack strength and will flow. In sum, the lithosphere is a 100-to-150-kilometer-thick layer of rock on the Earth's surface that is strong. At the present time we recognize twelve major plates (Figure 9-2); the modifier "major" acknowledges that some plates are very small in size and don't appear on a map at this scale.

How do the Plates Move?

The lithosphere is the layer of the Earth that includes the crust and upper 100 to 150 kilometers of the mantle. Below the lithosphere lies the asthenosphere, located in the upper mantle

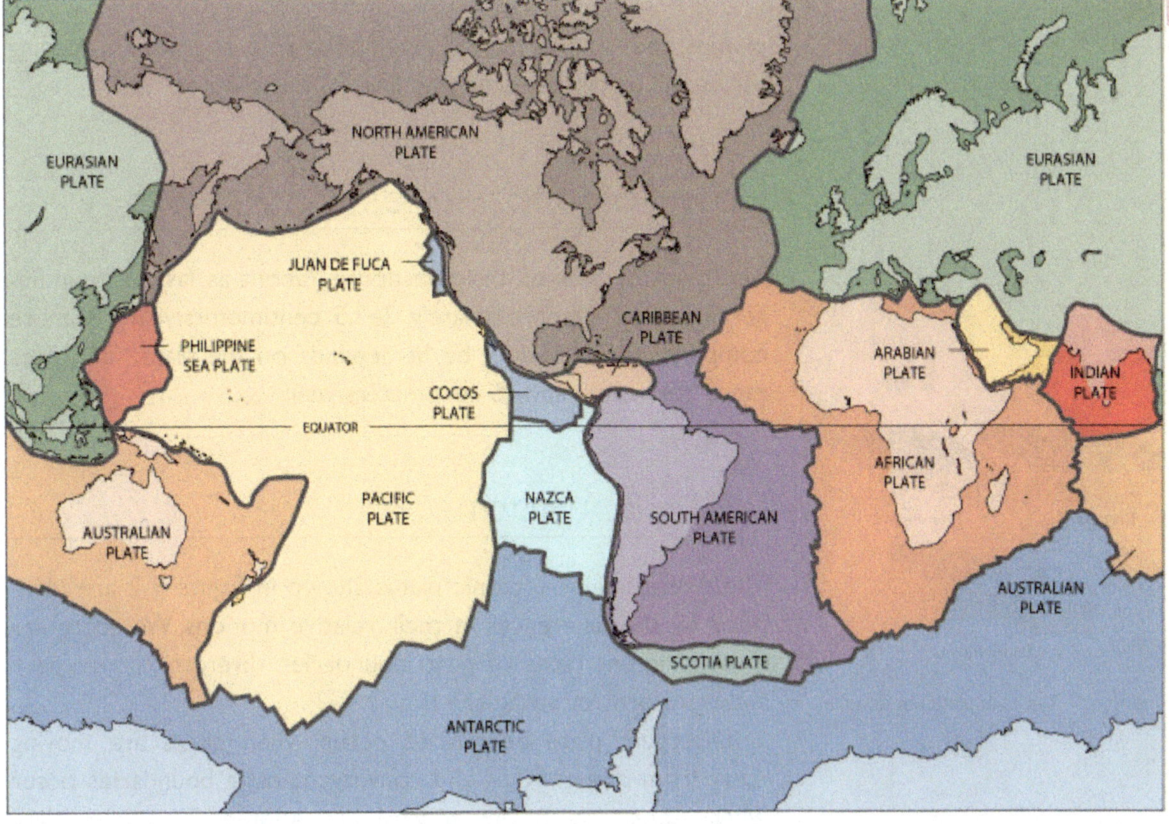

Figure 9-2 Map of modern plates.

Image from http://pubs.usgs.gov/publications/text/slabs.html.

between 150 and 450 kilometers in depth; the lower boundary of the asthenosphere is variable and may extend to depths of approximately 700 kilometers in some regions of the mantle. The upper boundary of the asthenosphere, at the contact with the overlying lithosphere, is the low-velocity zone (LVZ), a region where seismic velocities abruptly decrease. The asthenosphere is a slightly melted, mostly solid, ductile layer. Just as an ice skater can move across the ice on a thin film of water melted by the friction of his or her skate blade on the ice, the lithosphere can move on the more viscous asthenosphere below it. However, unlike the ice skating analogy, the asthenosphere is not created by the moving lithosphere above it but results from changes in mineral packing and physical properties in a region of elevated temperatures and reduced pressure within the mantle.

The energy required to move the lithospheric plates comes from the heat from within the Earth (convection cells and mantle plumes; see Chapter 8) and gravity. The role of gravity in moving plates will be discussed further later in this chapter.

How Fast do Plates Move?

The quick answer to this question is "about as fast as your fingernails grow" (approximately 2–2.5 centimeters/year). A more complete answer would be "it depends on the plate," as values range between 2 and 15 centimeters/year.

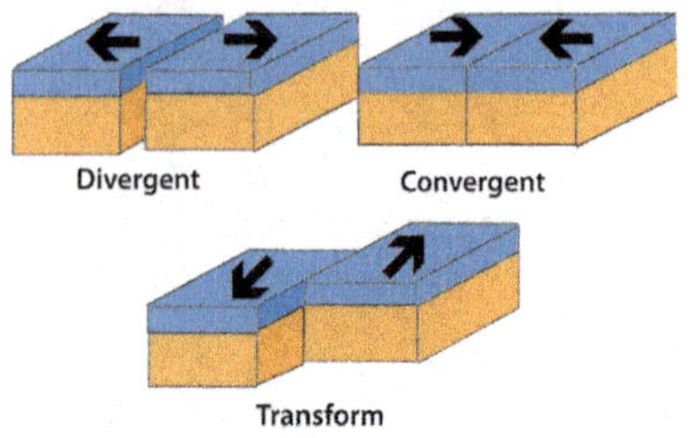

Figure 9-3 Types of plate boundaries.

Plate Boundaries

The twelve major tectonic plates shown in Figure 9-2 are identified by the differences in their relative motions. We recognize three different types of plate boundaries: divergent, convergent, and transform, or strike-slip (Figure 9-3).

Divergent plate boundaries occur when plates are moving apart from one another, and convergent plate boundaries occur when plates are moving toward one another. Transform plate boundaries are characterized by plates that are sliding by one another sideways. Seismic studies of earthquakes associated with plate boundaries indicate that divergent plate boundaries are characterized by stresses indicating the rocks are under extension, or "pulling apart." Similar studies indicate that rocks on both sides of convergent plate boundaries are experiencing compression, or squeezing, while those along transform boundaries are experiencing sideways shearing. This data is very useful in understanding the forces operating on the rocks involved in plate boundary interactions. Rocks at divergent plate boundaries are literally being pulled apart, while those at convergent plate boundaries are being pushed together. Rocks are shearing by one another sideways at transform plate boundaries. What processes are occurring at the plate boundaries that would produce these stresses?

Plate Tectonics, Volcanism, and Earthquakes

Figure 9-4 is a map of the world showing magnitude-5 or greater earthquake epicenter occurrences over a five-year interval of time (1990–1995). The different colors represent different focal depths.

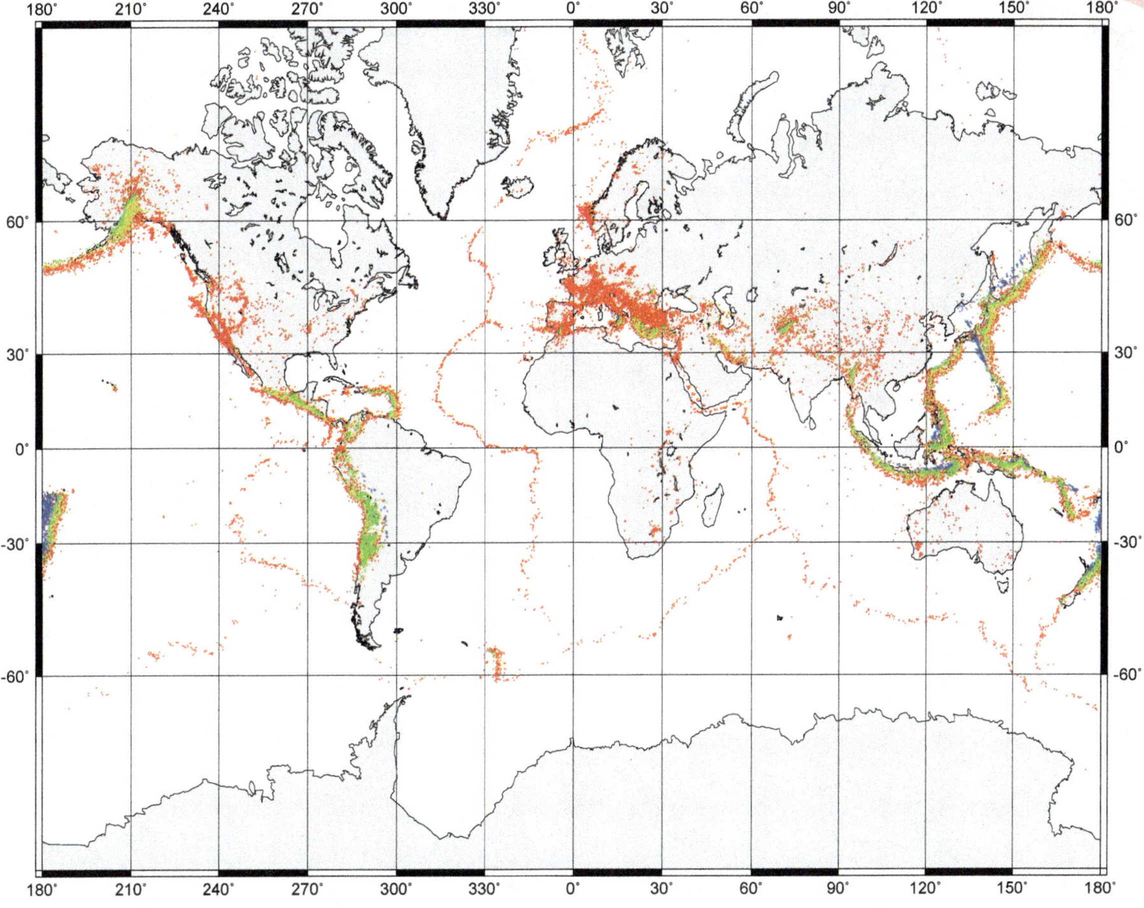

Figure 9-4 Seismicity around the globe. Red color = shallow focus (< 33 kilometers); orange = intermediate depth foci (33–70 kilometers); green = deeper intermediate focus (70-300 kilometers); blue = deep focus (300-700 kilometers).

Image from Sawyer, Dale, *Discovering Plate Boundaries.*

Source: http://plateboundary.rice.edu/quakes.11.17.pdf.

Scientists develop their theories by examining and interpreting data. Examine the data shown in Figures 9-4 and 9-5, the global distribution of earthquakes and volcanoes. Are these features randomly distributed across the globe? Do you see a pattern to the distribution? I'm sure you've decided that there are regions that appear to have more or less active volcanism than other areas do. The same is true for earthquakes. Once a scientist discerns a pattern, the next step is to try to describe it in more detail. You may try to convey the linear nature of some of the occurrences, for example. You might note the relationship of some of the data to the edges of continents. If you look carefully at Figure 9-4, you might see that the different colors used in the figure provide you with data on the depths of the

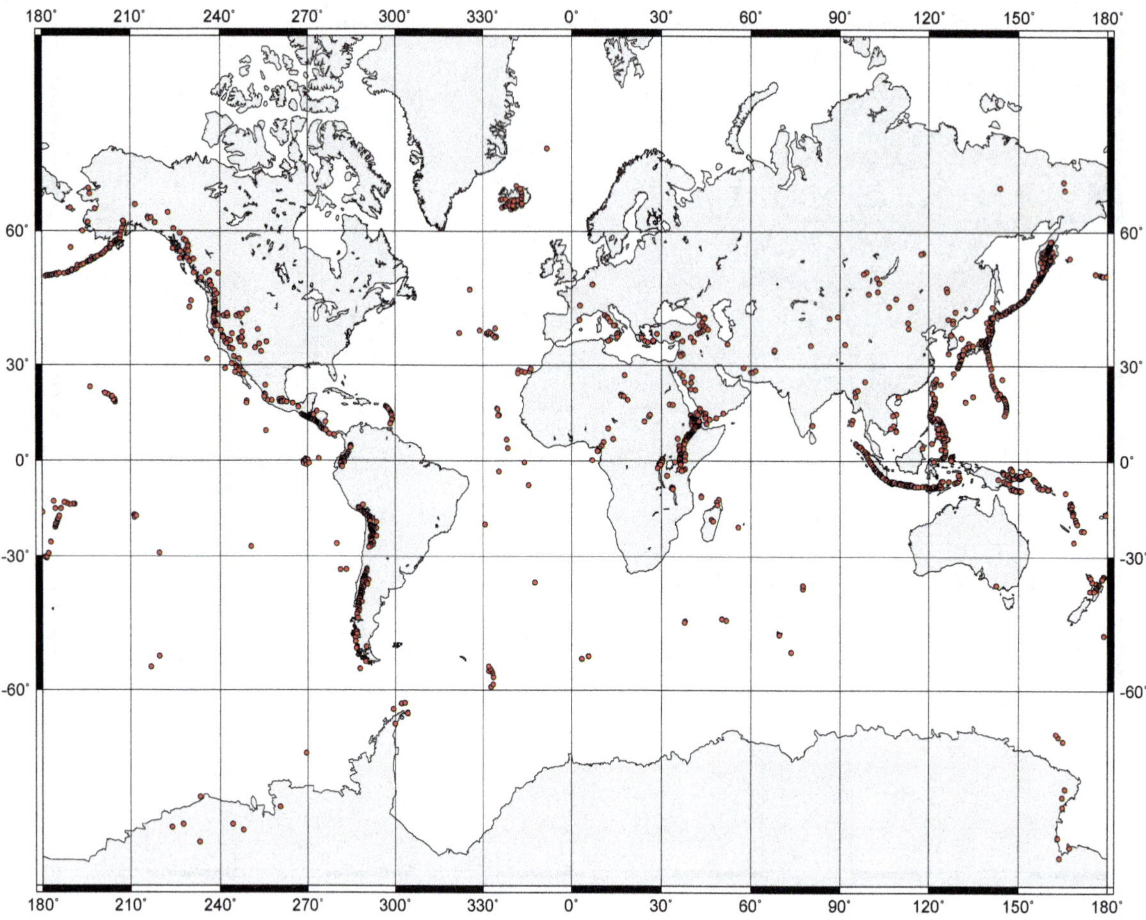

Figure 9-5 Red circles indicate global distribution of currently or historically active volcanoes.

Image from Sawyer, Dale, *Discovering Plate Boundaries*.

Source: http://plateboundary.rice.edu/volcano.11.17.pdf.

earthquake focus. Is there a color pattern present? The more descriptive you can be about data, the easier it will be to move to the next step and think about "why." Why are there parallel belts of earthquakes at progressively increasing depths in some regions of the Earth? Once a scientist crafts "why" questions, he or she can begin to plan out how to answer the "why." What do the data suggest to you? How could you test what you think is going on? Are there theoretical reasons why you would think that a particular process or event is occurring here? What data would you need to collect to test your ideas? All these questions describe the process of doing science, and they all start with making observations about the world around us.

Refer back in the chapter to Figure 9-2, which shows the major tectonic plates in today's world and their relative motions. Compare the locations of plate boundaries to the observations you made about the distribution of earthquakes and volcanoes. Do they appear to be related? You probably said "yes" in answer to this question, but now you need to go deeper. Do *all* plate boundaries have the same occurrences of volcanism and earthquakes? Earlier in this chapter we classified plates based on the relative motion at their boundaries. Factor this into your observation of volcano and earthquake global distribution; are the distributions of volcanoes and earthquakes different at divergent, convergent, and transform plate boundaries? You should be able to make a list of *at least* six more detailed observations about possible relationships.

Interpreting Volcano and Earthquake Distributions

Our first attempt at interpreting your observations will be based on what scientists term "first principles" or an interpretation based on the most "bottom-line" concepts of your discipline. Recall from an earlier chapter that an earthquake occurs when ______________ (fill in the blank). Hopefully, you said, "When rock breaks" or "When there is movement on a fault," because that's what an earthquake is, the released energy from a stressed rock. OK, now you can ask, "What stressed the rock?" or "What is occurring at this place in the Earth that caused stress to build up in the rock, ultimately causing it to fail and break?" Now things are getting interesting, as there are several processes that could generate stress in rocks. In order to explore these further, we need to revisit seismic wave interpretation.

Focal Mechanism Studies

When a seismometer records the arrival of an earthquake wave, geologists are very interested in the first motion exhibited by the P waves at this location. When the seismograms at several stations are analyzed and the geographic distribution of compression (the first motion is squeezing) and dilation (the first motion is expansion) is plotted, we can interpret whether the initial movement of the Earth at the focus is characteristic of normal, reverse, or strike-slip fault movement (examine Sidebar 7-1 to review these different types of

Figure 9-6 In these graphs, black quadrants represent regions under compression, while white quadrants are dilational for an earthquake focus at the center of the beach ball. Below the graphs are shown blocks of the Earth's crust with faults (red lines); from left to right: strike-slip, reverse (thrust), and normal (extensional) faults.

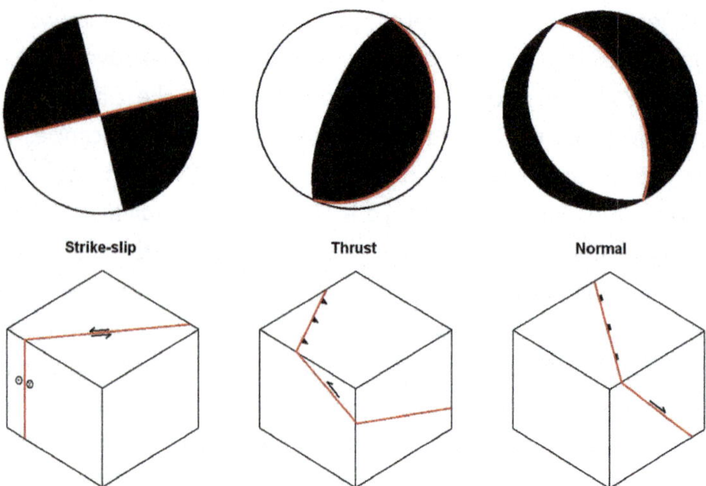

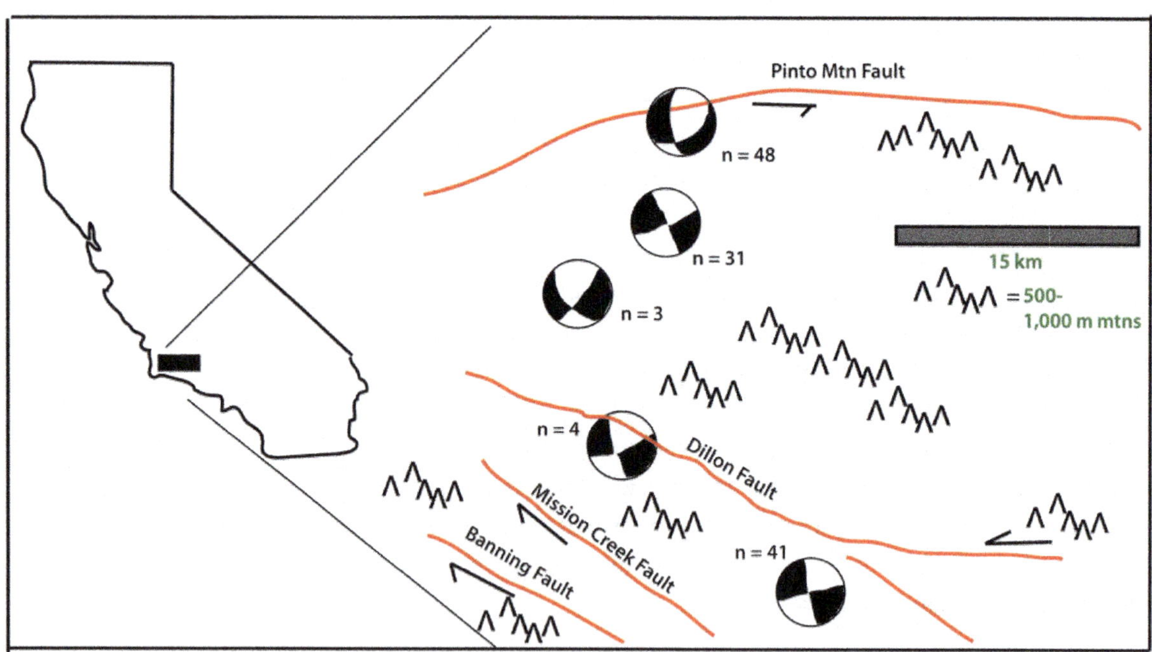

Figure 9-7 Focal mechanism data along a short segment of a fault in California.

faults). This is called a *focal mechanism solution*. Focal mechanism solutions are interpreted from seismic data plotted on circular graphs, which show, using black-and-white quadrants, the distribution of compression and dilation along a fault.

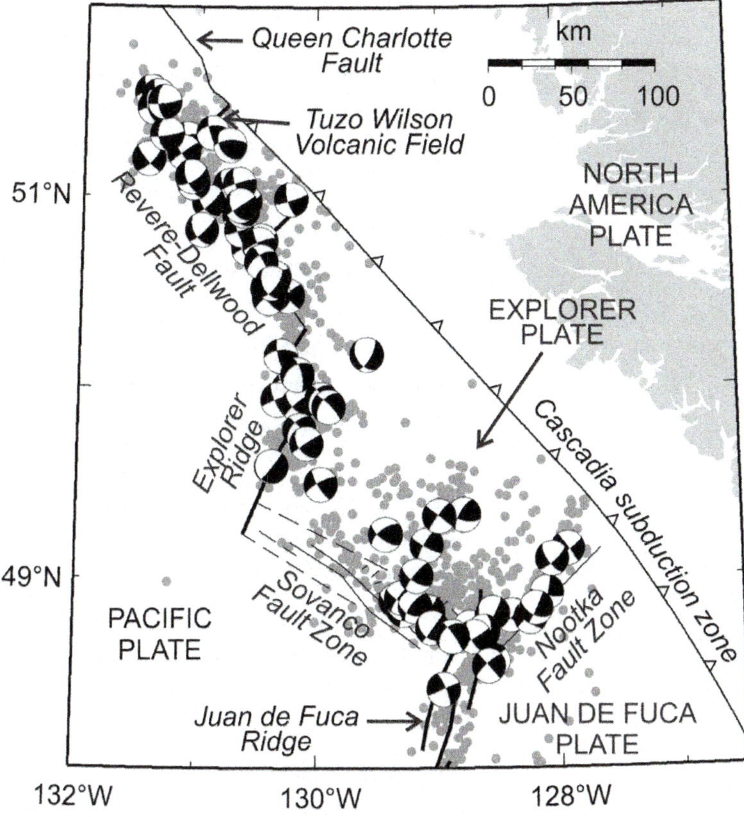

Figure 9-8 Focal mechanism solutions for shallow-focus earthquakes in the Pacific Ocean, west of British Columbia, Canada.

Source: http://www.bssaonline.org/content/100/5A/2002/F9.large.jpg.

How these graphs are produced is beyond the scope of this discussion; however, it is important to know that seismologists have a technique of analyzing seismograms that tells us what types of stresses are present when an earthquake is triggered.

Figures 9-7 and 9-8 both illustrate focal mechanism solutions for several magnitude-4 and -5 earthquakes: California and the Pacific Ocean west of British Columbia, Canada, respectively. Do these appear to have been generated from movement on strike-slip, reverse, or normal faults?

These two examples demonstrate that geologists can use these "first-motion studies" to determine the types of fault movement that trigger an earthquake. Because each type of fault is produced by specific stresses, we can use first motion data to tell us what stresses the rocks were under prior to the earthquake. The California data set in Figure 9-7 is an example of the horizontal

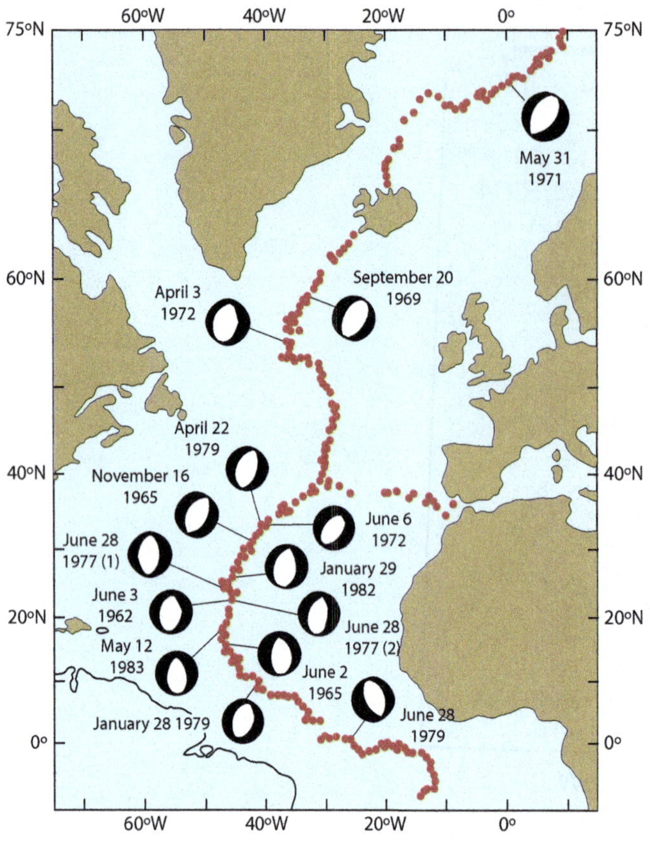

Figure 9-9 Mid-Atlantic ridge earthquake focal mechanism solutions.

Source: http://www.geosci.usyd.edu.au/users/prey/Teaching/Geos-2111GIS/FMS/Atlantique.gif.

fault motion associated with strike-slip faults. Not surprisingly, these faults are characteristic of transform plate boundaries. The northwestern Canada data set in Figure 9-8 is more complex, showing a combination of fault motions, both strike-slip and extensional. This region lies on a convergent plate boundary; however, you would probably predict that the fault motion should show compression. Why would we find extensional and sideways stresses here? This apparent incongruity is an example of why first-motion studies are so useful. They give us details about the processes in plate boundaries that need to be explained if we are to fully understand what happens when plates interact. We'll discuss processes at convergent plate boundaries later in this chapter, when an explanation for these types of faults will be more apparent.

Stepping back and looking at this data set from the perspective of "how science works" is also valuable. If you make a prediction that the earthquake first motions at convergent plate boundaries should show compression, and they don't, you have a few options. You could say, "The plate tectonic model must be wrong!" or "We don't really know what's going on here!" Alternatively, you can say, "What is it about motions at convergent plate boundaries that I don't yet understand? Let's look more closely." By reframing your confusion in the form of further questions to explore, science "advances." It is entirely possible that in the process of asking further questions, several anomalies arise and you really *do* have to reject your hypothesis. This outcome is what Kuhn would describe as a scientific revolution.

Observations of Divergent Plate Boundaries

If you examine the plate relative motion data in Figure 9-2, you can identify several plate boundaries that are characterized by divergent plate motion, for example, the plate boundaries between South America and Africa. If you study Figure 9-4, you will be able to make observations about the seismicity of this region. For example, what are the focal depths of these earthquakes? Hopefully, you were able to note that this plate boundary is seismically active and that the earthquakes are all shallow in focus (< 33 kilometers in depth). The focal mechanism data (Figure 9-9) indicate that the bulk of faults are normal (although strike-slip faults are also present). Synthesizing the earthquake data tells us that the lithosphere in this type of plate boundary is under extension at very shallow depths. Literally, the lithosphere is being pulled apart along this plate boundary.

Is there a characteristic volcanism at divergent plate boundaries? Examine the data shown in Figure 9-5. How would you characterize volcanism along the divergent mid-Atlantic plate boundary? For example, you may say that this region is nonvolcanic (volcanoes are absent or very rare), or volcanically active but not highly characteristic of this region, or possibly that volcanoes are common to abundant. Confirm your observation by examining other divergent plate boundaries (check Figure 9-2 to find these), such as the East Pacific Rise (between the Pacific, Nazca, and Antarctic plates) and east Africa (Kenya and Ethiopia). Because many divergent plate boundaries are within ocean basins (why is that?) and obscured from view by kilometers of water, the few that are on land are especially useful to examine. Iceland is a portion of the mid-Atlantic plate boundary that is above water, and the East African region is also accessible, so these are places where we can see what volcanism looks like along divergent boundaries (Figure 9-10, A–C). Because so many divergent plate boundaries lie under the oceans, they have been the site of research expeditions via submersibles, either manned or remote-controlled (Sidebar 9-1).

Based on your examination of Figure 9-10, what types of volcanoes appear to be characteristic of the divergent plate boundaries? (You may want to refer to Figure 5-11.) Be sure to include fissure eruptions in your description. Recall that the type of volcano that forms says something about the

(A) Gada Ale, Ethiopia (287m). Lava flow fills large fissure. Photo by Jean Fillipo.

(C) Cinder cones near Hayli Gubbi, Ethiopia. The cones have been cut by several faults. Yellow material is sulfur precipitated from toxic volcanic gas emissions. Photo by Marco Fulle.

(B) Krafla, Iceland.

(D) Krafla fissure. This fissure system lies directly on the mid-Atlantic plate boundary in central Iceland. While volcanoes, such as Krafla (Figure 9-10B) are common, much volcanism is from fissure eruptions. Lava is extruded along linear fracture systems that extend for kilometers. In the photo below, steam rich in sulfur can be seen emanating from the meter-wide fissure. This fissure lies within kilometers of Krafla volcano. The active fissure system in Iceland extends the length of the entire island (approximately 300 kilometers). Note: Fissure volcanism is not characterized by the positive topographic features that we call "volcanoes," so occurrences of these types of eruptions are not plotted in Figure 9-5.

Figure 9-10 Illustrations of divergent plate boundary volcanism.

viscosity of the lava that is being erupted. Based on the types of volcanoes you identified, hypothesize what the characteristic viscosity of the lava might be at divergent plate boundaries. What does this imply about the composition of magma in these regions?

Examine Figure 8-1, the global map of surface heat flux. Describe what these values are for divergent plate boundaries in general, and for our two examples (mid-Atlantic and East Africa). Do these values contradict or support your observations about volcanism along these types of plate boundaries?

The topography of the Earth's surface along divergent plate boundaries is also difficult to observe in many regions because of the presence of ocean basins, but a variety of remote sensing data on the Earth's gravitational field allows us to see topographic relief on the seafloor (Figure 8-2). With Iceland and East Africa as our land examples, use Figure 8-2 and the visual images of these regions (Figures 9-10 and 9-11) to provide more detail on their landscapes.

Figure 9-12 is a map of the age of the ocean basins. Compare your age observations to those of the other data sets we've discussed thus far (earthquakes, volcanism and heat flow,

Sidebar 9-1 Hydrothermal Vents on Mid-ocean Ridge Systems

Woods Hole Oceanographic Institution (WHOI) photograph of a black smoker, so named because of the dense concentration of suspended finely crystalline minerals that crystallize when the hot (> 300°C) water comes in contact with colder ocean water. The slurry, rich in iron and sulfur minerals, shoots out of the vent at velocities of more than 1 meter/second. The photo was taken from the WHOI submersible *Alvin*. In the foreground is an array of remote-controlled sampling equipment. The black smoker chimney is encrusted with a diversity of life forms adapted to these extreme conditions.

What we see at these sites is a fantastic array of life forms adapted to life in dark, warm-to-hot, mineral-rich waters adjacent to hydrothermal vents termed "black smokers," and volcanic fissures. These regions are examples of "extreme" environments on Earth, and they host uniquely adapted "extremophile" organisms. Because light does not penetrate to these depths in the oceans, there are no photosynthesizing organisms; the base of the food chain consists of microbes and bacteria that derive energy for metabolism from chemical reactions with elements ("chemosynthesis") dissolved in the hydrothermal water. The deep ocean hydrothermal vent regions are very important sites to study because they represent environments that might be present on other planets, such as Venus. They also might be the only present-day sites on Planet Earth that represent the same conditions that existed when life evolved on Earth. Finally, the chemical "soup" that erupts from these vents precipitates significant quantities of minerals, including gold and copper. Although it would be environmentally harmful to do so, might we be mining the seafloor around these plate boundaries someday?

(A) A lake fills a central valley in Iceland at the site of the first parliament held in this country in 930 AD. On the left side of the central valley is the North American plate; on the right side, the European plate. The mountain in the distance (left-center, capped by cloud) is a shield volcano. The photographer is standing near a cliff edge that represents a fault scarp (a cliff face formed along a fault).

(B) A fissure system runs parallel to the central rift valley in Iceland. The landscape is dissected by numerous faults. The rock visible is mafic igneous rock (basalt).

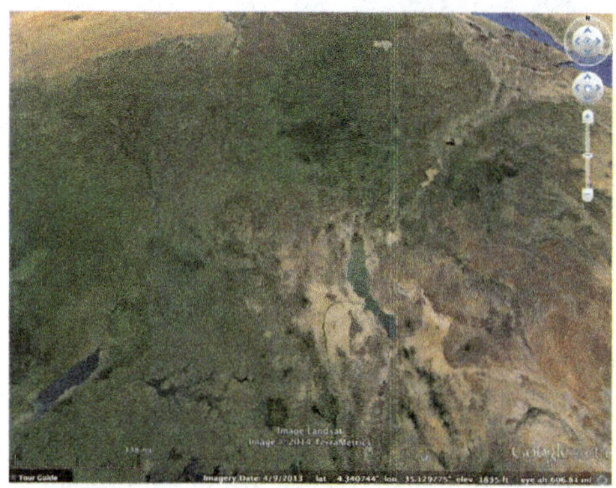

(C) The East African Rift Valley makes up the right half of this Google Earth image. Lake Turkana, Kenya, is in the center of the image. Ethiopia lies to the northeast. The India plate is to the right of Turkana, and the African plate is to its left.

Fig. 9-11C: Source: Google Earth.

Figure 9-11 Images of divergent plate boundary landscapes.

topography). Are young rocks associated with areas of volcanism, shallow-focus earthquakes, high heat flow, and elevated topography?

The magnetic anomaly record in the ocean basins is the final data set we will examine. Recall from Chapter 8 that magnetic anomalies are differences in the predicted strength or direction of the Earth's magnetic field at a location. Figure 8-27 presents the magnetic anomaly data for a portion of the Pacific Ocean floor to the west of British Columbia, Canada. Normal polarity rocks are colored, and reversed polarity rocks are white. In Figure 8-26, magnetic anomaly data are shown for the Atlantic Ocean seafloor south of Iceland. What is the relationship between magnetic anomaly patterns and age distribution?

Summary of Observations of Divergent Plate Boundaries

You have been presented with a variety of data sets and asked to make observations about them. You should have been able to note the following features:

- Shallow-focus earthquakes are common.
- Focal mechanism solutions indicate that the stresses are extensional in nature, forming normal faults.
- Volcanism is present. In some areas, such as Iceland, there appear to be clusters of volcanoes. In other areas of these plate boundaries, volcanoes are widely spaced to absent. The volcanoes do fall along a linear array that coincides with earthquake epicenters. Volcanoes are of the shield and cinder cone variety. Shield volcanoes form from low-viscosity magma. Much volcanism, however, is of the fissure style, which explains why there are not as many volcanoes as you might expect on the global volcanism map.

- Divergent plate boundaries are characteristically areas of high heat flow and elevation. On the ocean floor, the plate boundaries are less deep; on continents, they appear to be part of broad plateaus. Looking at the topographic relief more closely, divergent plate boundaries are characterized by volcanic highlands and valleys. Often the valleys have water (river or lakes) in them. Rocks are offset by normal faults, so cliffs and valleys are common. Movement on these faults might be related to the numerous earthquakes in these regions.
- In the ocean basins, divergent boundaries are characterized by the youngest rocks on the ocean floor. The age of the rocks increases away from the boundary, and the trend in age is symmetrical on both sides of the boundary. The same symmetry is seen in the magnetic anomalies. The increasing

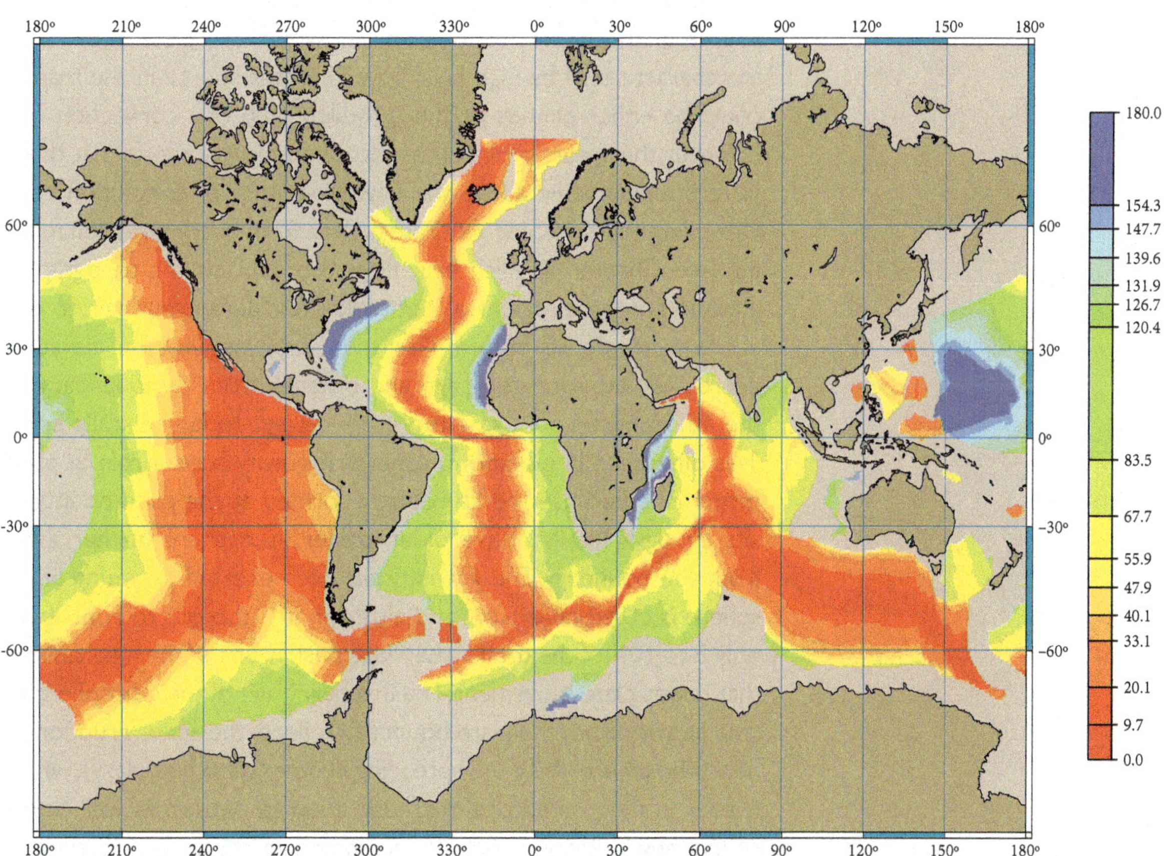

Figure 9-12 Age of ocean basins in millions of years (color-coded).

age of the ocean floor away from the boundary correlates with an increase in depth over the same distance.

Processes at Divergent Plate Boundaries

Now that you have a list of observations that you have made, the processes occurring at divergent plate boundaries can be inferred. The presence of shallow-focus earthquakes indicates that the uppermost lithosphere is being stressed, and the focal mechanisms indicate that normal faults, which form in conditions of extensional stress, are common. Literally, the lithosphere is being pulled apart, or rifted, here. Elevated heat flow suggests that there is magma present in the upper mantle, an interpretation supported by the presence of volcanism. Shield volcanoes are characteristic of low-viscosity magma, which would be the case if the mantle were melting to produce the magma. The chemistry of erupting lavas is also what we would predict for melted mantle. The presence of a large heat source in the upper mantle explains why these plate boundaries are characterized by high heat flow. The heat source in the mantle could be either plumes or the upwelling limb of a convection cell; however, the linear nature of the plate boundary suggests that these plate boundaries lie above the upwelling limbs of convection cells. Upwelling mantle would generate magma from decompression melting (see Chapter 5). High heat flow also explains the topographic elevation of these regions, as hot rock is less dense, causing it to rise above more dense mantle. This, then, explains why the elevation decreases with increasing age and distance from the boundary as the lithosphere cools, gets more dense, and topographically sinks. This difference in elevation means that gravity plays a role in plate movement, because the lithosphere formed at the plate boundary sits topographically above older, colder lithosphere further away from the boundary. The gravitational push exerted by young lithosphere is called *ridge push*, and although it is thought to contribute only 5 to 10% of the total force, ridge push is one of the mechanisms that drives plate movement. The increasing age of the lithosphere as you go away from the plate boundary reflects the active eruption of new lithosphere along the spreading center (plate boundary), which is seen in the symmetrical magnetic anomaly pattern as well. Some of this newly formed rock at the spreading center gets pushed in one direction, and some in the other, forming the symmetry on opposite sides of the central rift. All these processes are summarized in Figure 9-13.

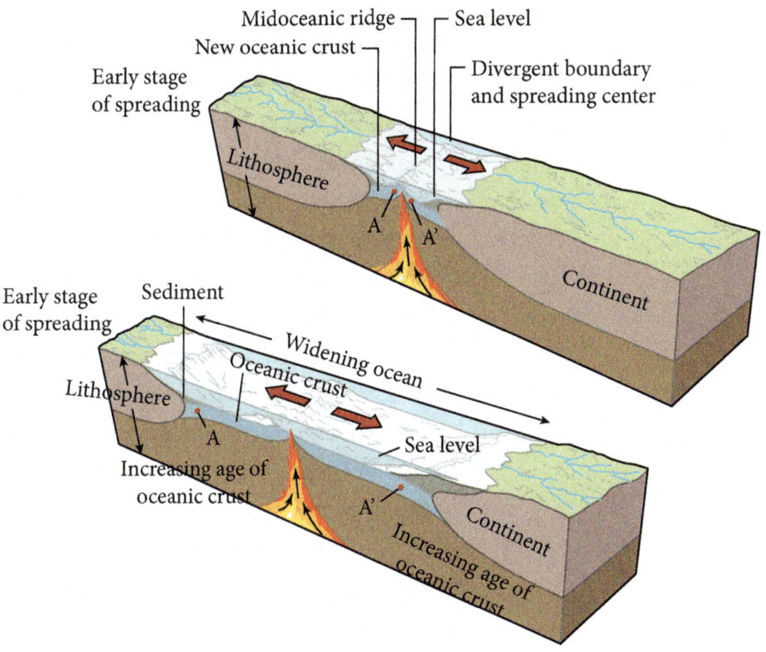

Figure 9-13 Divergent plate boundary processes. The upper figure shows the early stages of extension that rift a continent apart and form a new ocean basin. The lower figure shows a more mature divergent plate boundary where the ocean basin continually widens, as shown by the increasing distance between points A and A′. The East African rift seen in Figures 9-10 and 9-11 is in a stage of divergence where the central valley is not yet flooded by seawater.

Implications for the Earth System

The geologic processes active at divergent plate boundaries are very important for the recycling of Earth materials. The upwelling and eruption of mantle material (chemical species: elements, compounds, and molecules) "resupplies" the Earth's surface. The concentration of ores (economically valuable minerals) at mid-ocean spreading centers is one example of this. The creation of new ocean basins from the rifting of continents also has implications for the Earth's climate. We will see in Chapter 11 that ocean circulation patterns are a major control of the distribution of heat over the Earth's surface.

Observations of Convergent Plate Boundaries

If you examine the plate relative motion data in Figure 9-2, you can identify several plate boundaries that are characterized by convergent plate motion, for example, the plate boundaries between the South American and the Nazca plates or the Pacific and

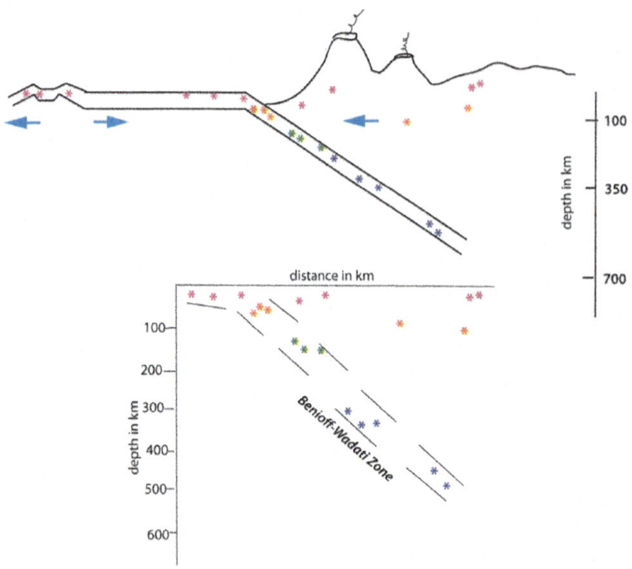

Figure 9-14 Schematic cross-section of a Benioff-Wadati zone. Focal depth key: red dots: shallow focus (<33km); green dots: intermediate focus (33–70km) ; purple dots: deep focus (>300km).

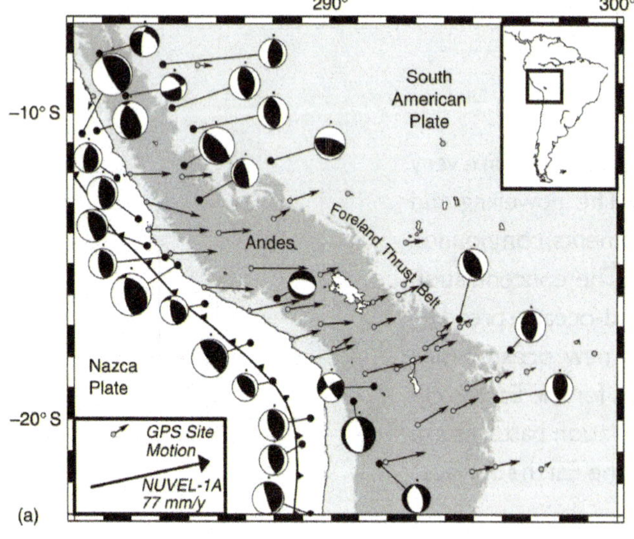

Figure 9-15 Focal mechanism data from a portion of the Nazca–South American convergence zone.

Source: http://www.earth.northwestern.edu/people/seth/Texts/iaspeiChap07.pdf.

North American plates south of Alaska. Examine Figure 9-4 and make observations about the seismicity of this region. For example, what are the focal depths of these earthquakes? The focal depths are more complicated than what you observed about those along divergent plate boundaries. Are most focal depths from one depth? Is there a pattern to the focal depths? Hopefully, you recognized that there are parallel color bands at convergent plate boundaries, and because each color represents different depths to the earthquake focus, this means that there is a series of different focal depths present. Examining a two-dimensional map, this pattern appears as parallel belts, but when projected into a third dimension, this means that the focal depths progressively increase (Figure 9-14).

From Figure 9-14, you can see that the parallel belts of color in Figure 9-4 represent earthquake foci. On the distance-versus-depth graph in Figure 9-14, you can see that the earthquake foci define a region, termed the *Benioff-Wadati zone*, where rocks are being stressed; this region defines the presence of the descending plate in the convergence zone. This tells us that at convergent plate boundaries, one plate is subducted beneath the other. *Subduction* refers to the process of slab descent into the mantle at convergent plate boundaries. A *subduction zone* is the region within a convergent plate boundary where one plate is being subducted beneath the other. The Benioff-Wadati zone tells us where the descending slab is within the mantle.

The focal mechanism data for magnitude-6-or-greater earthquakes along the Nazca–South American zone is shown in Figure 9-15. Compare the graphed data to the "beach balls" in Figure 9-6 to determine the nature of faulting in this region. Earthquakes at convergent plate boundaries are more complex than are those at divergent plate boundaries, as the data in Figure 9-15 show. Based on relative motions, you would be correct in predicting that all earthquakes would be the result of convergent stresses. The data, however, suggest that there is spatial variation in the type of fault motion at convergent plate boundaries. In Figure 9-15 the black line with "teeth" on it (small black triangles) marks the subduction zone (the boundary between plates, where one plate, in this case the Nazca, is being forced under the other, the South American plate). What type of faults are present to the left of this boundary? You would be correct if you replied, "Mostly reverse faults." To the east (right) of this, the faults are a combination of both normal and reversed. Even further to the east, the faulting is primarily thrust. We know from Figure 9-14 that the spatial distribution of earthquake data at convergent plate boundaries represents changes in depth as well, so our pattern of fault motion on a map really reflects changes that are a function of depth. The takeaway point here is that the nature of stresses in a subducting plate changes relative to its distance from the subduction zone. In other words, the subducting lithosphere at a convergent plate boundary is under compression at shallow depths, under both compression and extension at intermediate depths, and under compression at deeper depths.

Recall from Figure 9-8, which showed earthquake focal mechanisms off the western coast of Canada, that the nature of stresses at the Cascadia subduction zone were complex and not all compressional, as we might predict. The epicenter location of these earthquakes is on the subducting slab before it enters the subduction zone. It is literally bending and breaking. Thus, the focal mechanisms indicate extension ("opening" from the bending). The subducting slab must also be resisting downward motion into the mantle, and there is some sideways slippage (strike-slip fault motion) as the bent and broken slab begins descent into the mantle. As our discussion of the scientific process predicted, we now know more about the detailed geologic processes that occur at various places within the convergent plate boundary.

Is there a characteristic volcanism at convergent plate boundaries? Examine the data shown in Figure 9-5. How would you characterize volcanism along the convergent Nazca–South American plate boundary? For example, you may say that this

(A) Andes Mountains along the Argentina-Chile border. Two snow-covered volcanoes of tertiary age are visible. These volcanoes were active about ten to fifteen million years ago.

(B) Mt. Rainier in the Cascade range of the Pacific Northwest.

(C) Mt. Etna with Naples, Italy, in the foreground.

Figure 9-16 Illustrations of convergent volcanoes.

region is non-volcanic (volcanoes are absent or very rare), or volcanically active but not highly characteristic of this region, or possibly that volcanoes are common to abundant. Confirm your observation by examining other convergent plate boundaries (check Figure 9-2 to find these), such as the Asian-Filipino plate or Pacific-Australian plate boundaries. Figure 9-16 illustrates convergent plate boundary volcanic features (see also Figure 5-11B).

Based on your examination of the photos in Figure 9-16, what types of volcanoes appear to be characteristic of the convergent plate boundaries? Recall that the type of volcano that forms says something about the viscosity of the lava that is being erupted. Based on the types of volcanoes you identified, hypothesize what the characteristic viscosity of the lava might be, as well as the relative abundance of pyroclastic debris, in volcanic eruptions at convergent plate boundaries. What does this imply about the composition of magma in these regions?

Examine again the global map of surface heat flux (Figure 8-1). Describe what these values are for convergent plate boundaries in general, and for the Nazca–South America example in particular (the north Pacific plate is hard to see on this map projection). Do these values contradict or support your observations about volcanism along these types of plate boundaries? Are heat flow values as easy to characterize at convergent plate boundaries as they were for divergent? What does this imply?

Let's reexamine some seismic tomography data for what they might tell us about heat flow in subduction zones. Figure 8-7 is a map and cross-section of the western United States that illustrates seismic wave velocity anomalies. Earlier in the text we discussed what these data could tell us about the presence of mantle plumes. These data can also tell us whether there are areas of colder, more dense rock within the mantle. The black arcuate line on the map of Figure 8-7 shows the position of the cross-section of the upper 1,000 kilometers of the Earth. The Yellowstone plume is shown in bright red, but call your attention to the area to the left, under northern California. Here a region of bright blue is located, extending to a depth of about 200 kilometers. This indicates the presence of an anomalously cold region within the mantle. As we think about the processes occurring along convergent plate boundaries, we need to think about how to interpret these data.

Figure 8-2 and the photographs in Figure 9-16 might help you characterize the topography of the Earth's surface along convergent plate boundaries. Don't forget that part of a convergent plate boundary, the subduction zone, lies hidden from view beneath the ocean (Figure 9-14, although schematic, shows the position of a trench seaward of the volcanic land area). What can you infer about the depths of the ocean floor seaward of the continental margin of South America or Alaska? In order to estimate the maximum topographic relief (total elevation change) at a convergent plate boundary, you will need to estimate elevation change. You can do this by using the color-coding on Figure 8-2 and applying your skill at translating a two-dimensional map to three-dimensional perspective. For example, across the Filipino-Asia plate boundary at Japan, the topographic relief would be estimated to be approximately 9,500 meters (8,500 meters in depth in the trench to an average 1,000 meters of elevation on land). Not all modern convergent plate boundaries involve ocean-continent transitions. You should have noticed in Figure 9-2 that the boundary between the Indian and Asian plates is convergent. What is the elevation of Mt. Everest? Your topographic relief estimates and your answer to the "height of Mt. Everest" question should make it clear that topographic relief at convergent plate boundaries is among the highest on Earth.

Summary of Observations of Convergent Plate Boundaries

You have been presented with a variety of data sets and asked to make observations about them. You should have been able to note the following features:

- Shallow- to deep-focus earthquakes are very common.
- Focal mechanism solutions indicate that the stresses vary, and include both tension (forming normal faults) and compressional (forming reverse faults) stresses.
- Volcanism is common. The volcanoes fall along a linear array that coincides with earthquake epicenters. Volcanoes are of the stratovolcano variety, which indicates that both lava and pyroclastic material erupt. The abundance of pyroclastic material reflects explosive magma, formed from a silica-rich composition. High-viscosity magma forms in the presence of volatiles (gas and/or water) or if the composition is rich in silica.
- Convergent plate boundaries are characteristically areas of low heat flow; in fact, some regions, such as the Filipino-Eurasian boundary, involve some of the coldest lithosphere on Earth. Cold rock is more dense than hot rock of the same composition, and dense rock sits topographically lower in the mantle.
- Convergent plate boundaries are characterized by extreme topographic relief, from the tallest mountains on Earth to some of the deepest trenches.

Processes at Convergent Plate Boundaries

Now that you have a list of observations that you have made, the processes occurring at convergent plate boundaries can be inferred. The presence of shallow- to deep-focus earthquakes in the Benioff-Wadati zone indicates that the lithosphere is being stressed as it is forced to descend into the mantle under an overriding plate. The presence of earthquakes at depths of 300 kilometers or more indicate the depths to which the subducted lithosphere still may be identifiable within the mantle. This observation is confirmed by seismic wave velocity anomalies, which can identify the presence of old subducted lithosphere within the mantle, millions of years after its subduction. The seismic activity in the Benioff-Wadati zone also sheds light on material that is melting to generate magma fueling the abundant stratovolcanoes characteristic of convergent boundaries. Initially, it might be assumed that the descending slab itself is melting; however, seismic evidence clearly indicates that it is solid. Recall that in Chapter 5 we discussed several ways in which we can cause melting at different temperatures and pressures. If we add volatiles to hot rock, we lower its melting point (Figure 5-6). Ocean crust, formed at a divergent plate boundary, can accumulate kilometers of sediment on it before reaching a subduction zone. This material readily melts when heated, releasing volatiles under pressure into the subduction zone, in a process termed *slab dehydration*. This lowers the melting-point temperatures of the rock of the overlying plate, generating magma. Friction from the plates grinding past one another also generates heat and causes frictional melting. The composition of the magma will be highly variable, dependent on the composition of the rock of the overlying plate. Further melting is possible from conduction melting (magma transfers heat to the surrounding rock, melting it). As opposed to divergent plate boundaries, where magma was much more uniform in composition because of the common mantle source, the magmas at convergent plate boundaries are much more variable. While mafic igneous rocks such as basalt do form, intermediate compositions (andesite, diorite) and felsic rocks (rhyolite, granite) are common.

The tremendous pressures exerted on rocks within convergence zones make these prime locations for metamorphism, the alteration of preexisting rocks from heat and pressure. Rocks are deformed (folded and faulted) as they are metamorphosed.

Analysis of seismic velocity anomaly data tells us that subducted lithosphere is "negatively buoyant"; in other words, it is sinking into the mantle. Recent studies (ex, Conrad and Lithgow-Bertellon, 2002) have been able to quantify the force exerted by the descending plate, a term called *slab pull*, as a function of the density difference between the subducting lithosphere and the mantle below and the plate's size. Other geologists (ex, Schellart et al, 2010) have been able to model the relationship between the velocity of plate movement and the size of the subduction zone; larger subduction zones create higher plate velocities. This is strong support for the dominance of slab pull as the driving force to plate motion. The implications of this work are profound. The existing paradigm for plate motion focuses on the role of mantle heat flow. However, these new studies imply that the plates

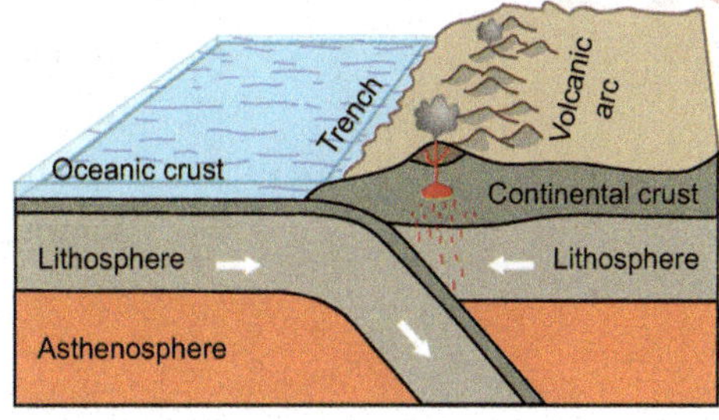

(A) Schematic diagram of processes at a convergent plate boundary where lithosphere bearing oceanic crust is being subducted beneath a plate bearing continental crust. The Nazca-South American plate boundary is an example.

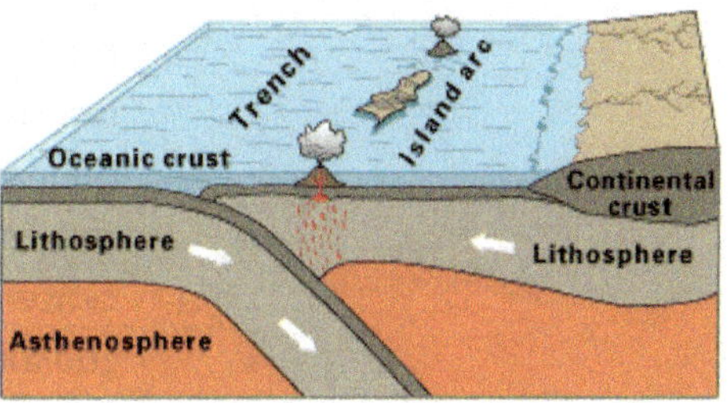

(B) Ocean-ocean convergence. Examples include the western Pacific: Australian and Pacific plate margins. The island arc describes the chain of volcanic islands that arise above the subduction zone.

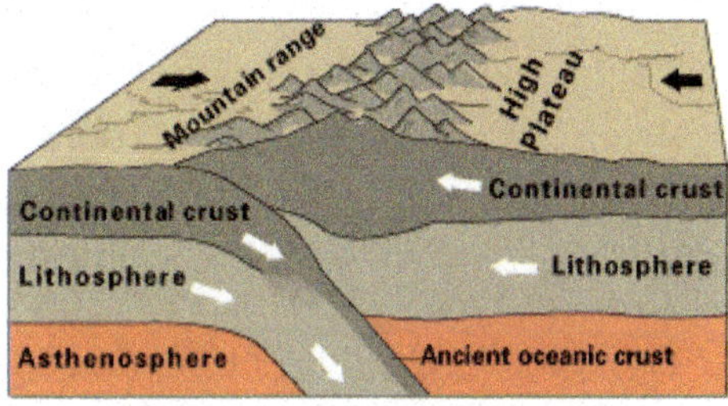

(C) Continent-continent convergence. The prime example of this type of collision is the ongoing Indian-Eurasian plate convergence, which is elevating the Himalayan Mountains. While lithosphere bearing continental crust is currently present on both plates, note that prior to this, the oceanic crust portion of the Indian plate was subducted beneath Eurasia. Now that this portion of the Indian plate is fully subducted, less dense, more buoyant continental material will not do so. The Himalayas have the height they do because of the "underplating" of Indian continental crust under Eurasian continental crust. The crust under the Tibetan high plateau is approximately 70 kilometers thick.

Figure 9-17

themselves exert a strong influence on their movement, and this in turn might influence heat flow within the mantle.

The processes occurring at a convergent plate boundary are summarized in Figure 9-17.

Implications for the Earth System

The geologic processes at convergent plate boundaries have a major impact on the Earth system. The subduction of lithosphere results in melting of materials that trigger volcanic eruptions that eject a variety of chemical elements. Chapter 5 summarized the tons of lead, copper, zinc, and other metals brought to the surface from the lower crust and mantle during convergent-plate-boundary volcanic eruptions. Metal deposits around hydrothermal vents indicate similar ore deposition at convergent eruptive centers. Volcanic eruptions impact atmospheric composition and alter the Earth's climate (Chapter 12). The melting of ocean floor sediments in subduction zones is one of the major ways in which carbon is cycled from the geosphere to the atmosphere. Hydrothermal fluids moving through metamorphic rocks precipitate a variety of important ore minerals, such as silver and gold. As we will discuss in the next chapter, the formation of mountains at convergent plate boundaries has important implications for the Earth's climate. Mountains alter local climates, but their uplift and weathering (chemical and physical breakdown) impact the hydrosphere and atmosphere.

Transform, or Strike-Slip Boundaries

If you examine the plate relative motion data in Figure 9-2, you can identify several plate boundaries that are characterized by transform, or strike-slip, plate motion, for example, the plate boundaries between the California (North American) plate and the Pacific plate. The network of faults along this plate boundary is part of the famous San Andreas fault system (several strands of parallel faults comprise the plate boundary). Strike-slip plate boundaries connect divergent and convergent plate boundaries and segments of both of these types of faults that intersect at an oblique angle. The lateral motion is how the change in relative motion between two plates is transferred laterally.

(A) Air photo of the San Andreas Fault in the Carrizo Plain region, approximately 160 kilometers north of Los Angeles. Because of the aridity of the region, the nearly 9 meters of offset from an 1857 magnitude-8 earthquake are still visible.

(B) Air photograph of Wallace Creek, near Bakersfield CA, offset by San Andreas Fault. The fault crosses the road and uplift along the fault is clearly visible. Radiometric dating in this region indicate that the SAF has been slipping approximately 34mm/yr for at least the past 13,000 years.

(C) Aerial view of the San Andreas fault in Coachella Valley, California. The fault scarp is clearly visible between the white arrows. Photo by Michael Rymer, USGS. The San Andreas fault is a right lateral fault, which means that the rocks that you see as you look across the fault are moving to the right.

Figure 9-18 Images along the San Andreas fault.

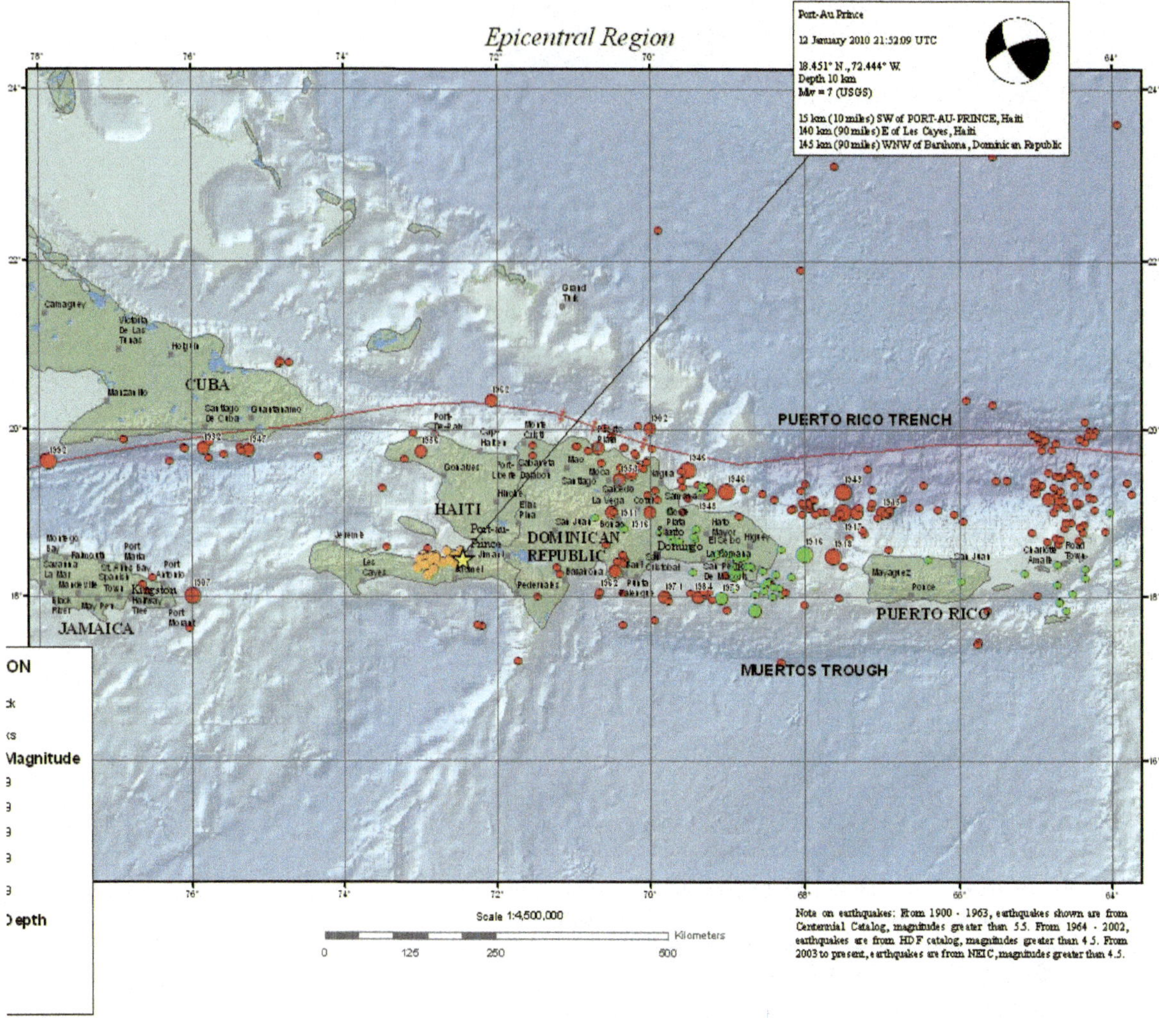

Figure 9-19 The January, 2010 Haiti earthquake represented strike-slip fault movement along the Enriquillo Fault Zone (EFZ) one of several strike-slip faults that links the convergent plate boundary between the North American and Caribbean plates (Puerto Rico Trench) and the convergent Cocos and Caribbean plates further to the west. Figure shows historical earthquake foci in this region. A summary focal mechanism solution for the 2010 earthquake is shown in the upper right. Like the 1700's earthquakes along this fault, the 2010 magnitude 7 earthquake (focal depth=13km) was also strike slip.

Examine Figure 9-4 and make observations about the seismicity of this region. For example, what are the focal depths of these earthquakes? Are most focal depths from one depth?

The focal mechanism data for a small portion of this boundary (Figure 9-7) indicate that the bulk of faults are strike-slip faults. Synthesizing the earthquake data tells us that the lithosphere in this type of plate boundary is under sideways shear at very shallow depths.

Is there a characteristic volcanism at transform plate boundaries? The volcano distribution data in Figure 9-5 are drawn at a scale that makes it difficult to distinguish volcanism occurring east of the plate boundary from any activity along the San Andreas fault system itself. From the data presented in Figure 9-5, you would probably say that volcanism is common; however, these features are actually well to the east of the present-day plate boundary. In the next section of the text we will discuss how plate boundaries can evolve over time, so perhaps the volcanism data reflect this. Recall from Figure 8-7 that we are able to image the presence of subducted lithosphere approximately 200 kilometers in depth beneath part of California and Oregon. This told us that this region used to be a convergent plate boundary. Perhaps some of the volcanic centers that we identify on the map of Figure 9-5 are related to a prior plate boundary. In general, transform plate boundaries are non-volcanic.

Examine Figure 8-1, the global map of surface heat flux. Describe what these values are for the San Andreas transform plate boundary. Another major transform plate boundary is the Anatolian fault system in northern Turkey. Do these heat flow values indicate a nearby source of mantle heat?

Figure 9-20 Google Earth image of the EFZ in southern Haiti. Much of the topography of the island is inherited from its former convergent setting; however, the location of the fault (faint red line) is a prominent topographic feature.

Source: Google Earth.

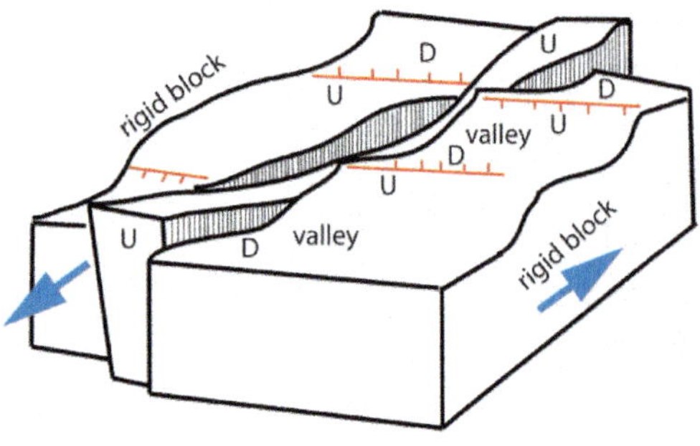

Figure 9-21 Schematic diagram of wrench faults and the formation of topography. Basins and uplifted blocks form when strike-slip faults tear open the Earth's surface.

The topography of the Earth's surface along the two transform plate boundary examples can be inferred from Figure 8-2; although, again, the scale of these maps doesn't allow you to see details.

As Figures 9-18, and 9-19, illustrate, the topography associated with strike-slip plate boundaries varies but always seems to include a valley where the fault axis lies. Water flows down the path of least resistance, which is often a fault surface. Consequently, the rocks around a fault surface easily erode away. As seen in Figure 9-18, ridges are also common topographic features; these are often the fault scarps themselves: the Earth's surface that has been moved upward from fault motion. You might say, "But wait—I thought that the motion on strike-slip faults was horizontal, not vertical; why is there an uplifted fault scarp?" Good question. Very few fault surfaces are vertical; usually they are inclined (notice that all our "beach ball" focal plane solutions in Figures 9-7, 9-8, 9-9, 9-15, and 9-19 are tilted). Even though the predominant motion on the San Andreas fault is right-lateral, there is a component of vertical offset.

Figure 9-21 shows a type of fault termed a *wrench*, or *tear fault*. These typically form where a strike-slip fault curves. The upthrown sides of the wrenched block would form hills such as those seen in the California landscape (Figure 9-18).

Much of the topography that you see in Figures 9-18, 9-19, and 9-20 reflects the geologic processes that occurred prior to the development of the strike-slip plate boundary. In both the California and Haiti examples, the prior plate boundary was convergent.

Plate Boundaries Change over Time

Plate boundaries evolve over time as the directions of motion of the plates change. This is very important when trying to interpret why the landscape looks the way it does: geology records ancient plate motions. For example, the geologic history of eastern North America records divergence between approximately 600 and 480 millions of years ago (Ma), followed by pulses of convergence from 480 to 300 Ma. Starting with the breakup of Pangaea approximately 280 Ma, this region has been bordered by a divergent plate boundary. Since that time, ongoing spreading about the mid-Atlantic ridge means that the continental margin of North America lies in the plate interior, so we call this a *passive margin.* A passive margin is a continent-ocean transition that does not correspond to a plate boundary.

The recent geologic history of California is shown in Figure 9-22. Approximately 30 Ma a spreading ridge on the Pacific Plate was subducted beneath a westward-moving North American plate. The San Andreas strike-slip fault developed as the mechanism to accommodate divergence to the south and convergence to the north. As subduction continued to the present day, the length of the San Andreas plate boundary increased.

The next chapter will examine the formation of mountains in more detail and also discuss how and why mountains are such important features in the Earth system.

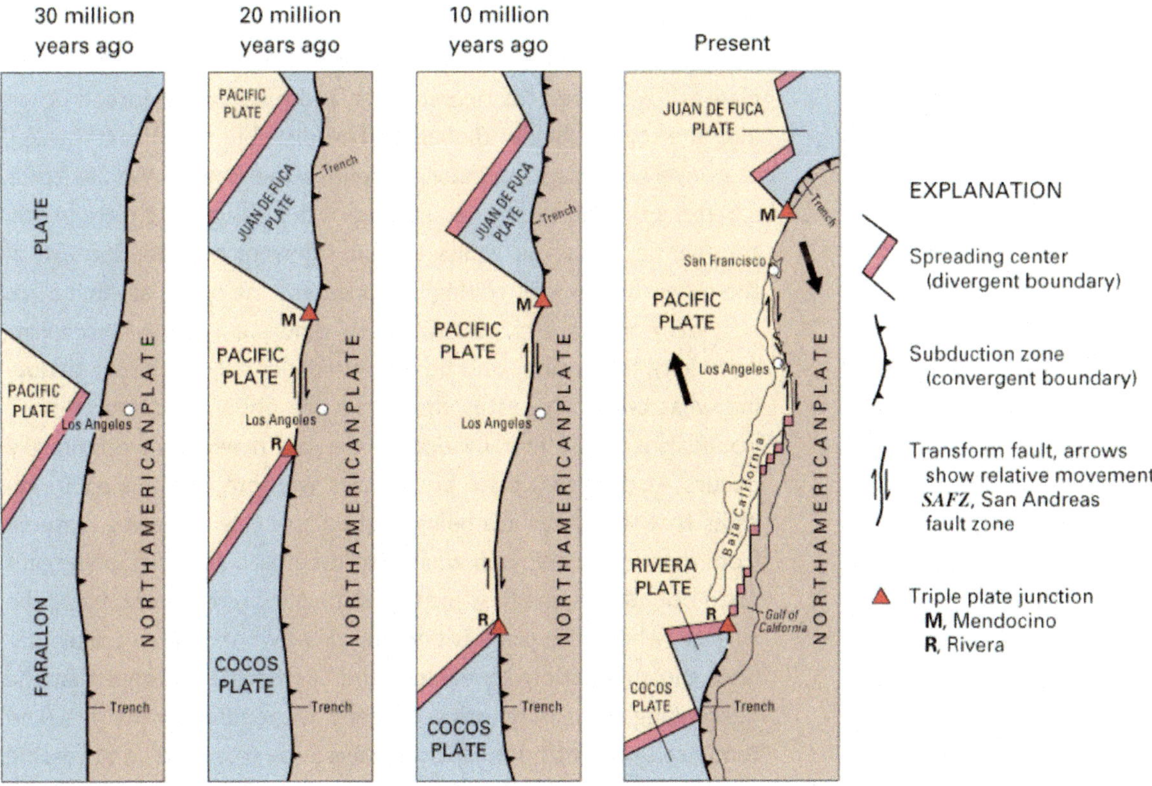

Figure 9-22 Evolution of the San Andreas fault.

Image from http://www.sanandreasfault.org/SAFBirth_Big.jpg.

Summary

Alfred Wegner coined the term "continental drift" in the early 20th C to describe his theory that the Earth's crust was once assembled into one large landmass, termed Pangea. Based on the "jigsaw puzzle fit" of the outlines of continents on either side of the Atlantic, as well as on the distribution of various rock types and fossils, the idea that the continents moved across the Earth's surface was rejected by geologists. Wegner had no mechanism to explain such movements. While they may seem similar, the theory of plate tectonics is actually very different from continental drift. The "plates" of plate tectonics do not consist of continental material, in fact they are composed of continent or oceanic crust plus nearly 100 km of upper mantle. The continents are just passengers on much larger moving plates. Plate tectonics also describes various ways in which these rigid slabs move relative to one another and why they move: heat flow from the Earth's interior. The boundaries between plates are characterized by their relative motions: divergent, convergent or strike-slip. Divergent plate boundaries are characterized by plates moving apart, with new crustal material erupting on the surface in a broad valley termed the rift zone. Convergent plate boundaries are characterized by plates moving towards one another, and one plate is forced down into the mantle below the other. The distribution of earthquakes at convergent plate boundaries defines the Benioff-Wadati zone, a pattern that describe the location of the descending plate within the mantle. Strike-slip faults exhibit sideways motion. Because of the different types of geologic processes that occur at each plate boundary, they are characterized by different types of volcanism, seismicity, topography and heat flow. Particularly useful for deducing what type of stresses the rocks are under at different plate boundaries is the first motion of the Earth when an earthquake occurs. At divergent plate boundaries we note that this data indicates that the plates are being pushed apart by upwelling material from the mantle below in what is termed "slab pull." At convergent plate boundaries the first motion data is more complex. Near the surface, where the place is bending downward, a deep trench defines the subduction zone. Here the earthquake indicate that the descending plate is resisting descent. Depending on its age (and temperature/density) it slips sideways away from the compression or even "pops upward." Deeper earthquakes however, indicate that the descending plate is being pulled downward into the mantle in what is termed "slab pull." The different geologic processes which occur at plate boundaries leave a distinct geologic record of rock types and structures. In this way we can study the geology of a region and determine its history of plate motions. These change over time, and a region that once may have been on a convergent plate boundary that uplifted mountains could currently be in a strike-slip or divergent setting. This suggests that the heat flow that drives plate motions also changes over time.

Chapter 9 Review Questions

- Define "plate tectonics" and describe how it differs from Wegner's theory of continental drift.
- What is the relationship between a "plate" and a "continent"?
- Use the analogy of an ice skate moving across ice to describe the relationship between the lithosphere and asthenosphere
- Examine Figure 9-4 and make at least 3 observations about the spatial distribution of earthquake foci data (do not suggest any interpretations of this data)
- Examine Figure 9-5 and make at least 3 observations about the spatial distribution of active volcanoes. (do not suggest any interpretations of this data)
- In Chapter 8 you made observations about the nature of heat flow data on the Earth's surface (Figure 8-1). Do you see any spatial similarity of the heat flow and volcano distributions and how might you interpret this? Recall that a similarity of data sets does not mean that there is a causal link. Can you hypothesize whether any relationship you noted in these data sets was "coincidental" or why you might think that they are related?
- In Chapter 8 you examined a map of the Earth's surface topography (Figure 8-2). Is there any relationship between either the volcano or earthquake data sets and topography? Can you hypothesize whether any relationship you noted in these data sets was "coincidental" or why you might think that they are related?
- Compare the earthquake foci and volcano data to the plate boundary map in Figure 2-3. Make at least three observations about relationships between both of these data sets and plate boundary location and type.

- Examine Figures 9-7 and 9-8 and interpret what the focal mechanism solutions indicate about the nature of the "first motion" of the Earth when earthquakes occur along faults in both of these regions.
- What types of volcanoes characterize divergent plate boundaries? Based on what you know about what controls the type of volcano that forms, what does this imply about the magma? What is the source of magma erupting at divergent plate boundaries?
- Examine the data for the age of the ocean basins in Figure 9-12. Describe the relationship between young rocks with areas of volcanism, shallow-focus earthquakes, high heat flow and elevated topography.
- Draw a cross section of the uppermost 250 km of the Earth at a divergent plate boundary beneath the Red Sea. Label the location of an upwelling limb of a convection cell, a magma chamber, volcanoes, ocean crust and continental crust (Egypt and Saudi Arabia)
- Examine Figure 9-4 and make observations about the nature of earthquakes at convergent plate boundaries (do not suggest any interpretations of this data)
- Examine Figure 9-5 and determine if there is a characteristic type of volcanism associated with convergent plate boundaries.
- What is a Benioff-Wadati zone?
- Figure 8-2 presented a map of the Earth's surface topography. Is there any relationship between either the volcano or earthquake data sets and topography? Can you hypothesize whether any relationship you noted in these data sets was "coincidental" or why you might think that they are related?
- Draw a cross section of the uppermost 250 km of the Earth at a convergent plate boundary on the western margin of South America. Label the location of the descending plate, an ocean floor trench, where melting and magma generation is occurring, the location on the surface of volcanoes and the magma chamber beneath it, volcanoes, mountains composed of folded and faulted rocks, and where metamorphism is occurring.
- Why is there no volcanism associated with strike-slip plate boundaries?
- Complete the following sentence, "Strike-slip plate boundaries connect….."

Chapter 10

Origin and Evolution of Mountains

At the end of this chapter you should be able to

- Describe the major geologic processes that are responsible for the uplift of mountains at convergent plate boundaries
- Draw a cross-section of a convergent plate boundary and label the geologic structures and major rock types present
- Describe the geologic processes that produced the features you can see in a cross-section of an orogeny
- Define “isostasy”
- Calculate the thickness of the root of a mountain, given its elevation

- Identify anticlines and synclines from a sketch or photograph
- Apply the principle of isostasy to explain why the continents sit higher in elevation than the ocean basins do
- Explain why rocks that formed at depths of 10 kilometers within the Earth millions of years ago can be seen today at the surface
- Explain what is meant by the statement "erosion causes the uplift of mountains"

Introduction

Mountains interest us for several reasons; for example, they are aesthetically pleasing, we recreate in them, and they host unique biological ecosystems. In the big picture of the Earth system, mountains are important because they modify climate in several important ways. Because of their elevation, mountains affect local climates, especially precipitation, and this in turn influences the distribution of water (the hydrosphere system), which of course, impacts the biosphere. In the broader picture, the uplift of mountains from plate tectonic processes is important to study for two reasons: first, because mountains represent increased surface area compared to lower-elevation landscapes. Second, mountains are composed of uplifted material from the Earth's interior, which through enhanced weathering and erosion associated with increased gradient and surface area, introduces material from the Earth's interior to its surface. Finally, mountains are the surface expression of processes going on within the Earth's interior, and as such they provide insight into what's going on "under there." This chapter will focus on how mountains form, why they rise to lofty elevations, and how they interact with the hydrosphere and atmosphere to impact the Earth's climate.

Formation of Mountains at Convergent Plate Boundaries

Mountain ranges are the most visible record of convergent plate boundary interactions, and the formation of mountains is called *orogenesis*. A specific plate collision that builds a specific mountain range such as the Appalachians is termed an *orogeny*. What geologic processes occur during an orogeny to create this elevation? In the text that follows we will review the roles played by volcanism, deformation, and crustal thickening. First, we will examine a cross-section through an idealized mountain belt (Figure 10-1), or *orogen*. This figure illustrates the parallel belts within a mountain range, each of which reflects different geologic processes associated with different regions within the plate collision.

From ocean crust landward (right to left), the cross-section illustrates the position of the *subduction zone*, composed of the trench and deformed ocean floor sediments in a region termed the *accretionary prism*. Between this package of folded and faulted marine sediments and the volcanic arc is a basin termed the *forearc basin*, which accumulates sediment eroded off the adjacent land area. Continuing landward is volcanic terrain and then a region termed the *foreland fold and thrust belt*. This region consists of folded metamorphic and sedimentary rocks thrust away (landward) of the collision zone.

Figure 10-1 illustrates a feature discussed in Chapter 9: magma feeding the volcanoes is generated from the melting of lower crust/upper mantle; the addition of volatiles from the subducted plate lowers the melting temperatures of these rocks. Note also that

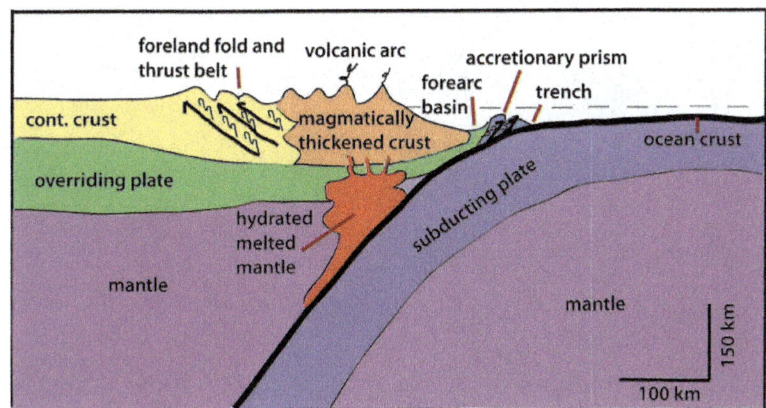

Figure 10-1 Schematic cross-section through an ocean-continent collision. The dashed line above the ocean represents sea level.

the ocean crust is a thin layer on top of the lithosphere, whereas the continental crust is thicker. The diagram also shows that the continental crust sits topographically higher on the mantle than does the oceanic crust. Figure 10-1 also illustrates the elevated mountainous terrain in the orogen. In addition to being elevated, the continental crust in the orogenic zone is thicker than the continental crust outside of both the mountainous area and the ocean. This reflects the relationship between crustal composition (density), thickness, and elevation that is *isostatic equilibrium*. The principle of *isostasy* says that the elevation of crust on the mantle is a function of its density and thickness.

We will now examine the processes that occur in the regions identified in Figure 10-1. These process are responsible for uplifting the crust to form mountains.

Volcanism

Stratovolcanoes are a feature of convergent plate boundaries because of the melting processes localized here. Sediment carried by the subducting plate causes melting of the rocks in the overlying plate. This occurs because of frictional heating generated from plates moving past one another and from the addition of volatiles (water and gas), which lowers the melting point for these rocks. The composition of the magma generated at subduction zones will vary as a function of the rock types present in the upper plate plus some uppermost mantle. Intermediate to felsic magmas are generated as a result. This composition produces a more viscous

Figure 10-2 Thrust fault in the Iglesia Basin, Argentina. The foreland basin of a portion of the Andes Mountains is shown in this photograph. Much of the landscape adjacent to the Tertiary volcanic field is alluvial sediments from erosion of the mountains. In the foreground of the photograph is an emergent thrust (red line; "teeth" are on the top layers of rock on the stack, which is moving toward the viewer). These rocks are being thrust eastward out of the zone of convergence to the west. Movement along this thrust is about 1 to 3 millimeters/year. For scale, the yellow circle encompasses a transportation bus.

magma; stratovolcanoes form from eruptions here. As Figure 10-1 illustrates, the addition of magma and the igneous rocks that form when it cools thickens the continental crust in this region. Essentially, we are adding material to the crust from below, raising the land surface on top.

Fold and Thrust Belt

The compressive stresses from the collision can extend landward hundreds of kilometers from the trench. Prior to the collision, the rocks in this region were most likely sedimentary rocks that accumulated in river or shoreline environments, as well as felsic continental crust. When compressed, these rocks will metamorphose and deform. We have already discussed how rocks might fault from compression (Chapter 7); reverse faults are especially common in this setting. Figure 10-1 shows thrust faults (shown as the heavy red lines with arrows); note that they are pushing rocks sideways and stacking them on top of one another.

The net effect of the emplacing thrust sheets on top of one another is to create a stack of rocks. This lateral displacement of rocks is an example of how the crust in this region gets thicker.

Figure 10-1 uses a squiggly-line symbol to denote that rocks in the foreland basin are also folded (rocks in the accretionary prism are folded, too). A *fold* is a geologic structure that describes rock layers that are bent or tilted. Sidebar 10-1 illustrates the terminology used to describe the two most common fold structures, anticlines and synclines. Folded rocks are an

Figure 10-3 Mt. Rundle, Canadian Rockies. A thick sequence of sedimentary strata of Devonian through Pennsylvanian age comprises a thrust sheet that has been emplaced eastward in the orogeny that uplifted the Canadian Rockies, approximately 175 million years ago.

Figure 10-4 Ductile deformation of a rock layer: the generation of folds from compression.

Figure 10-5 Google Earth image of Massanutten Mountain region, Virginia. Ridges (darker green) are anticlines; valleys (lighter green) are synclines.

Sidebar 10-1 Folds in Rocks

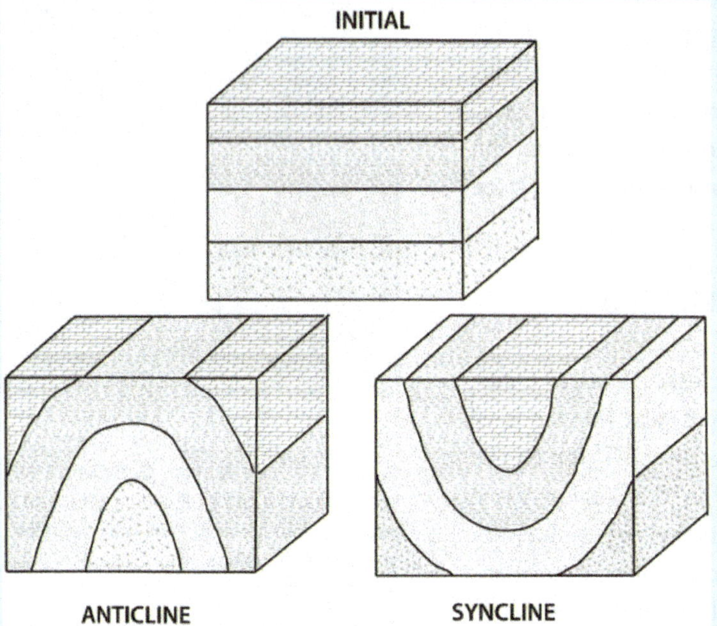

An *anticline* is defined as a fold when the *oldest* rock layer is in the core of the fold. A *syncline* is defined as a fold where the *youngest* rock layers are in the core of the fold.

Anticline

Syncline

SB 10-1B: USGS / Copyright in the Public Domain.
SB 10-1C: Mark A. Wilson / Copyright in the Public Domain.

example of *ductile* or *plastic deformation* (Figure 10-4); the stresses applied to the rocks allowed grains or crystals to slide and rotate instead of break.

As Figures 10-2 and 10-4 show, folding squeezes the rocks sideways, and as a result, the rocks are laterally shortened. Assume that the piece of paper shown in Figure 10-4 is 11 inches long. After squeezing, the paper might be only 8 inches long; the 3 inches of shortening are taken up by the up (anticline) and down (syncline) folds. In the real world, this translates to folded rocks elevating rock layers to make ridges, while synclines make linear valleys (Figure 10-5). Some of these folds can create mountains that attain heights of 1,000 to 2,000 meters in elevation.

In addition to folding and faulting, elevated temperatures from burial and adjacent volcanism combine with compressive pressure to metamorphose some of these rocks. Recall the concept of metamorphic grade from Chapter 5, Figure 5-19. From the collision zone landward, we might predict a transition from high-grade to low-grade metamorphism over a broad region in the orogeny. Rocks in many tectonic settings that are immediately adjacent to a hot magma might locally metamorphose (contact metamorphism), but geographically broad *regional metamorphism* is characteristic of orogens.

Hot fluids generated in metamorphism are one of the sources of ore formation and vein deposits. Recall from Chapter 4 that an ore is an economically valuable mineral, such as iron, gold, silver, copper, or nickel. These elements may be present in trace amounts in a variety of host rocks. The interaction of these rocks with hot fluids generated during metamorphism can concentrate minerals into ore deposits. Gold deposits in the Sierra Nevada Mountains, the Yukon, Colorado, and other areas are all associated with orogenic activity.

Crustal Thickening and Isostasy

Although it is rock, over geologic time the mantle flows like a viscous fluid (*viscosity* is a physical property that describes its "stickiness," or resistance to flow; water has very low viscosity, maple syrup has a slightly higher viscosity, and oil that you put in your car engine has a slightly higher viscosity). When topography is created on the Earth's surface—for example, through thickening the crust by piling stacks of reverse faults on top of one another—the crust is thickened. The mantle must respond to this new weight. How it responds can be examined through the analogy of an ice cube in water. *Buoyancy* is an equilibrium position between downward (gravitational) and upward (buoyant) forces. If you place an ice cube in water it will float at a level that represents the equilibrium between the weight of the ice cube and the weight of the water it displaces. This is what *Archimedes' principle* says: "A body immersed in a fluid is buoyed up with a force equal to the weight of the displaced fluid" (remember that weight = mass × volume).

How high will our ice cube float in water? Perhaps you recall that more than 75% of an iceberg is hidden below the water's surface (cue the music from the movie *Titanic*). Why isn't more iceberg floating above the water's surface? Or, to rephrase, how much iceberg has to lie below the water to hold the iceberg above the water? The answer to this question lies in balancing the downward forces of gravity against the upward forces of buoyancy.

$$\text{The depth of ice below the water} = \frac{(\text{Density of ice}) \times (\text{Height above water surface})}{(\text{Density of fluid}) - (\text{Density of solid})}$$

In English, the thickness of the root is equal to the density of the ice times its thickness above the water level divided by the density difference between the solid (ice) and the liquid (water).

You can see from this equation that the density difference between the ice and water is critical in determining how much ice is hidden below the sea surface (we can call this the "root" of the iceberg). The density of ice is 0.9 g/cm^3, and that of water is 1.0 g/cm^3. If an iceberg is sticking 10 meters out of the water,

$$\text{Depth of ice below water } (= \text{root}) = \frac{(0.9 \text{ g/cm}^3) \times 10 \text{ m}}{(1.0 - 0.9) \text{ g/cm}^3}$$

$$= 9 \text{ meters} / 0.1$$

$$= 90 \text{ meters}$$

Translated, the equation says that the thickness of the root is equal to the density of the ice (solid) times its thickness above water level, divided by the density difference between the solid and liquid. So, for an iceberg that sticks out of the water 10 meters, there is a "root" underneath the water's surface that is 90 meters thick! It's not the part of an iceberg that you can see that is so hazardous but the much larger portion below the water line that you can't see that can lead to disasters like the sinking of the *Titanic*, probably the most famous ship that sank because of a collision with an iceberg.

What does this have to do with elevation on the Earth's surface? The same principles that we use to discuss floating ice cubes applies to our analysis of why the continents stick out above sea level on the Earth's surface. For the average height of the continents (approximately 1 kilometer), there must be a corresponding thickness below the Earth's surface. Continents are composed of less dense granitic rocks that "float" like an iceberg on the more dense mantle beneath.

Granitic crust density (ρ) = 2.7 g/cm³ (rho, or ρ, is the symbol for density)

Mantle density ρ = 3.3 g/cm³

$$\text{Depth of continent's root} = \frac{(2.7\ \text{g/cm}^3) \times 1\ \text{km}}{(3.3 - 2.7)\ \text{gm}^3/\text{cm}}$$

$$= 2.7\ \text{km} / 0.6 = 4.5\ \text{km}$$

In other words, the average continent has about 4.5 kilometers of granite holding it up from below.

How deep does the root of a mountain 3.5 kilometers high have to be?

$$\text{Depth of the root} = \frac{(2.7\ \text{g/cm}^3) \times 3.5\ \text{km}}{(3.3 - 2.7)\ \text{gm/cm}^3}$$

$$= 9.45\ \text{km} / 0.6 = 15.75\ \text{km}$$

The root has to be nearly 16 kilometers thick. The 3.5-kilometer-high mountain has another 16 kilometers of root beneath it, for a total thickness of around 20 kilometers. This material is in addition to the "normal" thickness of continental crust.

How Isostasy Is Related to the Uplift of Mountains

Imagine that our iceberg changes over time; for example, if our iceberg extends 10 meters above the water, but strong sunlight causes it to melt, how will the root respond? Recall Archimedes' principle: the downward weight is balanced by upward buoyant force. If material (with its weight) is removed from the portion above the water line, the buoyant root will push up from below. This winds up replacing some of what is removed (some, not all, because of the density difference). So, if all 10 meters of exposed ice melt, the buoyant root, lightened of the weight of overlying ice, will push upward 9 meters. Thus, the iceberg will lose only 1 meter of height! Because the ice from the root continually pushes up as ice from the top is removed, it takes a while for an iceberg to completely disappear.

Mountains don't melt, but they *do* erode! Many mountains in the Andes are 4 kilometers high, above the average height of 1 kilometer of the South American continent. From the calculations that we did earlier, this means that the root is how deep?

$$\text{Depth of root} = \frac{(2.7\text{ g/cm}^3) \times 4\text{ km}}{(3.3\text{ g/cm}^3 - 2.7\text{ g/cm}^3)}$$

$$= 10.8\text{ km} / 0.6 = 18\text{ km}$$

So, the 4-kilometer-high Andes have another 18 kilometers of granite root beneath them (for a total of 22 kilometers of granite). How long will it take the Andes Mountains to erode to the level of the rest of South America (1 kilometer)? The erosion rate would be equal to the amount of elevation removed through weathering divided by the amount of time, or

$$\text{Erosion rate} = \frac{\text{Amount of elevation removed through weathering}}{\text{Amount of time}}$$

Erosion rates measured in some areas are about 2 millimeters per year, so

$$2.0\text{ mm/yr} = \frac{4{,}000 - 1{,}000\text{ meters}}{X} = 1{,}500{,}000\text{ years}$$

This equation says that it would take 1,500,000 (1.5 million) years to erode 3,000 meters of our Andean Mountains. *But* . . . our iceberg example showed that as you remove material from the top of the iceberg, it is replaced by material from the root below. *So*, the Andes won't erode away to become a more flat part of South America in 1.5 million years, because granitic material from the root will rise up to replace removed height! The Andes won't disappear, but the uplift response to erosion does explain how rocks formed deep within the Earth make it to the surface.

We determined previously that the Andes crust is an additional 22 kilometers thick (18 kilometers root and 4 kilometers tall mountains) on top of 30 kilometers of thick crust, for a total thickness of 52 kilometers. So, we don't erode away the 4-kilometer mountain; we erode lots of uplifted root that continually rises as the overlying mountain peak is removed.

$$\text{Rate} = \frac{\text{Change in thickness}}{\text{Change in time}}$$

The rate of erosion will equal the total thickness of the crust (22 kilometers, converted to meters and then millimeters) divided by the time (X):

$$2 \text{ mm/yr} = \frac{22 \text{ km}}{X}$$

$$2 \text{ mm/yr } (X) = 22 \text{ km} \times 1{,}000 \text{ m} \times 1{,}000 \text{ mm} = 11{,}000{,}000 \text{ yr}$$

So, it can take a long time to completely erode mountains! The key point here is that as material is eroded off the tops of mountains, it continues to be pushed up by compensating forces from below. This brings rocks formed deep within the Earth to the surface. Another way to look at this is to say, "Erosion of mountains causes uplift of the crust beneath."

As you drive east to west across the Green Mountains of Vermont, you pass the sequence of low- to high-grade metamorphic rocks. You should recall that metamorphic rocks form at depth in the lithosphere when plates collide. How did these rocks get to the surface so that we can look at them out of our car windows on the interstate? From our discussion of isostasy, you now know that these rocks have moved upward toward the surface as the overlying mountain peaks have been eroded away. We actually think that (based on the types of metamorphic minerals that are present) the metamorphic rocks making up the Green Mountains formed under temperature and pressure conditions of 20 to 30 kilometers in depth! You now get an idea of how we would figure out how high the original Green Mountains were: we can calculate how much rock would have to be eroded away to get rocks 20 kilometers down to rise toward the surface in response to Archimedes' principle of buoyancy! You might want to try this calculation to see how much crust was eroded away to bring 20-kilometer-deep root to the surface.

Thickening the Crust to Uplift Mountains

This first part of this chapter reviewed how we create mountains on the Earth's surface through the physical processes of faulting and folding that elevate the continental crust. Of course, as mountains rise, the crustal root below must thicken. How do we add material to the bottom of a continent? During plate collisions, much magma does not make it to the Earth's surface to erupt in volcanoes. Instead, it remains in the interior, in large magma chambers, or plutons. Also, the plate that is being subducted represents a thickness of oceanic crust that is present beneath the more buoyant continental crust, and both sit above the more dense mantle. So, mountains will rise from plate collision by crustal thickening. By analogy, it would be like watching the top of an iceberg rise because ice was being added below the water level (icebergs don't behave this way, fortunately!).

How far would the Earth's surface rise if the root of a continent thickened from 30 to 40 kilometers?

Step 1:

$$30 \text{ km} = \frac{(2.7 \text{ g/cm}^3) \times X}{(3.3 \text{ g/cm}^3 - 2.7 \text{ g/cm}^3)}$$

$$30 \text{ km} = \frac{2.7\text{g/cm}^3 X}{0.6} = 6.67 \text{ km height}$$

Step 2:

$$40 \text{ km} = \frac{(2.7 \text{ g/cm}^3) \times X}{0.6}$$

$$= 8.89 \text{ km height of the Earth's surface}$$

Thus, the Earth's surface rose by more than 2 kilometers (the difference between 8.89 and 6.67 kilometers) because of thickening of the root by 10 kilometers.

At this point you should be able to answer questions such as the following: (1) How high would mountains rise from crustal thickening of 15 kilometers? (2) Take your answer to the previous question. The crustal thickening occurred over a time interval of fifty million years. What is the uplift rate of the mountains in millimeters/year?

Summary

The most prominent characteristic of convergent plate boundaries are the mountains that form here. In general, the geologic processes that uplift mountains, termed orogenesis, thicken the crust by folding, faulting and "underplating" material at the base of the crust. Folding and faulting also "shorten" the crust; compression, or squeezing forces rock layers together, bending them into folds, or breaking and moving them by faulting. The two most common fold structures are anticlines and synclines. There are three types of faults: normal, reverse and strike-slip, all of which are distinguished by the relative motion that occurs on either side of the fault. Melting of the uppermost mantle and lower crust creates magma above the subducting plate and the magma injects into the crust, adding material to it. Thickening the crust causes uplift because of the relationship between the thickness and density of crustal rocks and the elevation that the crust seeks relative to the mantle below it. This relationship is termed isostasy. Just as the height that an iceberg can float in the ocean is a function of the amount of ice below the water line and the density difference between the ice and water, the elevation that the Earth's crust can attain is a function of its thickness and the density difference between the crust and mantle.

Chapter 10 Review Questions

- The Himalayan Mountains are still rising. Hypothesize on the processes which must be occurring below the Earth's surface that is causing their uplift.
- What is an "orogeny"?
- The principle of "isostasy" is based on "Archimedes principle." What does Archimedes Principle say?
- Apply the isostatic equation to determine the thickness of the root of a mountain 5 km high if the density of the crust is 2.7 g/cm^3 and that of the mantle, 3.3 g/cm^3. Once you have calculated the root thickness, what is the total thickness of the root plus mountain?
- What are the definitions of an anticline and syncline?
- Geochemical analysis suggests that the rocks found in central Vermont were once buried to a depth of 30km yet they are now at the surface? How can erosion cause the uplift of buried rocks and bring them to the surface?

Part IV

Water, Air, and the Earth's Climate

Chapter 11

The Hydrosphere

At the end of the chapter you should be able to

- Draw a concept sketch of the hydrosphere
- Describe and explain the major changes in the composition of water as it moves from rain to river to ocean reservoirs
- Give several examples of how the hydrosphere and atmosphere interact
- Examine a chemical reaction and identify the type of chemical weathering that has occurred
- Give examples of several physical weathering processes that involve water
- Describe the relationship between physical and chemical weathering
- Summarize the characteristics of a good aquifer

- Write a chemical reaction that describes how limestone buffers acid rain
- Describe the fate of the ions that enter seawater from rivers
- Give examples of water's role as an agent of chemical and physical weathering, as well as erosion
- Describe what the concept of a stream equilibrium profile means in terms of the erosion of an uplifted mountain
- Compare and contrast soil and sediment, including their relationship to bedrock composition
- Draw the major present-day ocean gyres
- Describe what is meant by "ocean acidification" and explain why it is happening
- Define thermohaline circulation and describe why it is a factor in heat distribution on Earth
- Calculate rates of sea level change
- Explain how predicted rates of sea level change are determined

Introduction

One of the objectives of Mars expeditions is to determine whether water was once present on Mars. Why? Water is a fundamental requirement for life on Planet Earth. If we find that water was once present on Mars, this means that the conditions were present for the formation of life. This chapter will focus on the hydrosphere and its critical role in a variety of geologic processes that strongly influence how the Earth works. Our focus will be on the hydrosphere's relationship to the geosphere and, somewhat, to the atmosphere. We will review the major reservoirs of the hydrosphere, how the composition of water changes as it moves between reservoirs, the role of water in some important geologic processes, and how the largest reservoir of water on Earth, the oceans, are responding to climate change.

The Production of Water on Earth

The iconic photograph of Earth taken by an *Apollo* astronaut (Figure 11-1) shows the amount of our planet that is water. Although 70% of the surface of the Earth is covered by water, water is less than 1% of its mass. This abundant light molecule is critical to life on Earth. How did we become the "blue planet"?

Figure 11-1 The blue marble from its moon.

Image from http://eoimages.gsfc.nasa.gov/images/imagerecords/3000/3020/apollo_lrg.jpg.

In Chapter 8 we discussed the early differentiation that generated the layered Earth. The liquid and gas layers that comprise the hydrosphere and atmosphere are the least dense of the materials that separated as a function of their density. The gaseous atmospheric envelope was prevented from escaping to space by the Earth's gravity. Based on the theory of density differentiation, it is logical to hypothesize that the surface reservoirs of water, such as the oceans, represent condensed water vapor released from volcanic eruptions in the first millions of years of Earth history. The water vapor present in volcanic eruptions was bound in the crystalline structure of many minerals in the mantle. The water content of eruptions varies between 37% and 97% (by volume concentration). Geologists have hypothesized that the major part of the hydrosphere formed about one hundred to two hundred million years after the accretion of the Earth; however, precisely when mantle degassing produced the approximately $1400 \times 10^6 km^3$ of surface and near-surface water on Earth is unknown. We do know that by 3.8 billion years ago, the Earth had come close to attaining its present volume of surface and near-surface water and that the total volume of water has not varied by more than 25% over its 4.6 billion–year history.

The rationale behind a 25% estimate in the variation of the Earth's total water budget lies in part in the possible role that ice-bearing comets (Figure 11-2) and meteor bombardment played in delivering water to the early Earth. Meteors that fall to Earth contain minute quantities of ice. Given that the rate of meteor bombardment in the first billion years of Earth history was estimated to be about two thousand times greater than the present-day rate of impacts, ice-bearing cosmic debris could theoretically have made a significant contribution to the Earth's water budget. Comet 67P/Churyumov-Gerasimenk, which we just landed on in 2014, is spraying about 2 cups of water per second off its surface. The variation can also be attributed to the loss of water from the uppermost atmosphere as a result of its photodissociation (stripping H off the H_2O molecule).

Figure 11-2 Comet 67P/Churyumov-Gerasimenk photographed from the Rosetta spacecraft. The comet is approximately 2.5 miles from tip to tip. Photo: European space agency.

While we consider the surface and near-surface water budget, we need to consider the "big picture" of water on Earth. Using analysis of the stable isotope ratios of hydrogen, geologists estimate that the mantle contains between 0.4 to 4 times the amount of water that we see on the Earth's surface and near-surface reservoirs. Plate tectonic processes drive the fluxes of mantle water through the incorporation of H_2O into mineral lattices during hydrothermal alteration processes at mid-ocean spreading centers, with its subsequent release from subduction of oceanic lithosphere at convergent plate boundaries. The huge mantle reservoir of water is not considered in surface process discussions of the water cycle. From a geologic perspective, however, the amount of water stored in the mantle is important; for example, water plays a major role in mineral transformations and the volatility of magma. In the remaining discussions of the water cycle in this chapter, we will not consider the mantle reservoir; rather, we will focus on the interactions of water with the atmosphere and surface geologic processes.

The Hydrologic Budget

Because the Earth is now a closed system, the water we have continually cycles between several reservoirs, which are shown in Figure 11-3. Table 11-1 lists the volume of water contained in the major reservoirs and the percentage of the total water on Earth that each represents. Note that this figure does not include the reservoir of water within minerals, discussed in the previous paragraph. We see the released water when steam erupts from a volcano. We see this water when steam erupts from a volcano.

In addition to the reservoirs, Figure 11-3 schematically illustrates the fluxes that move water from one to another.

An examination of Table 11-1 demonstrates that much of the concern regarding the melting of the polar ice caps and glaciers involves reducing and possibly eliminating the second largest reservoir of surface and near-surface water on Earth. It's also obvious where that water is going to go: the oceans!

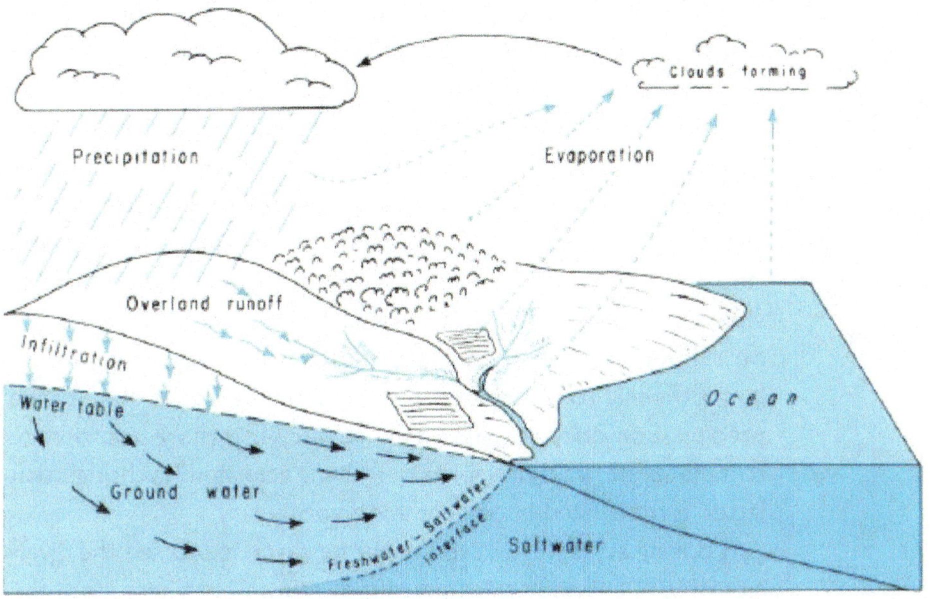

Figure 11-3 The hydrologic cycle.

Image from www.eoearth.org.

Ralph C. Heath / USGS / Copyright in the Public Domain.

Table 11-1 Size of the Major Reservoirs of Water in the Hydrosphere. Data from Various Sources

Reservoir	Volume ($km^3 \times 10^6$)	Percent of Total	Residence Time
total hydrosphere	1386–1460	100	
oceans	1338–1400	97	3000–3,200 yr
ice caps and glaciers	24–29	2	20–20,000 yr
groundwater	10.5	<1	100–10,000 yr
lakes	0.12	$\ll 1$	50–100 yr
soil moisture	0.065	0.005	1–2 months
atmosphere	0.013	0.001	9 days
rivers	0.0017	0.0001	2–6 months
biosphere	0.0006	0.0004	na

Because the largest reservoir is the ocean, there is a large surface area for evaporation of $H_2O_{(liquid)}$ into the atmosphere ($H_2O_{(gas)}$). Once in the atmosphere, water gas condenses into droplets onto small particulates such as dust; their composition can depend on the composition of the particulate material that they nucleate on and gases that dissolve in them once they form. Following droplet formation, water's composition can be affected by dissolution of gases, such as CO_2 and SO_2, that are present in the atmosphere. As a result, rainwater has a hugely variable composition around the world. This composition varies regionally because human and natural activities, such as industrial emissions and agriculture, place a wide range of different particulate compositions into the atmosphere as a function of regional land use (Table 11–2). Thus, rainwater can have pH values of weak acids ($pH < 7$). While most chemical data of this nature is on rainwater, some studies provide data on snowfall as well. It is worthwhile to remember that the liquid or frozen state of the precipitation does not alter its chemistry. It is more appropriate to talk about "acid precipitation" than "acid rain," although that latter term is fixed in our common vocabulary.

Looking at Table 11-2, you might be surprised to see the diversity of ions and molecules present in rainwater. Where does this material come from? In Chapter 12 we will discuss the composition of the atmosphere in more detail, but at this point we can summarize it as primarily composed of low-density gases. There are, however, small liquid and solid particles, less than 20 microns in size (a micron, 1 μm, is 1×10^{-6} meters), suspended in the fluid atmosphere. These particles are termed *aerosols*. Tiny aerosol particles are thrown into the atmosphere from winds moving across the land and sea (Figure 11-4), or the tiniest fraction emitted during volcanic eruptions. Burning fossil fuels emits aerosols, as do forest fires (natural and clear-cutting). Aerosols can also be particles that nucleate in the atmosphere. Many aerosols are water-soluble ions, such as SO_4^+ or NH_4^-; others are insoluble (SiO_2), and some are organic. Rainfall (and snowfall) are very effective at "scrubbing" the atmosphere of aerosols.

What are some observations that you can make about the data set in Table 11-2? How might you characterize the composition "world average precipitation"?

Gas Exchange with the Hydrosphere

Gases also dissolve in liquids, a process critical to interactions between the hydrosphere and atmosphere (but it also applies to the mixing of gases and lava!). The solubility of a gas in a liquid is primarily a function of temperature and the partial pressure of the

Table 11-2 Average Composition of Rainwater. Data in Milligrams per Liter. Data from Several Sources

Location Type of Environment	Central Amazon Tropical Rainforest	Pacific Island Ocean	Bermuda Island Ocean	SE Brazil Rural	Bejing Urban	Tokyo Urban	NC/VA Rural	CA US Urban	Alaska Polar
pH	4.9	4.92	4.79	4.4	6	4.52		5.5	4.96
H^+	12.6	12	16.2	36.3	10	29.9			11
Na^+	3.8	177	147	2.3	22.5	37	0.56	9.4	1
NH_4^-	3.7	2.1	3.8	18.6	236	40.4			1.1
K^+	1.5	3.7	4.3	2.5	13.8	2.9	0.11		0.6
Mg^{+2}	1	38.7	34.5	1.2	48.4	11.5	0.14	1.2	0.2
Ca^{+2}	1.8	7.4	9.7	2.3	209	24.9		0.8	0.1
C^{l-}	5.2	208	175	4.9	34.9	55.2			2.6
NO_3^-	5.4	1.7	5.5	15	106	30.5			1.9
SO_4^{-2}	3.4	30.6	36.3	17	314	50.2	2.2	7.6	7.2
Cd							0.65		
Cl–							0.57	17	
mm/yr	2300	1120	1130	na	441	1210			285

Figure 11-4 A 2000 NASA photograph of a large sandstorm blowing Saharan dust over the Atlantic Ocean. This dust storm covered more than 100,000 square miles in size. Updrafts in the atmosphere elevate this material to heights of approximately 15,000 feet, and winds can blow the dust so that it reaches the Caribbean and Florida. Recent studies attribute the decline in coral reef growth in the Caribbean to the increasing frequency and intensity of Saharan dust storms, and these events might also be related to the frequency and intensity of hurricane formation in the eastern Atlantic Ocean.

Image from http://eoimages.gsfc.nasa.gov/images/imagerecords/53000/53872/seawifs_canary_duststorm.jpg.

gas. You are very familiar with this relationship; every time you open a bottle of carbonated water, you see gas bubbles form and you hear the "fizz" of gas exsolving (coming out of solution). If the temperature stays constant, the pressure on the gas controls the amount of gas that goes into solution (higher pressure = more gas goes into solution; the cap on the soda bottle keeps the pressure high). If we lower the temperature, more gas goes into solution (this is why we chill our soda); conversely, the warmer the temperature, the more gas will leave the solution. This is why warm soda is "flat."

CO_2 and O_2 gases mix with the largest reservoir of the hydrosphere, the oceans at the sea surface. They can also dissolve in water droplets in the atmosphere. When the partial pressure of a gas is increased in the atmosphere adjacent to water, the gas will diffuse into that water until the partial pressures across the air-water interface are equilibrated.

$$CO_{2gas} + H_2O = H_2CO_3, \text{ or carbonic acid}$$
$$SO_{2gas} + H_2O = H_2SO_3, \text{ or sulfurous acid}$$

Because they involve bonding with hydrogen, the products of both of these chemical reactions are pH dependent (*pH* is a measure of the activity of the hydrogen ion). This means that both acids will dissociate into other molecules, depending on the acidity of the solution:

$$H_2CO_3 = H^+ + HCO_3^- \text{ } (HCO_3^- \text{ is the bicarbonate ion) and}$$
$$HCO_3^- = H^+ + CO3^= \text{ } (CO3^= \text{ is the carbonate ion)}$$

and

$$H_2SO_3 + H^+ + HSO_3^- \quad (HSO_3^- \text{ is the bisulfite ion) and}$$
$$HSO_3^- + H^+ + SO_3^= \quad (SO_3^= \text{ is the sulfite ion)}$$

Following the dissolution of these gases into water, they will dissociate into ion species until they all reach a pH-dependent equilibrium with one another. What is the significance of these reactions? The more ionic a molecule (such as $CO_3^=$ and $SO_3^=$), the more chemically reactive it will be with cations, such as those in many silicate minerals.

Once precipitation falls to Earth, its composition changes as a result of interaction with minerals, rocks, and soil. As Table 11-2 indicates, rainwater chemistry varies widely around the world as a function of air pollution, land use, and so forth. The same level of geographic variation is also seen in river waters around the world, but we can still compare the general concentrations of some dissolved elements (values in μg/ml) in rainwater to that of river water. This comparison should highlight the results of the interaction of rainwater after interaction with the geosphere.

River Transport of Ions

Rivers transport material in three ways: dissolved load, suspended load and bed load. *Dissolved load* consists of the elements in solution in the water. The *suspended load* consists of the fine-grained particles (usually less than 1 millimeter in size) that are carried in suspension in the moving water. The *bed load* consists of larger particles that bounce and roll along the stream channel bottom. These grains may be at rest a good portion of the time, only moving at very high discharge events, such as a flood following a storm.

Before we compare Tables 11-2 and 11-3 to try to discern how the composition of precipitation changes after contact with the Earth's surface, what are some observations that you can make about river water composition data in Table 11-3? How might you characterize the dissolved load of "world average river water"? What are the major differences that you see between precipitation and river water compositions? For example, you might observe the range in the concentrations of some ions (e.g., HCO_3^-), while others are generally about the same (e.g., K^+). We should explore what might explain the variation, or the lack of it. The data for the total dissolved solids (TDS) describes the total concentration of dissolved major ions. The values shown in Table 11-3 are hard to interpret until placed in context: the world average value for rivers is approximately 100 milligrams/liter, which is nearly twenty times the value for world average rainwater!

What is the takeaway point of this data? Chemical reactions of precipitation with rocks, minerals, and soil significantly change its chemistry, enriching it in most ions. The chemical processes that describe these interactions are the result of *chemical weathering* reactions. *Weathering* is a geologic process that describes the breakdown of rocks and minerals. Along with plate tectonics, weathering is the major factor that sculpts the Earth's surface. The energy that drives plate tectonics is the Earth's internal heat,

Table 11-3 Dissolved Load Composition of River Water, Corrected to Remove Effects of Pollution, in Milligrams per Liter. Tds = Total Dissolved Solids. Runoff Ratio = The Average Runoff per Unit Area/Average Rainfall.

Source: E.K. Berner and R.A. Berner, Global Water Cycle (1987).

	Continent					
	Africa	Asia	South America	North America	Europe	Oceania
Chemical Species						
Ca^{+2}	5.3	16.6	6.3	20.1	24.2	15
Mg^{+2}	2.2	4.3	1.4	4.9	5.2	3.8
Na^{+}	3.8	6.6	3.3	6.5	3.2	7
K^{+}	1.4	1.6	1	1.5	1.1	1.1
Cl^{-}	3.4	7.6	4.1	7	4.7	5.9
SO_4^{+}	3.2	9.7	3.5	14.9	15.1	6.5
HCO_3^{-}	26.7	66.2	24.4	71.4	80.1	65.1
SiO_2	12	11	10.3	7.2	5.8	16.3
TDS	57.8	123.5	54.3	133.5	140.3	120.6
Water discharge ($10^3 km^3/yr$)	3.41	12.47	11.04	5.53	2.56	2.4
Runoff ratio	0.28	0.54	0.41	0.38	0.42	n/a

but solar energy is the energy source for weathering. Weathering and *erosion* are examples of how the geosphere, atmosphere, and hydrosphere interact. What is the difference between weathering and erosion?

Weathering refers to the physical and chemical breakdown of rocks; erosion refers to the transportation of the detritus and its distribution across the Earth's surface. In chemical weathering, the reaction with water (hydrosphere), air (atmosphere), and organic material (biosphere) breaks down the minerals in a rock. In *physical weathering* the rock physically breaks apart. In other words, *rocks* can physically and chemically weather apart, but in the chemical weathering processes it is the *minerals* that decompose.

Chemical Weathering

There are three primary pathways for chemical weathering: *dissolution*, *hydrolysis*, and *oxidation*. The type that occurs is a function of the mineral's chemical composition and crystalline structure.

- *Hydrolysis:* the interaction of water (often as a weak acid) with a mineral to produce a new mineral and a water solution with lots of ions dissolved in it. For example,

$$KAlSi_3O_8 + 22H_2O = 4K^{+2} + 4OH^{-2} + Al_4Si_4O_{10}(OH)_8 + 8H_4SiO_{4l}.$$

In English, that's potassium feldspar plus water produces a free cation, a hydroxyl ion, the clay mineral kaolinite, and a solution.

- *Dissolution*: a weak acid completely or partially dissolves a mineral away, leaving a new solution. For example,

$$CaCO_3 + H_2CO_3 = Ca^{+2} + H_2O + 2CO_{2}g.$$

In English, this is calcite plus carbonic acid produces a free cation, water, and carbon dioxide gas. The ionic bond between the Ca cation and the CO_3 anion is weak, so weak carbonic acid *dissociates* the molecule and the mineral breaks down, with byproducts going into solution in water.

- *Oxidation*: iron-rich minerals "rust," which forms a new mineral, an "iron oxide," such as hematite. For example,

$$FeS_2 + 2H_2O = FeO_2 + 2H_2S.$$

In English, this is pyrite plus oxygen in water produces iron oxide and hydrogen sulfide gas. Oxidation and reduction involve the loss and gain of electrons, respectively. There is no actual exchange of electrons; the "gain" and "loss" are in the "oxidation state," which describes a hypothetical charge an atom would have if all bonds with other atoms were ionic.

Here are two applications of chemical weathering for you to think about. Limestone is less physically and chemically stable than granite is. Why? Limestone is composed of the mineral calcite, which has a weak ionic bond between the cation Ca^{++} and the ionic compound $CO_3^{=}$. The mineral can be easily dissociated (chemically "broken apart"). The cleavage plane in calcite also allows it to be easily physically broken apart. Why is basalt (fine-grained igneous rock composed of olivine and plagioclase feldspar) less physically and chemically stable than granite (quartz, potassium feldspar, and mica)? The minerals in basalt, which are iron-rich, are prone to chemical weathering through oxidation and hydrolysis because of the relatively weak ionic bonds attached to the silicate tetrahedrons in all these minerals.

How do you predict a mineral's tendency to weather? There are some generalities to start with: (1) a mineral containing iron is prone to oxidation. Thus, iron minerals such as pyrite (FeS_2) or olivine (Fe_2SiO_4) will be less stable than will minerals without iron. (2) Minerals that are ionically bonded are susceptible to chemical weathering because of their tendency to dissolve (i.e., dissociate) in water. Thus, halite (NaCl) or calcite ($CaCO_3$) more easily undergo dissolution than does quartz (SiO_2). (3) There is a relationship between the chemical stability of minerals and their sequence of crystallization in Bowen's reaction series. The minerals that form first in Bowen's reaction series are most likely to chemically weather, a relationship termed the *Goldrich series*. Why would this relationship exist? In a very basic sense, minerals that form at the highest crystallization temperatures are very out of equilibrium with temperatures and pressures at the Earth's surface. The late-forming minerals at the end of Bowen's reaction series, quartz (SiO_2) and potassium feldspar ($KALSi_3O_8$), are complex crystalline structures whose covalent bonds make them more stable. They crystallize at lower temperatures and are therefore less out of equilibrium with surface temperatures and pressures than, for example, olivine.

A

B

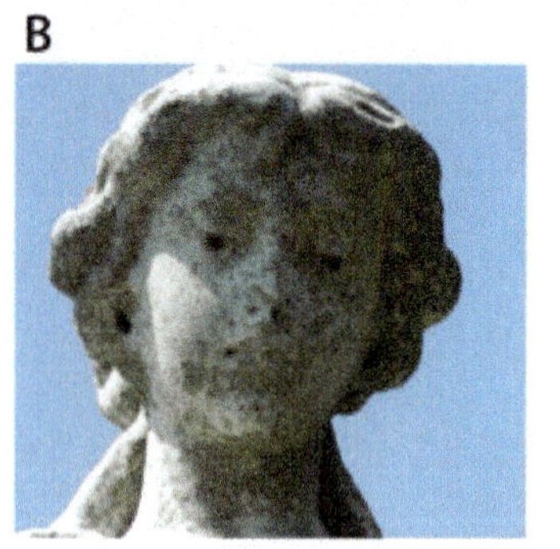

Figure 11-5 Dissolution of marble statues in a wet climate. Details on the heads of the sheep (A) and angel (B) are lost through dissolution of the carbonate rock.

Chemical weathering rates are enhanced by warm temperatures and moisture. *In general*, physical weathering rates are enhanced by fluctuating temperatures and moisture. There are exceptions to this, such as wind erosion in deserts. Water is a catalyst that makes chemical reactions go faster; it allows for the solution of gases (to produce weak acids, or "acid precipitation"). All chemical reactions go faster at warmer temperatures.

Now that we have reviewed the three types of chemical weathering reactions, examine Table 11-3 again and hypothesize about what minerals could break down to liberate the major ions listed. You could apply your response to this question to another: Why might this explain the huge regional variation in river compositions?

Groundwater

The chemical weathering reactions just discussed are not happening simply at the Earth's surface. Water that infiltrates into the subsurface becomes *groundwater*, and these chemical reactions continue in this environment. Return to Figure 11-3 and examine the fate of rainwater that reaches the ground. Some is taken up by plants (evapotranspiration) and transferred to the atmosphere (transpiration). Much runs off on the surface (streamflow and overland flow), and some infiltrates into the soil and bedrock. Soil (discussed below) will retain moisture (soil moisture), but some precipitation will percolate downward to become groundwater. The *water table* is the name we give to a zone in the shallow subsurface that is saturated with water; the pore spaces between grains of sediment are filled with water. Shallow wells (tens of meters deep) often go down to the water table; however, the amount of water present is very variable as a function of precipitation. A more reliable aquifer is a deeper rock layer within the subsurface with high porosity (lots of pore spaces filled with water). It is usually overlain by a rock layer with low *permeability* (any pore spaces present are not connected, so a fluid won't flow through the material easily), which helps trap the water in the layer below. Because surface waters are so susceptible to pollution, we drill into aquifers to obtain drinking water.

Table 11-1 shows that the amount of water on Earth in the groundwater reservoir is quite small (<1%) compared to that in the oceans and glacial ice. Because of increasing population, agricultural irrigation, and industrialization, this reservoir is under increasing stress. As the recent headlines about droughts reminds us, this reservoir behaves as any closed system, and through use, we are increasing the outflow. What is the corresponding rate of recharge, or inflow? You might be able to predict that this would be a function of precipitation and infiltration rate in the recharge zone. There can also be a lag time in rates of recharge. The Ogalla Aquifer in the central United States is recharging at between 0.6 millimeters/year to 150 millimeters/yr. Compare this to a usage rate in the past century that dropped the elevation of the aquifer by more than 1.5 meters. To compound the current discrepancy between extraction and recharge, climate change is altering evaporation/precipitation rates of surface water, and we must be concerned that recharge rates for groundwater will only decrease.

Figure 11-6 Solution cracks in limestone. Joints and fractures that form in limestone concentrate small amounts of precipitation that dissolve the rock. Over time the small cracks become wider and deeper.

Karst Landscapes

Groundwater is important for human use, but it also serves to modify the landscape. A particular landscape that frequently develops in limestone bedrock is termed *karst*. Named after a region in Germany underlain by limestone, karst describes the dissolution of limestone and the formation of a variety of landscape features. Some of these are small, for example, solution pits and cracks (Figure 11-6). Other features are large and dramatic, such as the stone pillars of the Liu River Valley in China (Figure 11-7). Other features are not visible on the land surface because the dissolution is occurring from groundwater moving through limestone rock (Figure 11-8). In these cases, caves and caverns form. When the overlying rock collapses downward, a sinkhole forms. For some people, the collapse of a

Figure 11-7 Stone Forest, Yunnan Province, China. Karst topography is common in southeast Asia because of the abundance of limestone bedrock and the warm, humid climate. Several South China karst regions were named UNESCO World Heritage sites in 2007.

Figure 11-8 Mammoth Cave, Kentucky, is one of the world's longest cave systems (400 miles).

sinkhole is the first indication that the rock beneath their feet was dissolving away.

The interaction of limestone bedrock and precipitation also influences the quality of surface water composition. The *acidification of lakes* describes a phenomenon where acid precipitation lowers the pH of bodies of water such as lakes and ponds. This, in turn, negatively impacts aquatic life in this water, with the collapse of the food chain often resulting. Lakes suffering from acidification lack fish and other "normal" flora and fauna. Prior to the passage of the Clean Air Act in 1970, many lakes in Canada and the northeastern United States suffered from acidification from sulphur released by coal burning in the midwest creating acid rain, some so severely that the lakes were nearly sterile. Some lakes, however, were less impacted. Lake Champlain, bordering New York, Vermont, and Quebec, was one of those lakes. Why would Lake Champlain resist lake acidification when neighboring lakes in New York and New Hampshire were so hard-hit? The answer lay in the bedrock composition. Much of the rock in the Lake Champlain *watershed* (the area of land that all drains into the same basin) is composed of the carbonate rocks limestone ($CaCO_3$) or dolostone [$(MgCa)(CO_3)_2$]. When acidic precipitation fell on the land around the lake, a dissolution weathering reaction occurred:

$$CaCO_3 + H_2CO_3 = Ca^{++} + CO_{2g} + H_2O\text{, or}$$
$$(MgCa)(CO_3)_2 + H_2CO_3 = Ma^{+} + Ca^{++} + CO_{2g} + H_2O.$$

Table 11-4 Ocean chemistry composition in mM/l. Residence time represents the length of time to replace the component in seawater given average river loads and a water residence time of 36,000 years. Data from various sources

ion	ppm Oceans (10^3)	ppm Rivers (avg)	Ocean Residence Time (10^3 yr)
Cl^-	19.3	6	87,00–130,000
K	0.4	1	16,000
$SO4^=$	2.71	8	12,000
$HCO3^-$	0.14	52	100
Mg^{++}	1.3	3	17,000
Ca^{++}	0.4	13	1,000
Na^+	10.7	5	72,000

Like a landscape covered with an antacid tablet, the minerals in these carbonate rocks buffered, or neutralized, the acid precipitation. By the time the surface and soil waters ran into the lake, its pH was near neutral (pH = 7). So, while karst topography may indicate that dissolution of carbonate rocks is occurring, creating solution pits, caves, and caverns, it also indicates that acid precipitation is being neutralized.

The Effects of Weathering on the Ocean Reservoir

Full exploration of the ocean reservoir would constitute a separate course (on oceanography), so we will confine our discussion to a few critical interactions with the geosphere and atmosphere.

As Table 11-3 shows, many different ions are being delivered to the oceans through the world's rivers. So, isn't the chemistry of the ocean changing as a result? What is the fate of all these dissolved ions that enter the oceans through streamflow, glacial flow, and groundwater? Intuitively, you might think that the composition of seawater would have to be constantly becoming more ion-rich, but it is not. The composition of seawater (Table 11-4) has been generally constant for at least the past six hundred million years.

Examine Table 11-4 and make some observations about the differences between the ion concentrations in average river water and seawater. You should be noting that seawater can have as much as approximately 10^3 times higher concentrations of some ions.

Salinity is one way to describe the number of some types of dissolved ions in a given volume of water. Seawater has a salinity of about 35 parts per thousand (ppt). Drinking water salinity is less than 1 ppt. The salinity of brackish water environments, such as estuaries, generally ranges on average between 1 and 30 ppt.

What happens to all those ions as they enter seawater? Clearly, river water is loading the oceans with dissolved chemical species, and they accumulate, which explains why concentrations of seawater are so much higher than those of river water. Will this accumulation continue indefinitely? Self-regulation of the Earth system would tell us that if ocean water chemistry is not changing, these ions must be leaving ocean water at a rate roughly equal to their input. Is there evidence to suggest that seawater chemistry stays generally constant over hundreds of thousands or millions of years? We have no way to sample the ancient ocean water directly; however, we can sample shells of marine organisms and, by analyzing their chemistry, make some interpretations about seawater composition. Although there has been variation about a mean, the oceans have been at their present composition for nearly half a billion years. Support for this interpretation also comes from marine biology: the physiologies of organisms that evolved for them to live in fresh or marine water has remained constant. The abundant fossil record of marine life forms, such as clams, coral, and snails, has been present since approximately six hundred million years ago.

Where do the Ions go When They Enter the Oceans?

The ions that come into the oceans from rivers, groundwater, and melting glaciers suffer three fates. First, some ions precipitate into minerals that form on the seafloor (Figure 11-9). Is ocean floor mining a future resource? Some entrepreneurs think so. Second, many ions are incorporated into the shells and soft body parts of marine life. This is particularly true for calcium (Ca^{++}) and bicarbonate (HCO_3^-) ions, which combine with Ca^{++}, also delivered to the oceans from runoff, to make calcite ($CaCO_3$). Much of this precipitation occurs within the skeletal tissue of marine plants and animals (i.e., making their shells), and some is precipitated onto the seafloor. Both processes make the sedimentary rock limestone: the type of rock type you can see around the Champlain Valley. Some of the fossils are large enough to see, but the entire limestone rock is composed of tiny crushed pieces of shell material. The final fate of the river-supplied ions is their

Figure 11-9 Manganese nodules on seafloor, photographed by a submersible operated by Woods Hole Oceanographic Institute. The nodules range from golf ball to bowling ball in size.

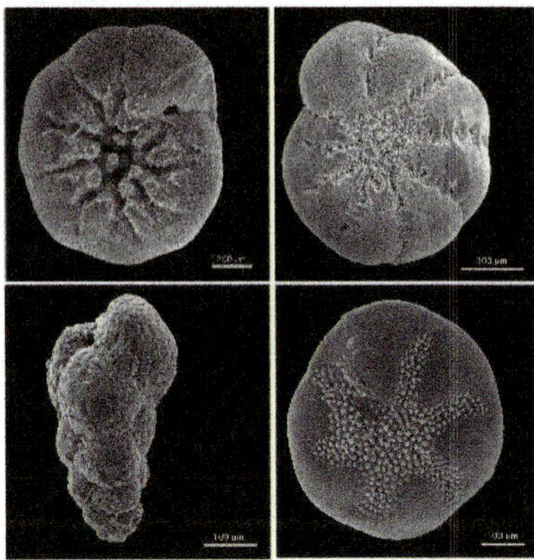

(A) Foraminifera (single-celled protozoa).

(C) Coral.

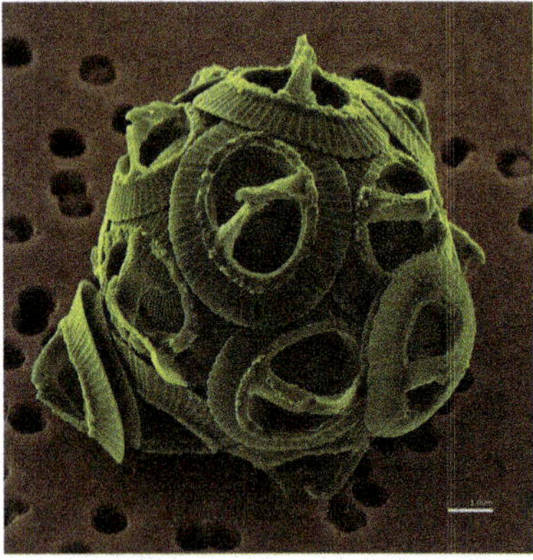

(B) Coccolithophorids (single-celled protozoa).

(D) Clams.

Figure 11-10 A selection of organisms that secrete $CaCO_3$ skeletons. Photos A (foraminifera) and B (coccolithophorids) are tiny animals that are part of the base of the marine food chain. Scale bars on A range between 100 and 300 microns. Scale bar on B is one micron. Photo C shows modern colonial coral building rigid structures on the shallow, warm seafloor. Photo D shows a modern pelecypod, or clam, one of many invertebrate marine organisms that make their shells from calcite or aragonite.

loss from seawater from their evaporation into the atmosphere as aerosols. Ions are lost from seawater from their incorporation into the atmosphere as aerosols and also by precipitation in minerals from chemical reactions with magma on the sea floor (see Chapter 10).

Chemical weathering as a process is an excellent example of the wide range of interactions between the hydrosphere, biosphere, and geosphere that occur on the Earth's surface. There are multiple examples of self-regulating feedbacks, some of which we will discuss in the next chapter, as they impact global climate.

Heat Capacity of Water and Its Role Thermo-regulating Earth

In Chapter 4 we discussed the chemical and physical structure of the water molecule and its effect on water's properties (see Table 4-2). One of these properties is critical to our discussion of the role of hydrosphere in controlling Earth's climate: the heat capacity of water is four times greater than that of air. The property of *heat capacity* describes a material's ability to store heat. Because water covers nearly 70% of the planet's surface, heat on the Earth is moved by water and the atmosphere; the relative contributions of these is a matter of debate among scientists. The oceans play a major role in moderating global temperature because ocean waters are not stationary, and ocean currents transfer solar energy from the equatorial regions poleward. We should examine ocean circulation to see why this is the case.

Surface Circulation in the Ocean

Surface ocean currents are created by the global wind belts. Because the Earth rotates, the Coriolis effect causes an apparent

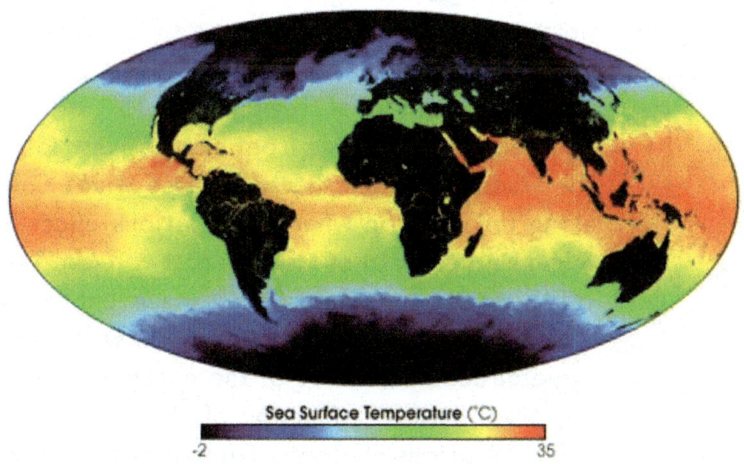

Figure 11-11 Present-day sea surface temperatures (SST) reflect the latitudinal zonation in temperature from the equator to the poles that is a function of the angle of incidence of solar radiation.

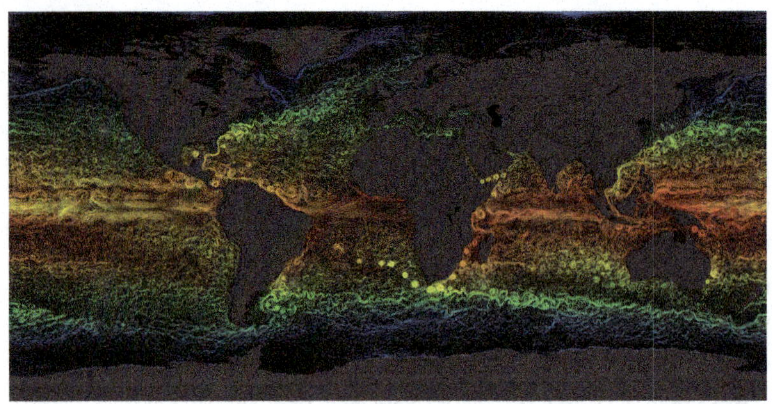

Figure 11-12 Ocean currents and sea surface temperatures from a joint NASA/MIT/Jet Propulsion Lab project. This visualization of ocean circulation and temperature is based on oceanographic circulation models that synthesize satellite and buoy data of sea surface temperatures, which are color-coded (blue = coldest; red = warmest). Data seasonally averaged. Note the Gulf Stream off the east coast of North America, which originates as a warm surface current in the Caribbean and sweeps up the Atlantic coast of the United States before swinging east across the Atlantic, cooling as it does so, before reaching northern Europe. The presence of warm Gulf Stream waters keeps colder Labrador Current waters from descending into the Atlantic and cooling Europe.

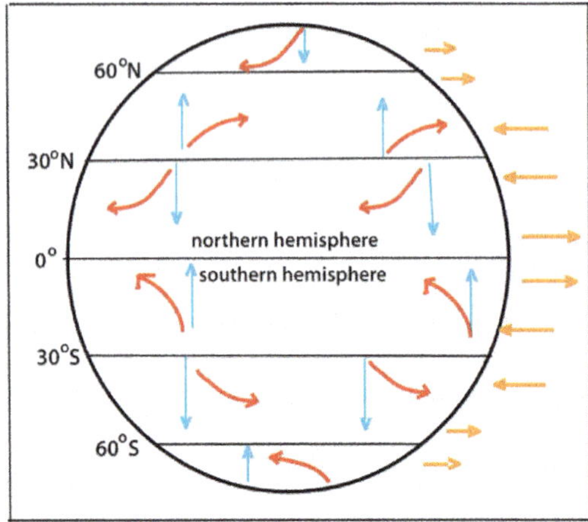

Figure 11-13 Yellow arrows on the side of the Earth indicate vertical motion of air heated by solar radiation. When the descending currents impact the Earth's surface, wind is created. Blue arrows represent air movement if the Earth was not rotating. The red arrows show the paths of the major wind belts. Because of the rotation of the earth, the Coriolis Effect deflects these winds, which are responsible for generating surface currents in the ocean. Note that winds are named for the directions they blow from.

deflection in the movement of fluids across the Earth's surface. Instead of winds blowing in a north-south direction in the northern hemisphere, they curve to the right (Figure 11-13). Likewise, north-south blowing winds in the southern hemisphere are deflected to the left.

In the equatorial region, +/−15° of the equator, winds blowing from east to west form the Trade Wind belt. Because these winds flow along the equator, where the Coriolis effect is minimal, their path of travel is east to west. The trade winds set up a surface ocean current called the Equatorial Current that also flows in the same direction. Because of the configuration of the ocean basins (a function of plate tectonics) and the Coriolis effect, the Equatorial Current is deflected northward along the east coast of the United States. We call this boundary current the Gulf Stream. The Gulf Stream travels northward, bringing warm water as high as 45° latitude before the eastward-blowing wind belt termed the

Prevailing Westerlies pushes it eastward (Figure 11-14). This current arrives at the British Isles, warming them. Some warm waters move northward as the North Atlantic Drift Current, and some are deflected to the south-moving Azores Current flowing along the west coast of Europe, where this coastal current ultimately joins with the Equatorial Current. We call a loop of ocean currents a *gyre*. Gyres also exist in the other ocean basins as well. These large surface current systems are responsible for moving heat from the equatorial regions to higher latitudes, and because water can so efficiently store heat, these currents are "heat pumps" for Planet Earth.

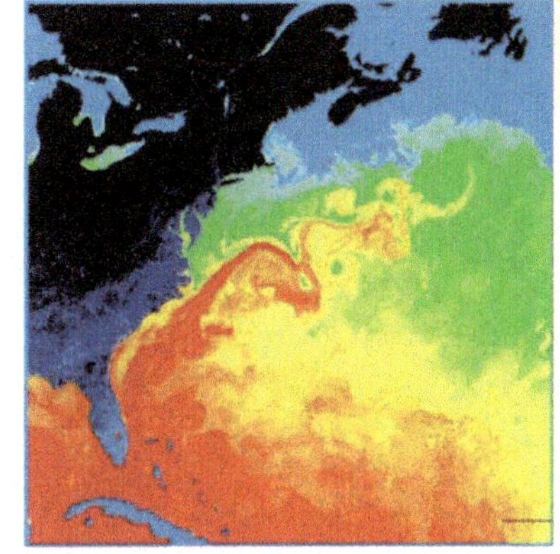

Figure 11-14 Gulf Stream waters move warm, salty equatorial waters northward, as shown in the NASA satellite image. Red colors are warm; pale blue colors are cold.

Thermohaline Circulation in the Ocean

Surface circulation tells only part of the story of water movement within the oceans. Water not only moves horizontally as a function of friction with blowing wind; it also moves vertically as a function of its density. The density of water is a function of its temperature and salinity. A *water mass* (or *water body*) is a segment of ocean water

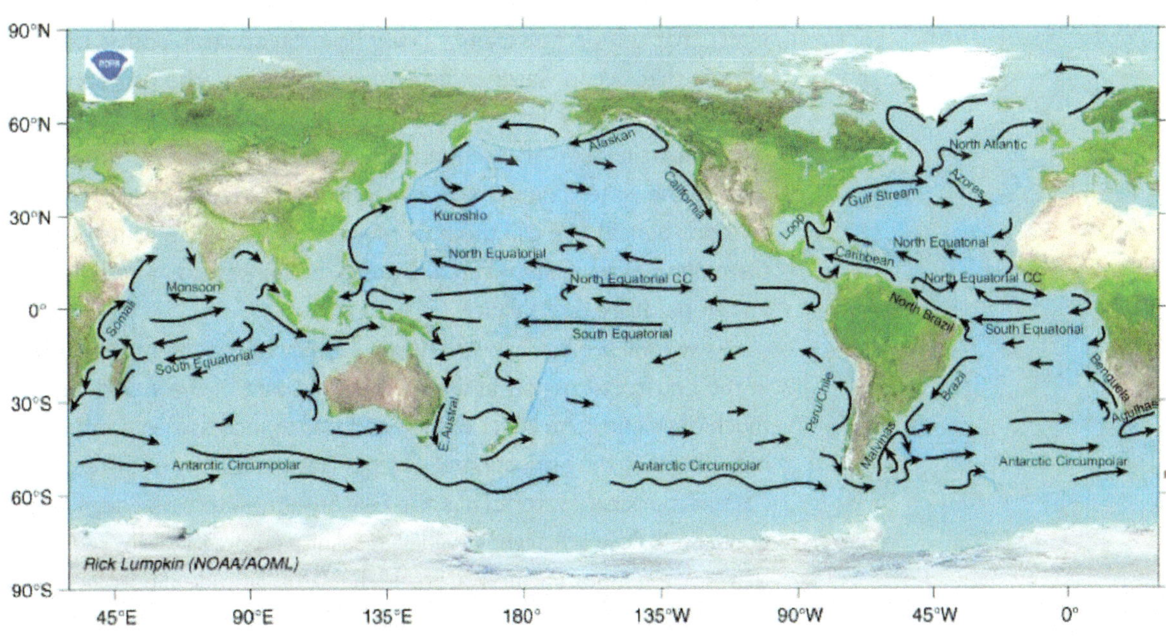

Figure 11-15 Major surface ocean currents are established by global wind belts with north-south flowing boundary currents constrained by the positions of continents. These two factors create surface gyres, or loops, of circulating surface water.

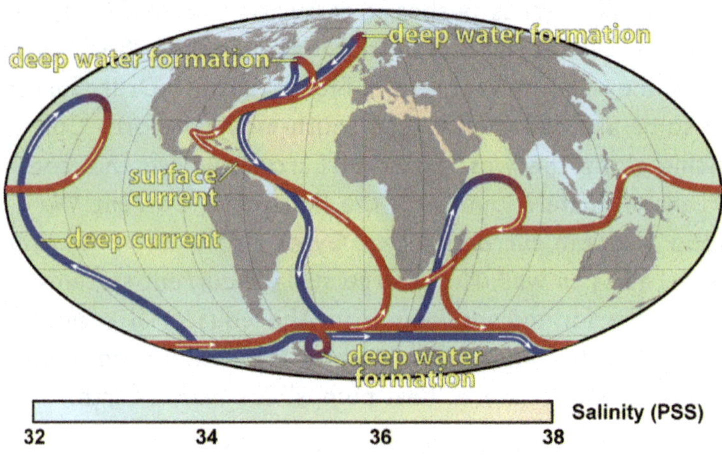

Figure 11-16 The great ocean conveyor system in the world's oceans is shown. Surface currents are in red, and flow within the water column is shown in blue; arrows show direction of flow. Average annual salinity values for surface waters are shown in color; PSS expressed in "practical salinity scale," a dimensionless ratio expressing the electrical conductivity of the solution.

Robert Simmon / NASA / Copyright in the Public Domain.

characterized by the same temperature and salinity; so, for example, we can consider the Gulf Stream to be a water mass. Because of the effect that temperature and salinity have on a water body's density, we see that ocean water is highly stratified: less dense on top, more dense below. As a water mass moves horizontally, if its density is different from that of surrounding water, the position of the water mass in the water column will change; it will rise (if less dense) or sink (if more dense). We call the flow of water in response to its density *thermohaline circulation* (THC). Density of seawater is a function of its temperature and salinity.

As we will see in Chapters 12 and 13, most direct solar energy falls in the equatorial region, and waters in this part of the world absorb much of this heat. An effect of this heating is to increase evaporation of moisture into the atmosphere, which raises the salinity of the equatorial surface ocean waters. Thus, the Gulf Stream waters that begin their northward flow are slightly higher in salinity (approximately 36.6 ppt) than, for example, Pacific waters at the same latitude (approximately 34 ppt). When the warm, salty Gulf Stream water encounters the saline, cold Arctic waters south of Greenland, the two water masses don't mix. Instead, the cold (i.e., high-density), less saline water sinks and flows southward at depth (Figures 11-16 and 11-17). This sinking establishes a complex three-dimensional ocean circulation pattern termed the *great ocean conveyor* (Figure 11-16).

At about 30 degrees south latitude, the water mass turns and flows into the Indian and Pacific oceans. Cooling and dilution result in the water mass losing its unique density, so it mixes with surface waters, driven by wind, to begin its circulation again. The sinking of cold surface waters drives the movement of surface waters, which are warmed by solar energy. The sinking of the cold, fresh Arctic Ocean water allows warm Gulf Stream waters to reach western Europe, warming it. Thermohaline circulation is estimated to move approximately 1.2PW (1 PW = 10^{15} watts) of heat across the globe. It's estimated that it would take slightly less than 1,000 years for a water molecule to travel the "conveyor belt" loop in the oceans.

Oceanographers believe that THC is responsible for maintaining the climate of northern Europe, and a stable climate in such a large landmass helps stabilize all of Earth's climate. There are several lines of evidence that suggest that disruption of thermohaline circulation has occurred in the past; the evidence for this includes geochemical data of sediments in the world's oceans as well as computer models of ocean circulation.

Is there reason to think that thermohaline circulation is at risk of collapsing in the future? Many oceanographers believe so. Because the high latitudes of northern Europe and Greenland are presently cold, much freshwater is trapped in ice or permafrost. Thermohaline circulation is driven by the density contrast between Arctic and Gulf Stream waters. The

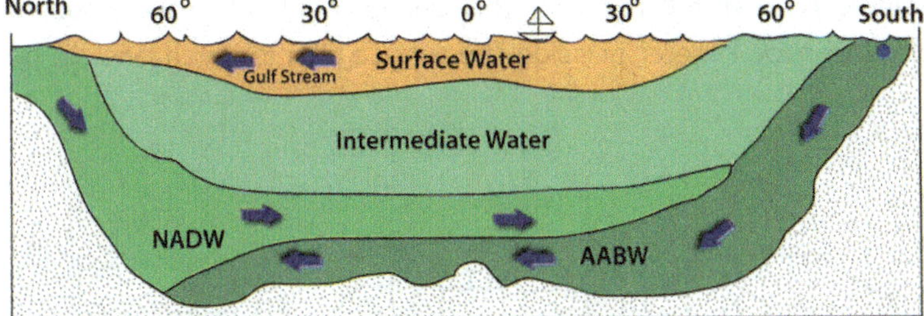

Figure 11-17 Thermohaline circulation in the Atlantic Ocean basin. NADW = North Atlantic Deep Water, the water body formed when salty Gulf Stream water cools and sinks. AABW = Antarctic Bottom Water, the very cold water that forms from surface waters isolated in a surface gyre around the Antarctic Continent (blue dot represents water flowing perpendicular to the blue arrows, or in and out of the page)

formation of sea ice in the Arctic creates a cold, salty surface layer of Arctic water that easily sinks beneath the warmer, salty Gulf Stream water mass. If global warming either limits the formation of sea ice in the Arctic Ocean or increases freshwater runoff from melting ice and permafrost in the northern seas, the magnitude of input of cold freshwater will interact differently with the warm, salty Gulf Stream water, and models suggest that surface mixing will cause the collapse of subsurface density-driven flow. The great ocean conveyor in its present configuration will shut down.

The climate response to an even partial collapse of THC will be significant. Climate models predict that Europe will be plunged into longer, colder winters as a result of cold ocean surface water sitting to the west. "Downstream" effects of THC include precipitation changes in Africa and Asia, with increasing drought likely. Thermohaline circulation really is a heat pump for Earth, with the movement of heat from the sun strongly influencing regional climates.

Physical Weathering by Water

Water is involved in much physical weathering. Because water expands about 10% when it freezes, liquid water that seeps into tiny cracks expands and cracks rocks apart. This can happen repeatedly in temperate climates; the process is called the freeze-and-thaw *cycle*. Over geologic time, an apparently solid rock can crumble apart. This breakdown of a rock into smaller pieces increases chemical weathering rates. Why? The physical breakdown alters the surface-area-to-volume ratio of the rock. Try this calculation: If you have a square block of rock 1 meter per side, the surface area for each face will be

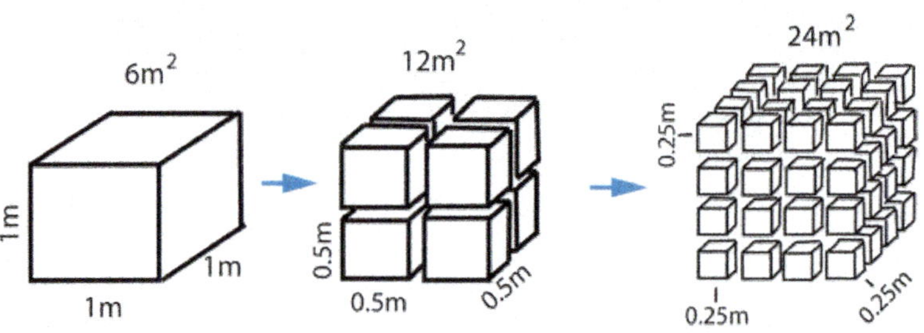

Figure 11-18 The increase in surface area as a result of physical breakup.

1 m × 1 m = 1m^2. This block has six faces on it, for a total surface area of 6 × 1 m^2, or 6 m^2. You hit the block with a hammer and it miraculously breaks into eight equidimensional blocks. The surface area of the face of each of these blocks is now 0.5 m, so the surface area of each smaller block is 0.5 m × 0.5 m, or 0.25 m^2. There are six faces of eight blocks with this dimension, for 6 × 8 × 0.25 m^2 = 12 m^2 surface area. If we hit each of these smaller blocks with a hammer and perfectly halve each of these, we'd produce sixty-four blocks. Each would have a surface area of 0.25 m × 0.25 m = 0.0625 m^2, so the total surface area is now 64 × 0.0625 m^2, or 24 m^2. The total volume of the rock stays the same, but the surface area changes. To summarize, each time you halve the surface area of each block, you double the total surface area.

On a small scale, the physical breakdown of a rock increases the surface area available for attack by water and the resulting onset of chemical weathering. This is why an increase in surface area increases chemical weathering rates and the breakdown of rocks. On a very large scale, the uplift of the Earth's' surface by the formation of mountains increases surface area (Figure 11-19). We can predict an increase in chemical weathering as a result.

Other processes of physical weathering involving water as the agent include glacial scouring, wave action, and stream dissection.

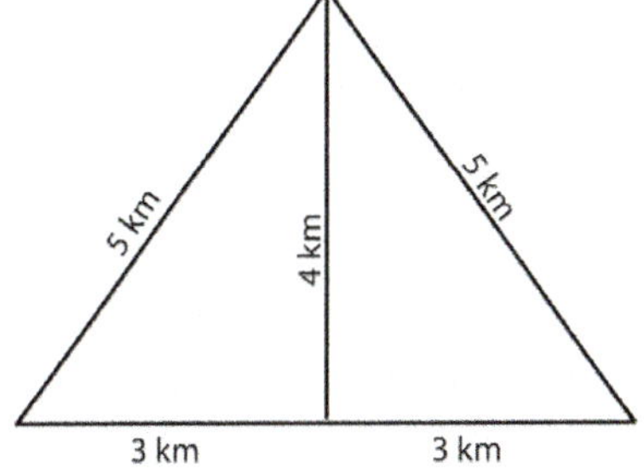

Figure 11-19 The effect of mountains on surface area of the Earth. On the isosceles triangles drawn here, the bottom legs are both 3 kilometers long. The total length is 6 kilometers. If a mountain is uplifted 4 kilometers high, the new slopes of the mountain will each be 5 kilometers long. Thus, there are now 10 kilometers of surface area of rock available to weather versus the original 6 kilometers. There is also an increase in slope, which represents an increase in potential energy to move weathered material down the slope.

Waves as Agents of Physical Erosion

The physical action of ocean waves hitting the coastlines is an example of the role that water plays in the physical breakdown of rocks (Figure 11-20). Like seismic waves, water waves bend (refract) and reflect, but in the case of water waves, this happens not because of a change in the density of the medium, but because of the depth of the water. Wave refraction serves to focus wave energy in a narrow region of a coastline, and significant erosion occurs as a result. Erosion from wave reflection is a byproduct of "hardening" a shoreline by dumping rip-rap, because waves erode twice, once as they come in, and again following reflection.

(A)  (B) 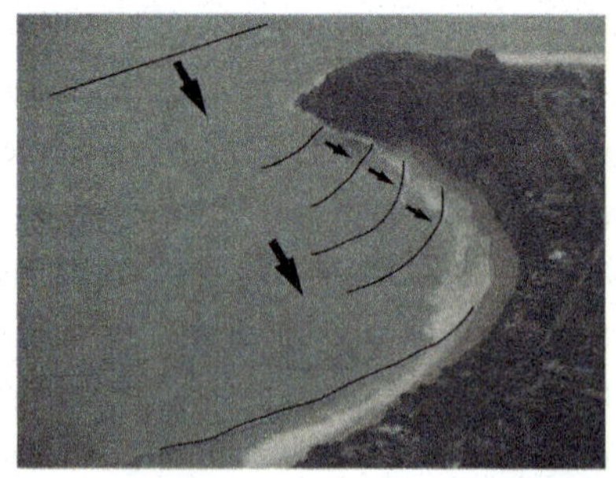

Figure 11-20 (A) A sea stack on the coast of Scotland. (B) Refracting waves on the California coast. The straight line indicates the direction of incoming waves. The waves that interact with the peninsula of land slow down, while the remainder of the wave front continues at the same speed. The light black lines show the bending of the wave front. The relative sizes of the arrows emphasize the difference in speed.

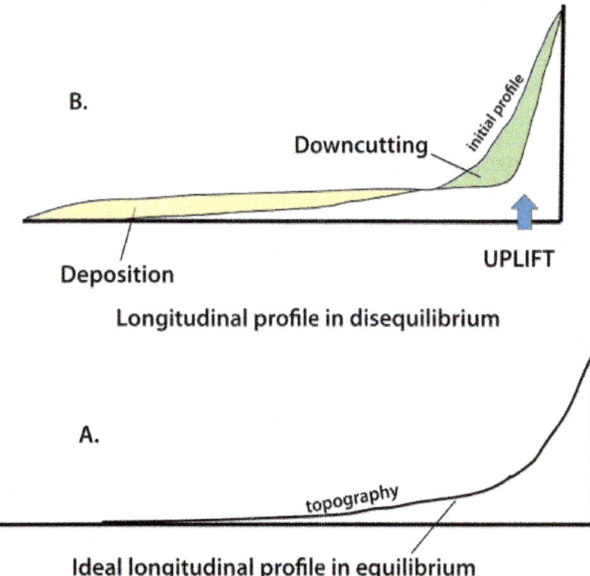

Figure 11-21 Equilibrium profile of a stream. (A) Initial state in equilibrium. (B) Uplift event triggers downcutting near the source area and deposition downstream. Over time the stream channel will adjust to move toward the equilibrium profile, where erosion and deposition are balanced.

Stream Dissection

Rivers and streams flow in channels, so perhaps it's obvious that the flow of water is responsible for scouring the surface that the water flows over. The concept of an equilibrium profile for streams says that a stream will evolve over time to a state of balance between erosion and deposition. The concept of an equilibrium profile (Figure 11-21) for streams says that a stream will evolve over time to a steady state between erosion and deposition. This balance results in a change of slope in the drainage area.

The controlling factors for whether a stream could achieve equilibrium include width, depth, velocity, slope, discharge, size of sediment, concentration of sediment, and bed roughness of the channel. A stream is a system in that a change in one of these components induces an adjustment in one or more of the other variables. This is a form of physical erosion of the landscape. The Grand Canyon is the most spectacular example of the efficiency of flowing water in physically weathering a deep channel.

Of course, the river also transports the detritus, so it is also an agent of erosion. Why do rivers and streams dissect, or cut down, a channel?

Water flows downhill in response to gravity. Thus, it represents the kinetic energy of moving water molecules. This energy acts on the rock or sediment substrate that it's flowing over. Rock will wear away from constant water flow but will do so even faster because of abrasion by the sediment that the river is carrying. A pothole is an example of this process (Figure 11-22).

Figure 11-22 Two potholes viewed from above. These circular structures form when pebbles or cobbles carried by the water get trapped in a depression in the rock and rotate around, bouncing off the wall and enlarging the hole.

How efficient are streams and rivers in physically weathering the landscape? Recent studies on the age of the Grand Canyon (Figure 11-23) indicate that it is much younger (five to six million years old) than previously thought (seventy million years old). How rapid has the rate of downcutting been? If we assume an average depth of about 1,800 meters and an age of six million years, this produces a rate of 3 millimeters/year. We might say that river erosion is not fast, but it is constant. The origin of the downcutting of the Colorado River into the Colorado Plateau is related to the uplift of the plateau. This is an example of the linkage between the geosphere and the hydrosphere: an upper-mantle heat source is responsible for the uplift in this region, and the hydrosphere's response is downcutting by the Colorado River.

Figure 11-23 2014 International Space Station photograph of the Colorado Plateau and the Grand Canyon.

Glacial Weathering

We might think of physical weathering by ice as similar to that done by water. Both flow downhill and are thus exerting kinetic energy on the Earth's surface. Ice, however, is a solid substance with greater strength than liquid water, so you would be correct if you predicted that this would make ice a more powerful agent in the breakdown and movement of Earth materials. Glacial ice is capable of grinding, pulverizing destruction of rock because the base of the ice often has rock material embedded in it. The glacier acts like very coarse sandpaper scraping the underlying bedrock as it flows. As Figure 11-24 shows, large valleys can be carved from this scraping behavior. A full discussion of glaciers would be worthy of a separate course in geology, so we can only briefly touch on glacial ice's ability to sculpt the Earth's surface. There are several features that we can observe in the world around us that demonstrate this (Figure 11-24).

The glaciers that covered much of the northern hemisphere in the past 2.5 million years left a record of scouring by ice on land and in the ocean. The tops of mountains in New England more than 4,000 feet in elevation have glacial striations on them, implying that the glacier was at least this thick. Large sandy islands such as Long Island, Nantucket, and Martha's Vineyard are terminal moraines and outwash plains, recording the huge amounts of sediment transported by the advancing ice.

An excellent example of the combined effects of flowing water and glaciation in eroding the landscape is recorded by the lowering of sea level with the onset of Pleistocene glaciation, 2.5 million years ago. With the flux of water into the glacial ice reservoir, the amount of water in the ocean reservoir decreased. Sea level dropped nearly 100 meters along the shorelines of the continents. The broad continental shelves were exposed. This represents another version of disruption of the equilibrium profile of rivers, but instead of uplift near a source, the distal endpoint dropped. Rivers responded by dissecting the exposed shelves. Figure 11-25 is a computer image of a depth chart of the continental shelf southeast of New York City. The Hudson Canyon, the seaward extension of the Hudson River channel, is shown. This drowned river valley was incised, or cut, during the last glaciation.

An excellent example of the interaction of the glacial reservoir of the hydrosphere and the geosphere is the effect that the weight of ice has on the crust-mantle isostatic equilibrium. We discussed isostasy when we were examining the mechanism by which mountains are

(A) Moraine in Iceland: the ridge of sediment (between red arrows) at the termination of a glacial advance.

(B) Moraine sediments in Iceland: glacial deposits are called *till*, a word that describes unlayered sediment of a wide range of grain sizes. The black color reflects the basalt rock being weathered by the glacier.

(C) Outwash: streams draining off the front of the glacier deposit sediment over a broad plain.

(D) U-shaped valley at the head of Leh Valley, Ladakh. Glaciers flowing down narrow river valleys widen them.

(E) Striations: grooves in bedrock from the scraping of boulders incorporated into the ice at the bottom of the moving glacier. The orientation of the striations tells us the flow direction of the ice.

Figure 11-24 Evidence of glacial processes on the landscape.

Figure 11-25 Computer image of the Hudson Canyon, cut by the Hudson River on the continental shelf, when the sea level was 100 to 150 meters lower than it is today.

Image from http://coseenow.net/online-community/groups/broader-impacts/forum/topic/looking-for-education-partners-for-a-hudson-submarine-canyon-project/.

Source: Center for Ocean Science and Education Excellence.

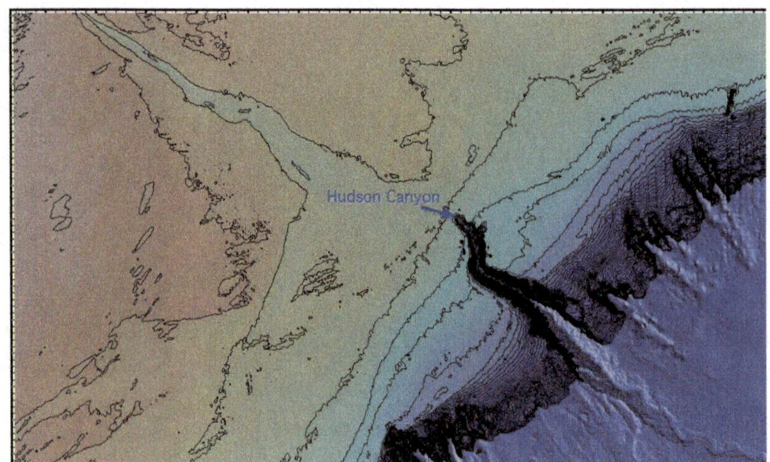

raised. The weight of kilometers of glacial ice had the reverse effect; it caused the crust to subside into the mantle. With the melting of the ice, the mantle "rebounded" to its former equilibrium position (Figure 11-26). With global warming and melting of the ice sheets of Greenland and Antarctica, we can expect a crustal response. The uplift of the crust from the release of the weight of ice might also trigger small shallow-focus earthquakes.

C. crustal rebound
mantle

B. glacial ice
crustal subsidence
mantle

A. crust
mantle

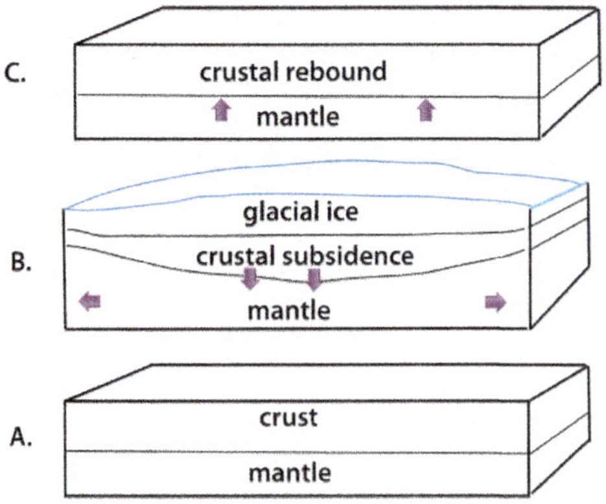

Figure 11-26 Isostatic response of the crust and mantle to glacial ice load. (A) Initial state. (B) Weight of ice depresses crust into mantle. (C) Melting of ice; removal of weight; mantle and crust rebound to original position. There can be a lag time of several thousand years between B and C.

Soil

In Chapter 5 we described the difference between sediment (detritus) and soil (detritus plus organic material plus liquids and gases). Soil is produced over time by the addition of these materials to the mineral fragments, so it very specifically reflects local bedrock geology, water availability and composition, and local organic material. Because plants extract atmospheric gases such as carbon dioxide and nitrogen, we can more correctly look at soil as the intersection of all four of the Earth system spheres.

The mineral content of soil is a function of the rock type available to weather apart, and the weathering reactions that occur. The organic content is a function of the flora and fauna, both dead and alive. Nutrients to support plant life are derived from the chemicals released from weathering reactions as well as from the decay of organic matter. Plant tissues store elements such as nitrogen or calcium, and they also exchange gases with the atmosphere through a process termed *transpiration*.

The Soil Profile

You could predict that if bedrock weathers apart to produce soil, there must be a transition zone between the two. You would be correct. A *soil profile* describes the various layers through this transition zone from solid rock upward (Figure 11-27) to organic material.

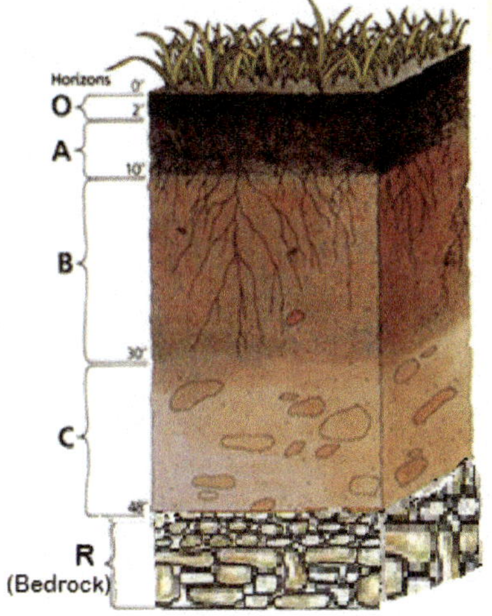

Figure 11-27 Idealized soil profile. The various layers in a soil profile are identified by letters. Not all layers will be present in all soils. Layer C is the unaltered bedrock. Layer B represents weathered minerals, usually rich in oxidized minerals and clay (products of hydrolysis and oxidation reactions) as well as organic material infiltrated from above. Layer A contains decaying organic material in greater abundance than the B layer; many more minerals are dissolved or intensely oxidized. Other layers may be present that reflect specific water saturation states or types of plant life that are not shown on this figure.

Soil Types

Because of the dependence of soil composition on bedrock type, you could predict that soil will vary from place to place (Figure 11-28). You could also predict that if the various soil layers represent the products of weathering reactions, such as hydrolysis, the intensity of weathering—and thus soil composition—would be influenced by climate (both temperature and precipitation). Additional factors that control soil composition would include the types of vegetation in an area, and the length of time that

Figure 11-28 USDA soil map a portion of Addison County, Vermont. Different letters correspond to different types of soils, for example designations starting with the letter "A" are varieties of "Adams loams" while the letter "V" represents the "Vergennes rocky clays." Different types of soils are suitable for different uses.

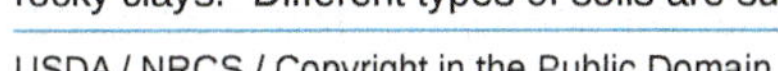

chemical weathering has been operating; the latter factor would be influenced by topography.

Weathering Summary

Weathering represents the intersection of several of the Earth system's spheres. The mineral transformations are in the geosphere; the water source is in the hydrosphere; the gases that dissolve in water molecules are in the atmosphere; and soil formation involves the biosphere. Soil can be viewed as the ultimate example of the interactions between all the spheres in the Earth system. We could predict that if any one of the spheres were perturbed, soils would be impacted. We are currently seeing this play out in some of the world's agricultural areas as climate changes.

Changes to the Reservoirs in the Hydrosphere

Can you predict how the hydrologic cycle might change with changes in global climate? Wondering where to start? Look at the hydrologic cycle graphic (Figure 11-3) and think about which of the reservoirs might be impacted by global warming. What would you predict would happen if global average temperatures increased, which effectively means that we would put more energy into the system? (Remember that temperature drives chemical reactions, including changes of states of matter).

Global Sea Level

You probably predicted that if global temperatures rise, there will be a flux of water from the glacial ice reservoir to the ocean reservoir. This will raise global sea levels. The graph in Figure 11-29 illustrates a range of model predictions about this possible outcome. Before we evaluate this model's, or any model's, predictions, we should understand what the assumptions are that went into its construction.

A *climate model* is a computer-generated prediction that uses mathematical algorithms to code all the variables that contribute to the Earth's climate. Obviously, this is a very complex subject, and in order to write equations that represent real-world relationships, you have to understand what all these are. You also need to be able to accurately quantify the fluxes; in other words, it's

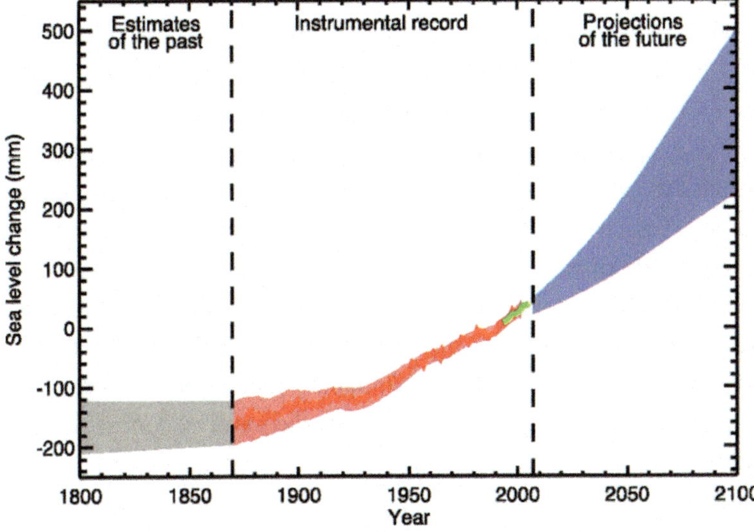

Figure 11-29 Predictions for global mean sea level change (IPCC). The data indicated by the red line represent the portion of the data set (1870 to 2009) based on actual measured values from tide gauges; the red shaded area represents the statistical variation of the measurements. The short green line indicates data measured from satellite telemetry. Data to the left of 1860, in gray, represents sea levels interpreted from climate proxies. The blue interval on the right of the graph represent sea level predictions for different model assumptions, including future CO_2 emissions. All models for global sea level change include the thermal expansion of water.

Source: Intergovernmental Panel on Climate Change (IPCC).

not enough to know that the evaporation of seawater into the atmosphere stores latent heat—you have to be able to say "how much heat" and "how much evaporation." An initial response might be, "This is ridiculously too complicated to be able to do," but it's what thousands of atmospheric scientists, oceanographers, and geologists have been doing over the past one hundred years. We actually know a lot about the ocean-atmosphere circulation relationship, and our models, usually abbreviated AOGCM (atmosphere-ocean general circulation models), are constantly under revision as we learn more. A challenge in these computer models is that the measured data for any variable (e.g., temperature, salinity) are collected from a specific place on the globe. How this data impact a neighboring place on the globe has to be factored into the model. In other words, the AOGCM needs to spatially integrate data in a three-dimensional world. All models have to clearly state what their input is, both measured values and assumptions about less well-constrained values. One of the latter is, of course, future emissions of CO_2 from anthropogenic activity. Will we keep introducing increasing levels of CO_2 into the atmosphere, or will we reduce this? The nice thing about a model is that it can tell us what might happen in each case. So, we can see that if we reduce CO_2 emissions to level X, Y will be the result. We could then have an informed "cost-benefit" discussion about whether or not we want to do this. The international team of hundreds of scientists working on climate predictions is called the Intergovernmental Panel on Climate Change (IPCC). They periodically issue reports that summarize the state of our knowledge about the Earth's climate, including predictions for how it is changing. The IPCC scientists have created seven different models (seven sets of different assumptions) using ranges in maximum and minimum values and rates of change. They are then able to take the mean values for all these outputs for their predictions. Thus, they are not the "gloom-and-doom" or "rosy-scenario" outcomes, but something in the middle.

In addition to assumptions about future CO_2 emissions, climate models make assumptions in two broad areas. The first of these are assumptions about the range of impacts from feedback mechanisms. Because there are several variables that influence the Earth's climate, we need to consider how changes in one variable might trigger changes in another. *Positive feedback* is a change that amplifies the original direction of change. An example of a positive feedback for global warming would be the following scenario: the melting of snow and glacial ice from increased heating will result in lowered albedo levels, which will lead in turn to decreased reflectance and more heat absorption. *Negative feedback* is a change that counteracts the original direction of change. For example, an increase in reflectance as a result of increased snowfall would be an example of a negative feedback for global warming, because an increase in albedo would result in decreased solar radiation and further cooling.

Second, discussion about changing global climates is complicated by regional and seasonal differences in the amount of solar irradiance and albedo, by reemitted long wavelength energy and its absorption by greenhouse gases in the atmosphere, and, finally, by variations in the distribution of heat by fluid circulation. The effects of feedback loops on changing climate further complicate modeling.

For example, glacial ice and snow have high albedo values. With global warming and melting of these surfaces, how will albedo change? A computer model will need to account for different rates of melting and how this will alter feedback. Another example might be how warming might change rates of flow of a glacier. Finally, climate models need to make assumptions about heat exchange between the oceans and the atmosphere. We have discussed the high heat capacity of water compared to that of air, and there is still ongoing debate about how much heat is transferred and at what rate. Fortunately, we can test how good our climate models are by looking at how well the predicted model replicated the actual measured historical data (Figure 11-30) and then derive confidence in the correlation.

Now let's return to the most recent IPCC sea level data (Figure 11-29). Take the midpoint for sea level estimates from 1800 to 1860. What would the rate of change for this interval be? Use the heavy red line to estimate the rate of sea level change between 1860 and 2000. Take the midpoint of the future sea level predicted change to calculate the rate for 2010 to 2100. How would you summarize what the range of models say about future global sea level change?

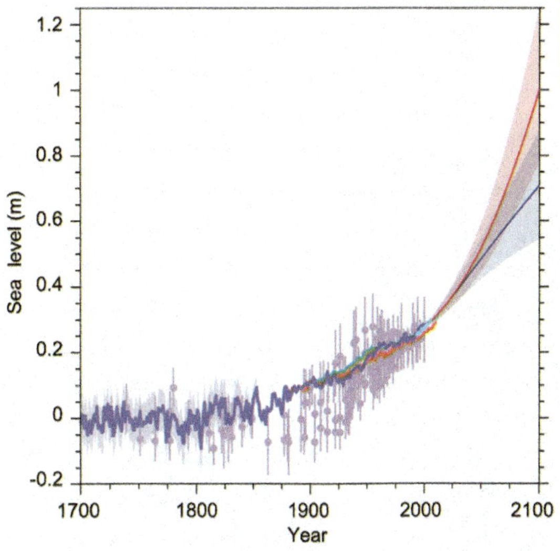

Figure 11-30 Comparison of measured and predicted values of sea level change.

Source: Intergovernmental Panel on Climate Change (IPCC).

Effects of Sea Level Rise on Coastlines

The flood levels from Hurricane Sandy, which hit the east coast of the United States in 2012, occurred during a part of the annual tidal cycle that elevated water levels 3 to 4 meters above a normal high tide. This storm made it clear that with increasing sea level, many large metropolitan areas in the United States were going to be susceptible to catastrophic flooding (Figure 11-31).

Effects on Precipitation

A second change in the hydrosphere as a result of climate change involves changes to the fluxes between the ocean surface and atmosphere (evaporation and precipitation). Studies suggest that an increase in global average temperatures will result in increased precipitation intensity, a process described as "the intensification of the hydrologic cycle." This would occur because rainfall amounts are influenced by atmospheric water capacity, which

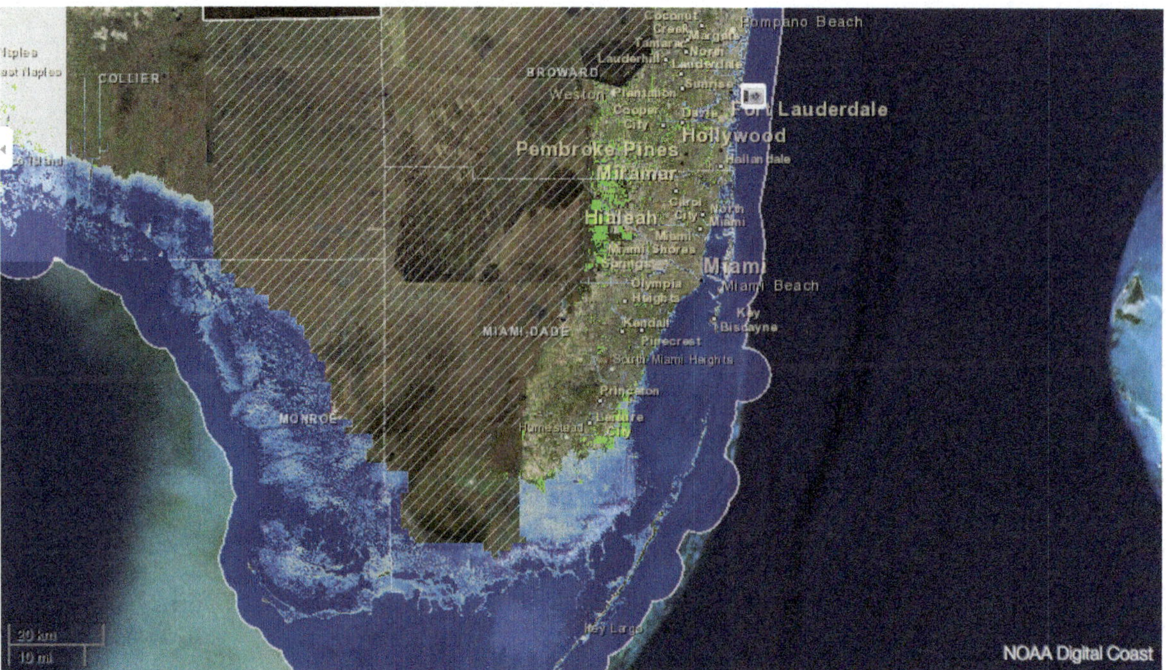

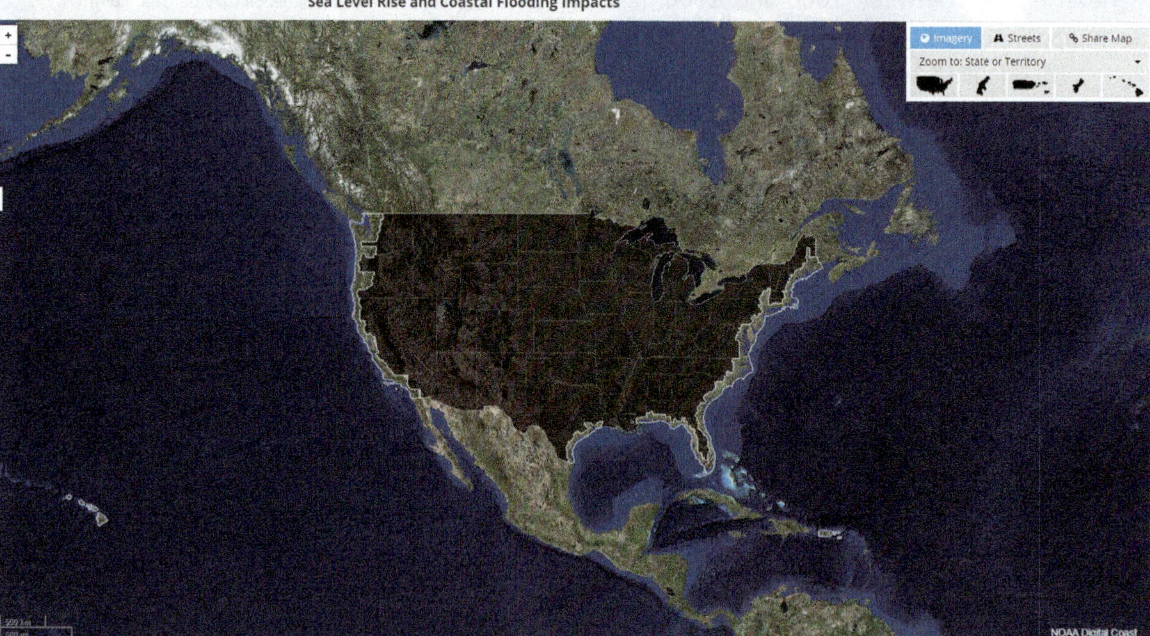

Figure 11-31 National Oceanographic and Atmospheric Administration (NOAA) maps show in pale blue shading the predicted areas of flooding from a sea level rise of 0.5 meter. Map A illustrates southern Florida and the Miami metropolitan area. Map B illustrates a portion of the Long Island, New York, south shore. Green regions are not inundated during fair-weather conditions. Unmapped areas are shown in a green diagonal hatch pattern.

would increase with elevating temperatures, triggering extremes in precipitation intensity and duration of dry spells.

In addition to more extreme and more intense weather events, studies predict changes in the geographic variation of precipitation and evaporation in the oceans, which in turn produces geographic variations on land: dry regions become drier and wet regions wetter.

Changes in Ocean Chemistry: Ocean Acidification

The next chapter, on the Earth's atmosphere, will more fully discuss the nature and extent of linkages between the atmosphere and hydrosphere reservoirs. In brief, excess carbon dioxide (CO_2) in the atmosphere leads to elevated dissolution of this gas into seawater, creating more H_2CO_3 (carbonic acid). Increased levels of carbonic acid results in lowered pH of ocean water, a process termed *ocean acidification*. The average pH of ocean water has decreased from 8.2 to 8.1 (7 = neutral) since the Industrial Revolution of the late nineteenth century (Figure 11-32) and is predicted to drop by 0.2 to 0.4 pH by 2100. This may not appear to be a significant decrease, but the pH scale is logarithmic, so a decrease of 1.0 represents a tenfold increase in acidity. Thus, it's possible that the oceans could be about five times more acidic by the end of the century than they are today.

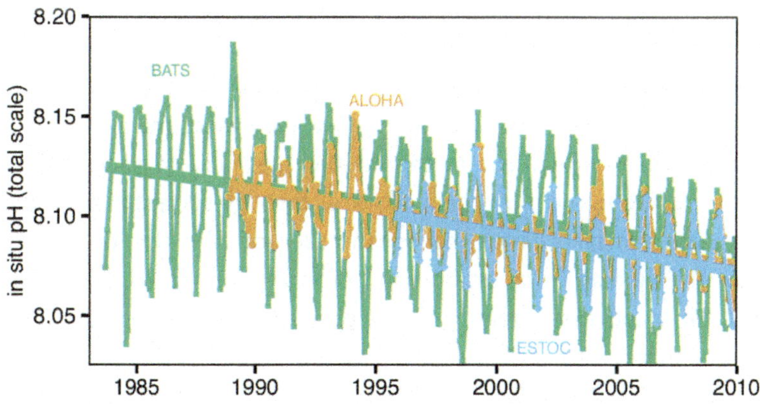

Figure 11-32 Ocean pH measurements from three localities shown in different colors. Solid lines represent seasonal averages.

Source: Intergovernmental Panel on Climate Change (IPCC).

Acidification of the oceans has been identified as a major cause of the phenomenon termed *coral bleaching*, which describes the mortality of coral because of the death of symbiotic algae that live within the coral colony. Additional studies have demonstrated that many organisms that build shells or skeletons by the extraction of Ca^{++} from seawater are negatively impacted by ocean acidification.

Methane Release from Freshwater Reservoirs

Chapter 12 will discuss the significance of methane (CH_4) as a greenhouse gas whose concentration in the atmosphere impacts global climate. One source of methane are bacteria, who release the gas when they decompose organic material in the anoxic settings, such as tidal flats, wetlands, and marshes. The rate at which methane is released increases as a function of temperature. New studies show that methane production in freshwater and shoreline ecosystems will accelerate as global temperature rises. Figure 11-30 suggests that sea level rise over the next decades will flood increasing areas of continental coastlines, and this might impact methane production, as it alters ecosystems where methane-producing bacteria dwell. There is also predicted methane release from melting permafrost.

Summary of Future Changes in the Hydrosphere

As discussed in the introductory chapters, a fundamental characteristic of a system is that it is "self-regulating," a term that describes the fact that "when one component changes, the others also change." We just examined how increased average global temperatures would lead to the melting (i.e., flux) of glaciers (i.e., reservoirs) and the movement of water through surface reservoirs (streams and lakes) into the ocean reservoir. We also discussed how the fluxes of evaporation and precipitation would change as a result of increased heating. These processes are examples of the hydrosphere's self-regulating, or adjusting, to changes.

These fluxes, as they changed, did not all impact the volume in reservoirs the same amount; as glacial ice melted, some water flowed through streams and some into lakes (where it resided for a period of time). In other words, there is not a one-for-one change in glaciers and change in the ocean reservoirs, and there is not a one-for-one change in glaciers and streams or glaciers and lakes. In the previous example, because there are a variety of pathways (fluxes) that the water molecule can follow, melting glaciers will *ultimately* result in rising sea level, but not immediately. This delay is termed the *lag time*. Takeaway point: there is not always an *immediate* impact in other reservoirs when you change one of them. Because of the lag-time effect and the variation in the rates of fluxes, we often do not see the immediate effects of changes in a reservoir. The dramatic correlation of changes in glacial ice volume and ocean water levels is shocking to scientists because we are *not* seeing the delay! There are several other aspects of global climate change where there is nearly no lag time, and there are others where there *is* a lag effect. This is one of the reasons why the public is confused about climate change predictions and daily or annual weather.

Climate models predict that as the Earth warms and polar ice melts, in addition to sea level rise, there will be an increase of 5 to 25% more water vapor in the atmosphere. This increase won't be uniform across the globe, and some regions will experience increased drought (southern United States, Australia, central Africa), while others will see an increase in precipitation (polar latitudes), and this will be highly seasonal.

The nature of precipitation is also predicted to change. For example, a region that experiences a one-day precipitation event once every twenty years will see this frequency increase to one every ten years or fewer. This translates to increased flood frequency.

Because of the flux of both energy and material, the hydrosphere and atmosphere are very closely linked to one another, or, as the climate models describe, they are coupled. The next chapter explores the structure of the atmosphere and its role in the Earth system.

Summary

Water is arguably the most important molecule on Earth. Necessary for life, its presence also controls the Earth's temperature and influences many chemical reactions. Planet Earth is unique because water exists on the Earth's surface in all three states of matter. The Earth's surface water originated within the Earth, trapped in the crystal lattices of minerals, and as the Earth differentiated, this water escaped through volcanic eruptions and "degassing" of the

Earth in the first billion years of its history. Much water, perhaps more than by degassing, was also delivered to Earth through collisions with icy comets in the earliest history of the Earth. Because the Earth is a closed system, the water on Earth cycles from one reservoir to another. This cycling of water is termed "the water cycle." While much water remains trapped in the crystal lattices deep within the Earth, the largest reservoir of surface water are the oceans, followed by the polar ice caps. Evaporation of water from the ocean surface transports water vapor into the atmosphere where condensation creates precipitation. Water falling back to the Earth's surface can runoff into streams, infiltrate into the soil or be intercepted by plants and animals. Ultimately, all of this water makes its way back to the oceans. During this cycle the composition of water changes significantly. Because of the dissolution of various gases in water molecules, there is no such thing as "pure" rainwater; all rainwater is slightly acidic (at best). As this water moves through surface reservoirs, such as groundwater or streams, it chemically reacts with soil and rock. The breakdown of rocks occurs through the chemical and physical breakdown of the minerals that comprise them. There are three primary types of chemical weathering of rock: hydrolysis, dissolution and oxidation. Each of these produces water by-products that are enriched in ions in solution. Ultimately this dissolved material is delivered to the oceans where it is removed through mineral precipitation onto the sea floor. Chemical weathering is one of the primary fluxes for moving Earth materials from one reservoir (minerals within rocks) to another reservoir (water). In addition to chemically weathering rocks, flowing water also serves to physically weather rocks, abrading them through river, ice and wave processes. Physical weathering can accelerate chemical weathering because smaller particles have a larger surface area available for chemical reactions to occur. Soil is the product of chemical weathering of rock and the introduction of organic material to the regolith that is produced. Soil composition varies geographically because of the variation in the type of rock available to weather, the climate and the type of vegetation available. Because of water's high heat capacity, ocean currents are critically important in influencing the distribution of heat around the world. The major ocean currents are produced by winds blowing across its surface. Because of the current shapes of the oceans, these wind-generated currents are deflected north and south to form large circular gyres that deliver warm equatorial water to higher latitudes. Ocean water also circulates in a vertical orientation, and differences in temperature and salinity control the density of seawater, causing it to sink and flow north and south kilometers below the surface. This thermohaline circulation is also important for global climate as the sinking of cold waters at high latitude draws warm surface waters northward. Without thermohaline circulation the temperature gradient from the equator to the poles would be much more severe. As global climate change proceeds through this century we will see the transfer of water from its storage in the polar ice caps to the oceans, with sea level rise resulting. An effect of increasing atmospheric carbon dioxide levels is the increase in gas exchange with the oceans, with resulting increased rate of production of H2CO3, carbonic acid, making the oceans more acidic and damaging marine life.

Chapter 11 Review Questions

- Using blocks and arrows, draw the hydrologic cycle
- What are the three largest reservoirs of water on Earth?
- What are general trends that describe how the composition of water changes from precipitation through to its return to the oceans
- What is the difference between erosion and weathering?
- What are the two types of weathering? How are they related?
- What are the three main types of chemical weathering and what are the general products of each?
- Describe why the chemical interaction of rain and limestone bedrock "buffers" the rainwater.
- What is the fate of ions that enter the oceans?
- Sinkholes are a characteristic of a karst landscape. What characterizes the bedrock geology of a karst region and why do sinkholes (caves and caverns) form?
- What properties characterize a good aquifer?
- The residence time of water in an aquifer can be 100's to 10,000's of years. What is the *primary* process by which aquifers get recharged? What is the *primary* process by which they are depleted? Are these in balance and if not, why not?
- What is the difference between soil and sediment?
- What is "regolith"?

- Because the bedrock of Hawaii is *mafic igneous rock*, which of the 3 primary pathways of chemical weathering would you expect to be dominant here and why?
- Why does the composition of soil evolve over time?
- Give examples of how water is an agent of physical weathering.
- Give several examples of processes that link the atmosphere and the largest reservoir of the hydrosphere
- How does the property of water termed "latent heat" help move heat around the Earth's surface?
- Examine Figure 11-11 and summarize the relationship between temperature of the sea surface and latitude?
- What property of water makes ocean currents so effective at moving heat poleward from the equator?
- What is the relationship between atmospheric circulation and the major ocean surface currents?
- What does thermohaline circulation describe? What role does it play in moving heat around the Earth's surface?
- Using Figure 11-20, calculate the amount of maximum and minimum sea level rise predicted between 2000 and 2100 in feet.
- Go to: http://www.coast.noaa.gov/slr/ and explore the map. Determine what sea level rise will flood the Jacksonville Jaguars home field (Everbank Field). Determine the sea level rise magnitude that will inundate most of Boston. (Note: these values are for non-storm conditions).
- Ocean acidification describes what process?
- Examine Figure 11-32 and summarize at least one observation you can make about ocean pH measurements over a 25 year interval.

Chapter 12

The Earth's Atmosphere and Energy Budget

At the end of this chapter you should be able to

- Communicate why volcanic emissions are not responsible for the past 150 years of increasing CO_2 levels in the atmosphere
- Describe the major layers of the atmosphere, their elevations, and their major characteristics
- Explain why the atmosphere is layered
- Define what the electromagnetic spectrum is and describe how it is related to the Earth's energy budget
- Draw and label the Earth's energy budget
- Define "albedo" and its role in solar radiation of the Earth

- Communicate why solar irradiance is not responsible for the current increase in global average temperature
- Draw the reservoirs and examples of fluxes in the carbon cycle
- Describe what makes a gas a "greenhouse gas"
- Explain why O_2 is not a greenhouse gas but O_3 is

Introduction

Study of the Earth's atmosphere and its processes is such a large subject that it would take an entire course to cover, even at the introductory level. For our purposes, we will examine how the atmosphere, geosphere, and hydrosphere interact and what role this plays in controlling the Earth's climate (next chapter). Obviously, the biosphere also interacts with the atmosphere, but we will only very briefly discuss these interactions.

The gaseous envelope around the Earth is a continuation of the differentiation of the solid Earth that we discussed in Chapter 8. Like the solid Earth below it, the atmosphere is also layered as a function of its density, and also like the solid Earth, its density is controlled by its composition and temperature. We will discuss each of the layers and their properties later, but before this, let's examine the origin of the gases that comprise this thin envelope.

Origin of the Earth's Atmosphere

The Earth's atmosphere originated as part of the segregation of the solid Earth into the core, mantle, and crust, tens of millions of years after the formation of Planet Earth 4.6 billion years ago. Both the hydrosphere and atmosphere represent the most volatile components of the Earth, and their accretion represents the final end products of a process termed "degassing" of the solid Earth. "Degassing" might sound like "Earth burps," and in a way, they are. You see modern degassing examples when a volcano erupts. In addition to pulverized rock particles (ash), gases are the primary constituents of a volcanic eruption (Table 13-1). A less obvious example of degassing involves the incorporation of H_2O and CO_2 into the crystalline structures of minerals. A major reservoir of CO_2 in the Earth is within the mineral calcite (the mineral that comprises the rock limestone). A major reservoir of H_2O in the Earth's crust is found in minerals such as mica [e.g., $KAl_3Si_3O_{10}(OH)_2$)] or hornblende [$(Ca,Na)_{2-3}(Mg,Fe,Al)_5(Al,Si)_8O_{22}(OH,F)_2$]. When hydrous minerals are heated, for example, in the production of magma, the volatiles are released as gases. One of the major industrial processes that emits CO_2 gas into the atmosphere is the production of concrete, accounting for up to 5% of global CO_2 emissions every year! In nature, the subduction of ocean floor sediments (primarily made up of $CaCO_3$ of shells of marine organisms) accomplishes the same gas release (Table 12-1).

Sulfur aerosols in the atmosphere, although not as abundant as water or carbon dioxide, can impact global climate following major volcanic eruptions that spew material high into the atmosphere. Both SO_2 and carbonyl sulfide (C-O-S molecule) droplets reflect incoming solar radiation, producing global cooling. This will be discussed more fully later in this chapter.

Table 12-1 Volcanic Gas Emissions. Emissions from Three Volcanoes from Three Different Tectonic Settings, in Volume Percent. The Three Most Abundant Gases are Generally Water, Carbon Dioxide, and Sulfur Dioxide. Both CO_2 And SO_2 can Dissolve into Water Droplets to Form Acid Rain.

Source: R.B. Symonds, from Volatiles in Magmas: Mineralogical Society of America Reviews in Mineralogy, vol. 30, pp. 1–6.

Volcano	Kilauea Summit	Erta Ale	Momotombo
Tectonic Association	Hot Spot	Divergent Plate Boundary	Convergent Plate Boundary
Temperature	1170°C	1130°C	820°C
H_2O	37.1	77.2	97.1
CO_2	48.9	11.3	1.44
SO_2	11.8	8.34	0.5
H_2	0.49	1.39	0.7
CO	1.51	0.44	0.01
H_2S	0.04	0.68	0.23
HCL	0.08	0.42	2.89
HF	–	–	0.26

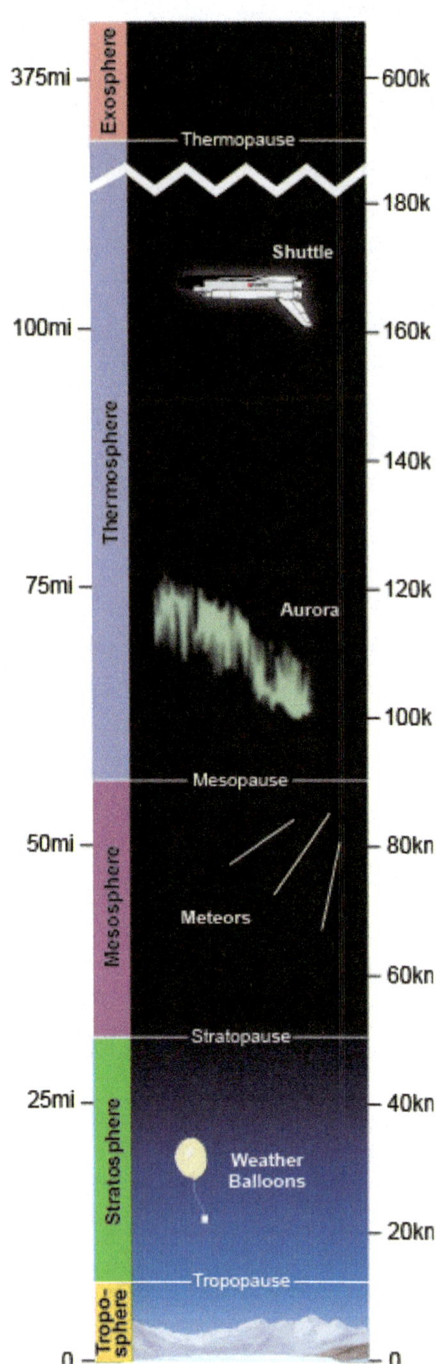

Figure 12-1 Layers of the atmosphere, with thicknesses.

Layers of the Atmosphere

The layers of the atmosphere are distinguished from one another on the basis of four properties: temperature, chemical composition, movement, and density.

Troposphere

The *troposphere* is the layer of the atmosphere that you are most familiar with, as all the weather that we experience occurs here. This layer ranges from the Earth's surface up to around 7 kilometers at the poles, and to about 17 to 18 kilometers in altitude at the equator. The troposphere is the densest of the layers of the atmosphere, comprising about 70% of the atmosphere's total mass; of this, 50% of the mass of the atmosphere is contained in the lowest 5.6 kilometers of the troposphere—density decreases with altitude. You've experienced the effects of this if you have ever climbed a mountain to above 3,000 meters. The air gets thin, and you find yourself short of breath. You've probably also experienced the drop in temperature with increasing altitude, a byproduct of the decreasing density. By the time you get to the top of the troposphere, the temperature drops to around –60°C. The rate of temperature change with altitude is approximately 6.5°C /kilometer.

The weather report that you read online or see on television is describing wind, pressure, and precipitation activity within the troposphere. A weather report also typically mentions the *jet stream*, a narrow air current (wind) that occurs in the upper troposphere, just below the tropopause boundary with the overlying stratosphere. The jet streams are responsible for driving the major high- and low-pressure systems below them in the troposphere. Thus, they are important "weather makers."

Composition of the Troposphere

Because the troposphere is the atmospheric layer that interacts with all the other spheres of the Earth system, we need to examine its composition. As Table 12-2 illustrates, nitrogen and oxygen comprise the bulk of the lower atmosphere, with the gases argon and carbon

Table 12-2 General composition of the troposphere (dry) in parts per million by volume. Values vary spatially and temporally. Note: the 2011 value for CO_2 concentration is 390.5ppmv. Data from various sources.

gas	vol %	ppmv
N_2	78.084	780,840
O_2	20.947	209,470
Ar	0.934	9340
CO_2	0.033	330
Ne	0.00182	18.2
He	0.00052	5.2
CH_4	0.0002	2
Kr	0.00011	1.1
SO_2	0.0001	1
H_2	0.00005	0.5
N_2O	0.00005	0.5
Xe	0.000009	0.09
O_3	0.000007	0.07
NO_2	0.000002	0.02

dioxide comprising a small percentage (by volume); all other gases are present only at trace levels.

The 2014 CO_2 level of 397 ppmv is the highest value attained for the atmosphere over the past eight hundred thousand years, and it will continue to rise because of fossil fuel consumption. Approximately one-fourth of this has been absorbed by the oceans.

Stratosphere

The *stratosphere*, extending from the top of the troposphere (7–18 kilometers) to around 50 kilometers, is separated from the troposphere by the boundary termed the *tropopause*. The stratosphere contains the ozone layer, a region where ultraviolet (UV) light interacts with the O_2 molecule to produce metastable O_3. Present in very small quantities (approximately 2–8 ppm), ozone serves an important function in the stratosphere, protecting the lower atmosphere from elevated levels of ultraviolet radiation, which can be hazardous to organic life. The ultraviolet radiation absorbed by ozone is then reemitted in the form of heat into the stratosphere, warming up the mid-upper regions of this layer. As a result of

Figure 12-2 Cumulonimbus cloud in the stratosphere. The sharp top to the cloud marks the upper altitude of convection.

this warming, the temperature of the stratosphere increases with height, creating a phenomenon termed a *temperature inversion layer*.

Unlike the troposphere below it, the stratosphere is very stable; the temperature inversion layer prevents the vertical convective motion of molecules characteristic of the lower troposphere. You can occasionally see evidence of the stratosphere's inversion layer. The anvil shape of the top of a thunderstorm cloud marks the base of the inversion layer (Figure 12-2).

Mesosphere

Above the stratosphere, the *mesosphere* extends to an altitude of approximately 80 to 85 kilometers, where the mesopause forms the boundary with the overlying thermosphere. Within the mesosphere, temperatures continue to drop with increasing altitude to about −100°C.

Thermosphere

The uppermost layer of the atmosphere is the *thermosphere*, extending from the mesosphere upward to altitudes of about 640 kilometers. The name of this layer reflects the importance of temperature changes that occur here. Intuitively, you might think that temperatures would continue to fall with increasing altitude; however, the opposite is the case. Solar radiation causes

Figure 12-3 The aurora borealis.

Figure 12-4 Photograph of the International Space Station in orbit above the Earth.

temperatures to rise to more than 1,000°C from only −120C at base; however, the extremely low pressure and density would make the temperatures feel much, much colder.

Within the thermosphere is a region termed the *ionosphere* (between 80 kilometers and approximately 550 kilometers). This region is aptly named for the abundance of charged particles (ions) and negatively charged electrons found here. You may have

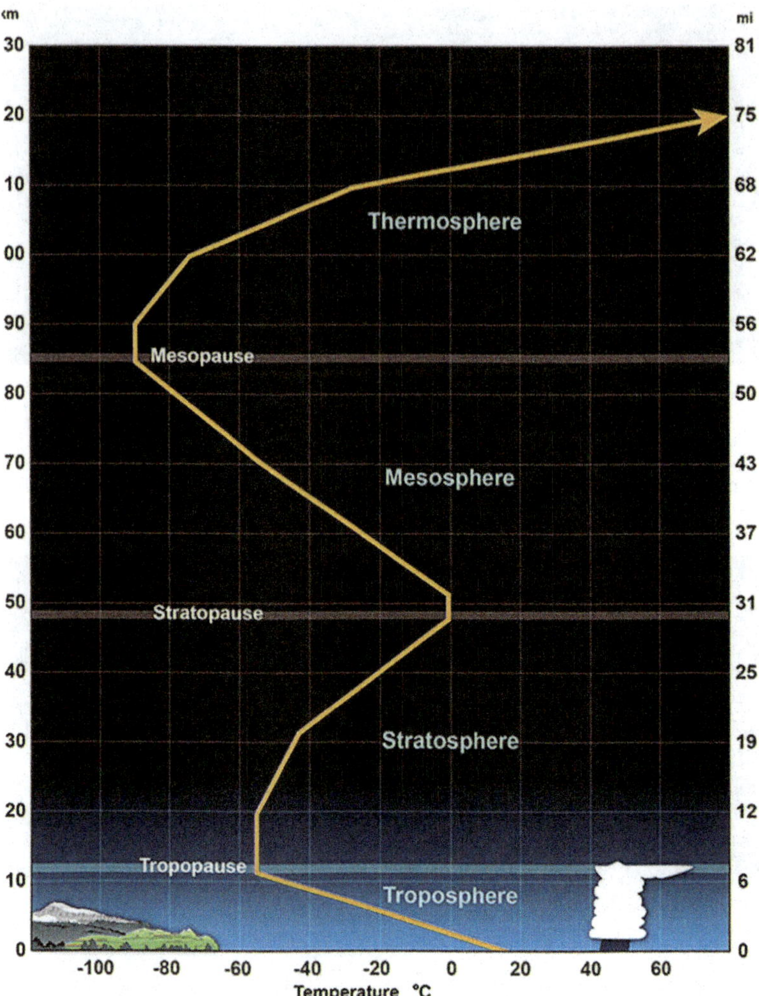

Figure 12-5 Temperature profile of layers of the atmosphere. A summary illustration of the vertical changes in temperature with altitude in the atmosphere. Note that temperature decreases with altitude in the troposphere and mesosphere but increases with altitude in the stratosphere and thermosphere.

NOAA / NWS / Copyright in the Public Domain.

seen the ionosphere "in action" if you have ever seen the aurora borealis (northern lights), shown in Figure 12-3. This visible light is produced by the interaction of the charged particles in the ionosphere and solar particles.

The International Space Station (Figure 12-4) orbits within the thermosphere, between altitudes of 320 and 380 kilometers.

Beyond the ionsphere, the outermost thermosphere is called the *exosphere*, which merges with "outer space." Satellites orbit the Earth in this region.

Heat Transfer and the Earth's Heat Budget

Heat Transfer

There are three ways in which heat energy is transferred: *radiation*, *conduction*, and *convection*. We discussed the process of convection within the mantle in Chapter 8. Recall that convection involved the physical movement of material as a function of its temperature-controlled density. Like a lava lamp, hot material rises, and cold, more dense material sinks. Convection also occurs in the atmosphere when warm air rises and cold air sinks. As warm air rises, air must move in to replace it, and these horizontal movements of the atmosphere are what we experience as "wind."

Conduction is the transfer of heat energy from one substance to another, from one atom or molecule to another. A pot on a stove will ultimately warm the water within it as heat is transferred from the pot to the water. Because of the low density of molecules in the atmosphere, the atmosphere does not efficiently conduct heat.

Radiation, a shorthand for "electromagnetic radiation," is the transfer of heat by a range of wavelengths and frequencies of energy. This is the primary way in which the atmosphere moves heat, so we must examine electromagnetic radiation in more detail.

The Electromagnetic Spectrum

How does the sun's energy travel through the vacuum of space? It can't do so by convection, as this depends on the transfer of energy of motion from one moving atom to another. It also can't do so by conduction, as this requires the transfer of energy from one atom to an adjacent atom. The only mechanism left is radiation. *Electromagnetic radiation* describes the group of self-propagating waves that result from the interaction of electric and magnetic fields. Electromagnetic radiation can pass through both matter and a vacuum, and the different types of waves are classified according to the frequency of waves. In order of increasing frequency (and decreasing wavelength) are radio waves, microwaves, infrared radiation, visible light, ultraviolet radiation, X-rays, and gamma rays (Figure 12-6).

Human eyes are adapted to sensing only a very small portion of wave frequencies; we call this the visible spectrum of light. A variety of filters can be worn that enable us to view electromagnetic

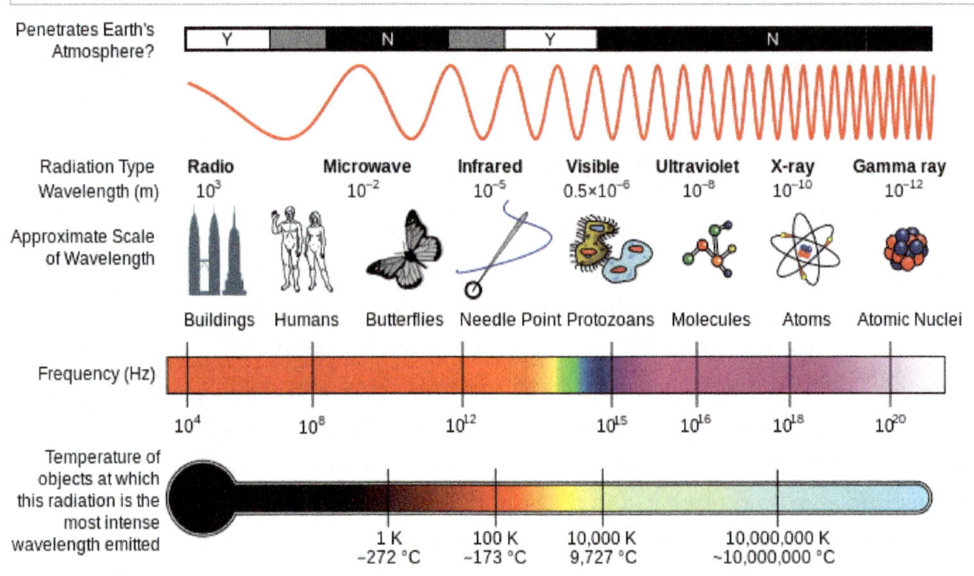

Figure 12-6 The electromagnetic spectrum.

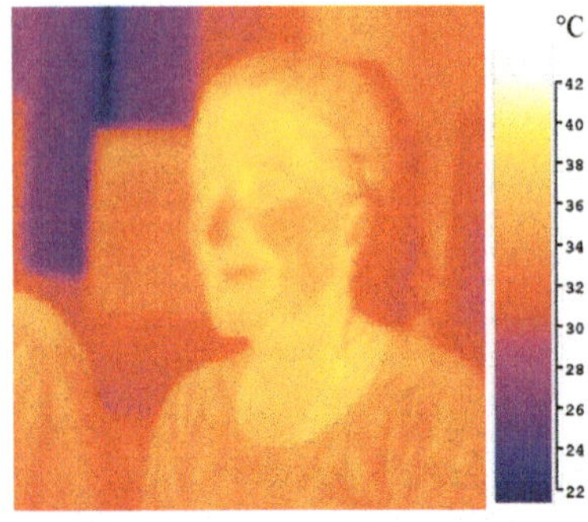

Figure 12-7 Infrared radiation has slightly longer wavelengths than visible light. Objects emit radiation at slightly different wavelengths as a function of their temperature. The image was taken with a filter sensitive to these wavelengths.

spectrum at another small window of frequencies (Figure 12-7). You also interact with the electromagnetic spectrum in many other ways, including using your cell phone, using a microwave oven to heat food, watching television, or listening to the radio (all various frequencies of microwaves). Visits to the emergency room might involve use of X-rays. The shorter the wavelength of electromagnetic radiation, the more damaging the waves are to living organisms. Thus, you can stand next to a microwave when preparing food, but exposure to gamma rays can be lethal.

We are discussing electromagnetic radiation because the Earth is heated by incoming solar radiation that interacts with the gases of the atmosphere as well as the surface of the Earth. The sunlight you see (visible light) is a form of *short-wavelength* radiation. Your eyes are unaware of other wavelengths arriving to Earth through space (Figure 12-8). As described earlier, several layers of the atmosphere play an important role in absorbing much other electromagnetic radiation, literally shielding the planet from harmful radiation. The Earth's heat budget, described below, describes the fate of the heat energy received by the Earth.

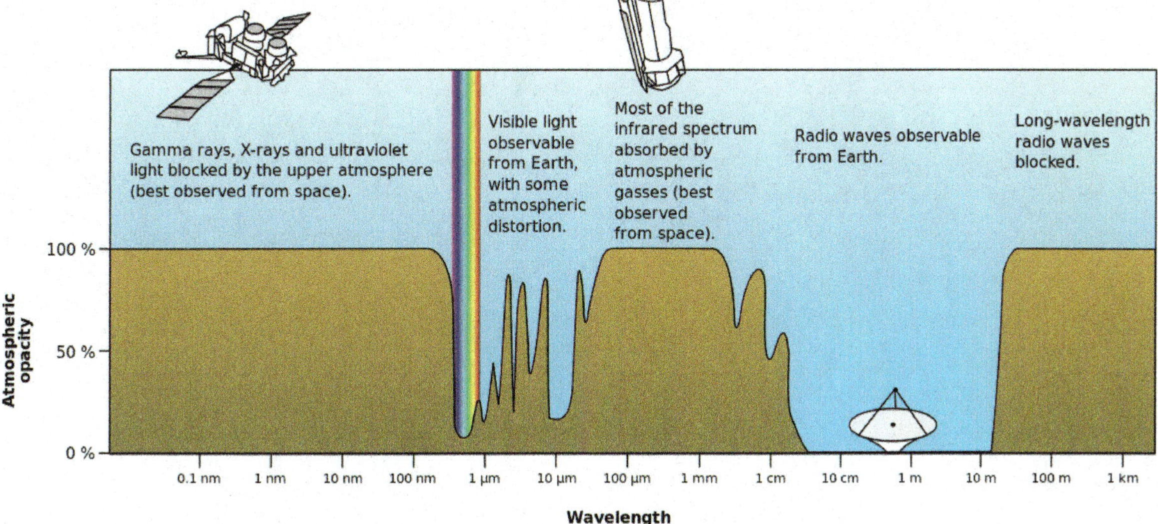

Figure 12-8 Most electromagnetic wavelengths do not reach the Earth's surface, but are absorbed by the atmosphere. Our eyes have evolved to be able to interpret the wavelengths that we term "visible light."

Heat Budget

Just like money flowing in and out of a savings account, the Earth's heat energy can be thought of as the net result of heat in, heat out. The Earth's average temperature would be stable when heat in = heat out. Global warming, then, represents either an increase in "heat in" or a decrease in "heat out." In order to figure out which of these is the case, we'll need to examine the heat budget in more detail. Figure 12-9 illustrates the Earth's heat budget. Solar radiation represents 100% of the heat received by the Earth. This equals "heat in"; approximately 1,370 watts/m^2 of solar radiation are received by Earth per year. Of this, nearly 30% of the energy reaching the upper atmosphere is reflected back into space; approximately two-thirds of this value is from the reflectance of aerosols (clouds and small particles), and the other one-third is from albedo of the Earth's surface. The processes by which heat goes out of the Earth system include 6% that is reflected back into space from the atmosphere (*albedo* is a term used to describe a surface's reflectance), 20% that is reflected back into space from clouds, and 4% that is reflected back from the Earth's surface. The atmosphere absorbs 16% of the incoming radiation, clouds absorb 3%, and the Earth's surface absorbs 51% of the heat energy. The total of heat in and heat out equals 100%.

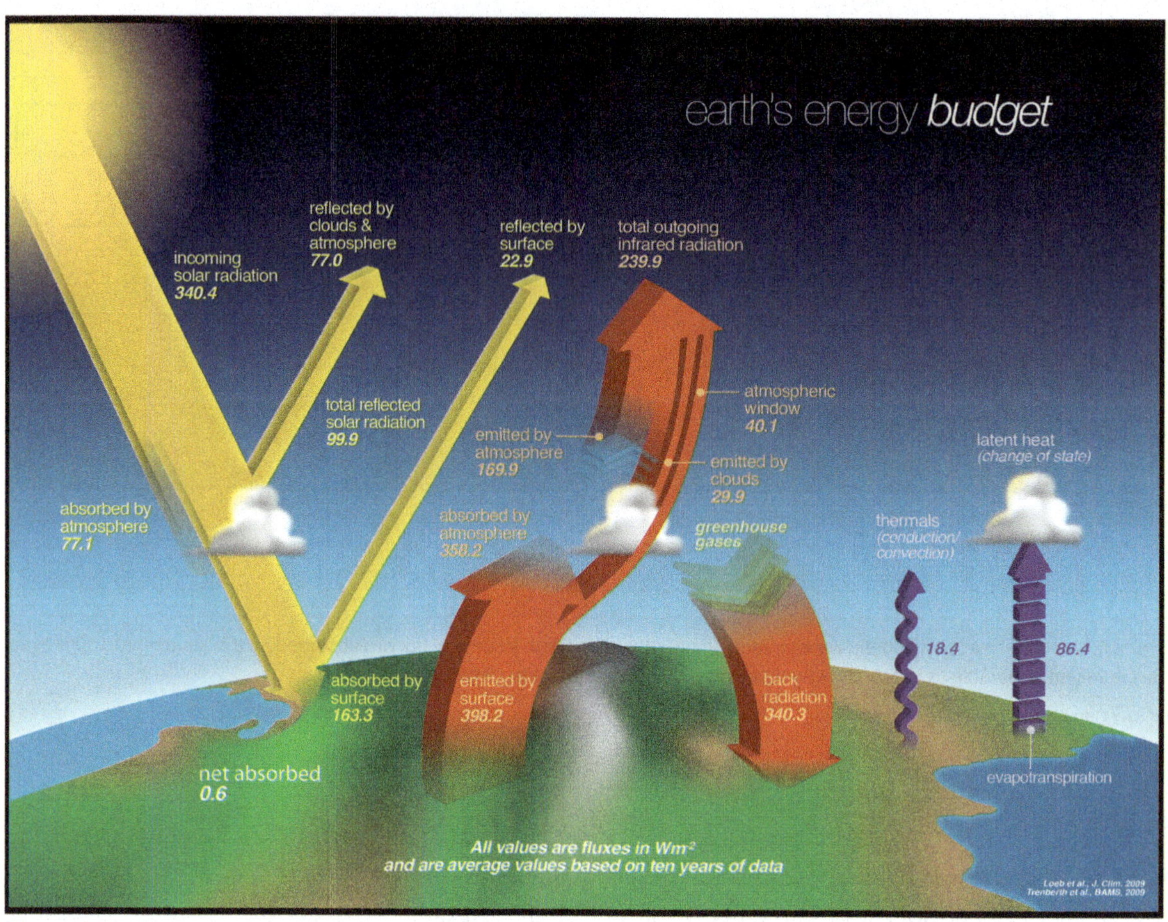

Figure 12-9 The Earth's energy budget. See text for discussion.

Image from http://science-edu.larc.nasa.gov/energy_budget/.

Albedo is a property that describes the reflectance of a material. Values describing albedo are presented as percentages; if 100% of visible light is reflected, a material has an albedo of 1.0. Light-colored materials have higher albedos than dark-colored substances do; therefore, snow, ice, and clouds have high albedos, while the oceans have low albedo values. Incoming short-wavelength solar radiation is reflected back into space also as short-wavelength radiation. This is a very different fate from that of the solar energy that is absorbed by Planet Earth, as we will see in the following text. The property of albedo essentially describes solar radiation that is reflected back into space before it can be absorbed to heat our planet.

What is the fate of the absorbed energy? Most of the energy (64%) is radiated back into space, as *long-wavelength* radiation,

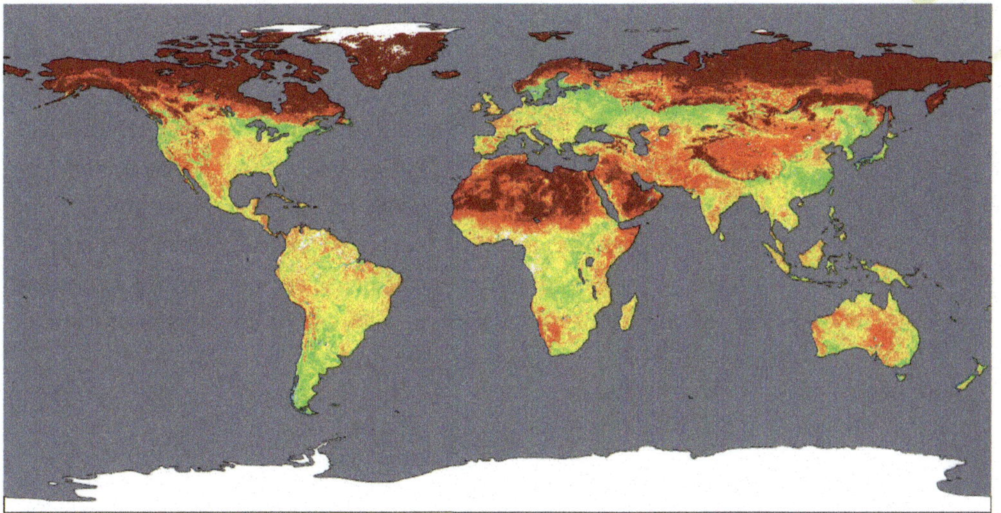

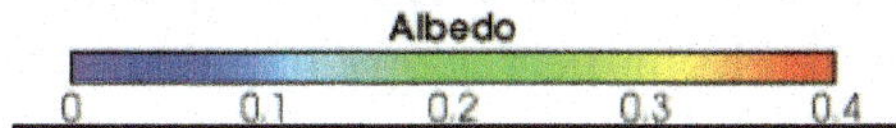

Figure 12-10 This Moderate-Resolution Imaging Spectroradiometer (MODIS) satellite image shows the difference in albedo over the Earth's land areas. The red-colored areas have the highest albedo (i.e., are the most reflective), yellows and greens are intermediate values, and blues and violets have low albedo (i.e., are least reflective). White indicates where no data were available, and no albedo data are provided over the oceans. The albedo of most land areas ranges between 0.1 and 0.4 (10–40%); the Earth's average is around 0.3. This image was produced using data compiled over a sixteen-day period, from April 7 through 22, 2002.

Image by Crystal Schaaf, Boston University, based on data processed by the MODIS Land Science Team. Image from http://visibleearth.nasa.gov/view.php?id=60636.

Source: Crystal Schaaf.

from clouds and the atmosphere. A much smaller percent (6%) is radiated back into space, also as *long-wavelength* radiation, from the Earth's surface. The Earth's surface heats the atmosphere so that 7% of the absorbed energy is conducted up to the troposphere. A much larger amount (23%) is carried aloft to the atmosphere by the *latent heat* of water vapor. (Latent heat represents heat stored and released with a change of state. In this case, heat is absorbed by water when it changes state from liquid to gas during evaporation.)

The terms *short-wavelength* radiation (incoming solar energy) and *long-wavelength* radiation (outgoing emitted energy) are in italics to stress their difference. The different wavelengths of electromagnetic radiation cause them to interact with gas molecules in the atmosphere very differently. The incoming *short-wavelength* radiation is absorbed by the Earth's surface and atmospheric gases and heated. An object that receives heat will reemit it, but it re-radiates as *long-wavelength* radiation. These wavelengths interact with the bonds between some molecules in the atmosphere, notably CO_2, CH_4, and H_2O. Most gases, such as nitrogen and oxygen, are "invisible" or transparent to longer-wavelength infrared radiation. Greenhouses gases are not. Instead, the bonds absorb infrared energy and vibrate (recall that temperature is a measure of the vibrational energy of an atom, so when atoms vibrate, or move, they are storing heat energy).

Fluctuation in Solar Irradiation

There are two heat energy sources that "run" Planet Earth. The first of these we discussed in Chapter 8, the Earth's internal heat generated from the decay of unstable isotopes, as well as the heat remaining from the formation of the Earth, 4.6 billion years ago. This energy source that drives plate tectonics is estimated to be approximately 47 +/− 2 TW (a terawatt, TW, is one trillion, or 10^{12}, watts) of energy. The second source, solar energy, drives all processes operating on the Earth's surface and in the biosphere and is significantly larger, approximately 173,000 TW of energy.

How much variation is there to the quantity of incoming solar radiation? Is there enough variation to explain why global average temperatures are rising? The Earth receives on average 1,366 watts/m^2 of energy from solar radiation per year (a ten-year average). When talking about financial budgets, the unit of measurement is the dollar. When talking heat energy, the common unit of measurement is a watt. A watt is a measure of energy transfer equal to 1 joule per second (a joule is a measure of work that is equivalent to the force of 1 newton over 1 meter, or 1 kg-m^2/sec^2). To simplify, watts measure the potential energy to get work done.

We know that solar energy output, or irradiance, varies in an eleven-year cycle related to sunspot activity (Figure 12-11). This variation produces a difference of approximately 0.2 W/m^2 of energy; in other words, fluctuations in solar output are not large enough to explain changes in global average temperatures (Chapter 13). It's been estimated that, at most, variation in solar irradiance would produce temperature changes on the order of tenths of a degree Celsius (0.1°–0.3°C).

While solar irradiation differences in the measurable past do not produce enough heat variation to explain current changes in global temperatures, if we look further back in Earth history, nearly half a billion years back in time, this was not the case. The early sun was significantly weaker than it is today, with estimates of as much as a 70% lower solar output at the time of formation of the Earth, 4.6 billion years ago. By half a billion years ago, solar output was estimated to be only 4% of today's value. When we discuss the geologic record of climate change (Chapter 13), we will discuss the significance of diminished solar output on global climate. At this point, however, variation in solar irradiance cannot explain current global temperature increase of about 0.5°C since 1900.

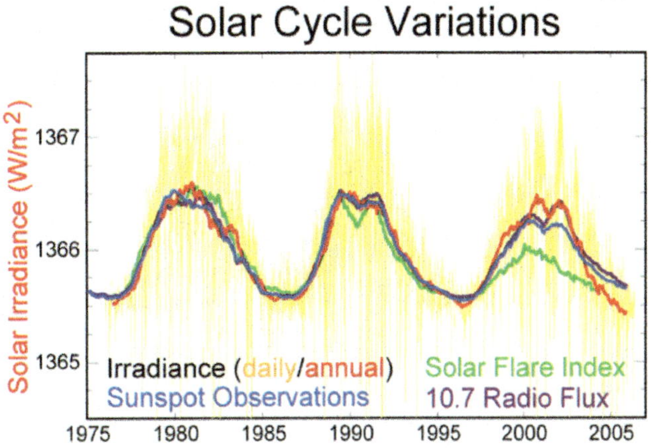

Figure 12-11 Cyclic variation in solar irradiation.

The Carbon Cycle

As we begin our discussion of atmospheric composition and its role in regulating the temperature of the Earth, we need to focus on the reservoirs and fluxes of one of the most important chemical elements in this process: carbon.

Figure 12-12 illustrates how the geosphere interacts with the carbon cycle. The six largest carbon reservoirs are shown: the biosphere and hydrosphere (the oceans and surface waters carrying the products of chemical weathering, geosphere (buried fossil fuels and the sedimentary rocks limestone and dolostone as well as unlithified ocean floor sediments, and the atmosphere. Figure 12-12 also shows processes that link the movement of carbon from one reservoir to another. For example, volcanic emissions. moves carbon from the Earth's interior into the atmosphere. Chemical weathering processes (hydrolysis and dissolution) move carbon dissolved in water droplets in the atmosphere (H_2CO_3) into the geosphere through the breakdown of minerals. The rates at which the fluxes shown on Figure 12-12 operate vary significantly. Carbon stored in sea floor sediments can take tens of millions of years before subduction and melting results in the release of carbon back to the atmosphere. Recycling of carbon from the biosphere into the atmosphere takes place at rates of minutes and days.

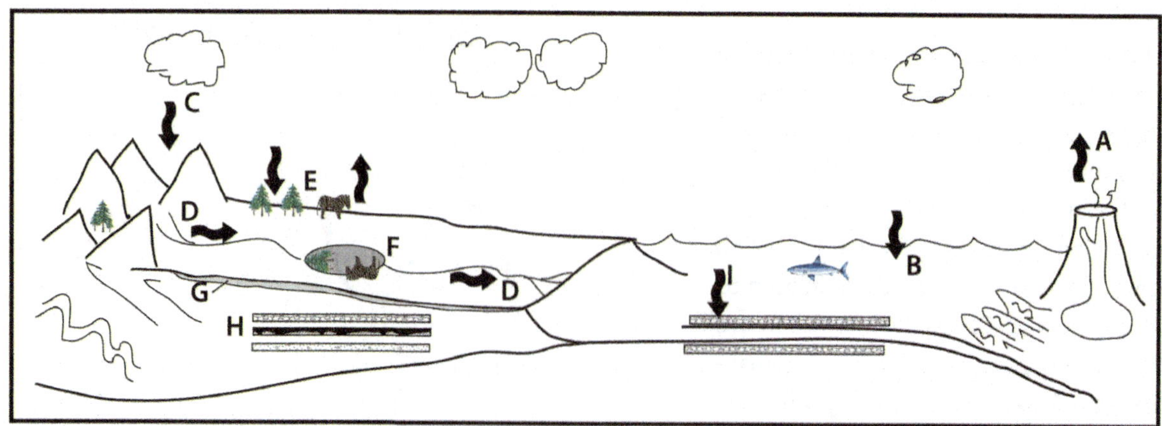

Figure 12-12 Cartoon illustrating the geosphere's interactions with the carbon cycle. A= degassing of CO_2 into the atmosphere from volcanic eruptions occurring at divergent and convergent boundaries. B = dissolution of atmospheric CO_2 gas into ocean water. C = dissolving atmospheric CO_2 into water droplets in the atmosphere, creates weak acid (H_2CO_3). D = chemical weathering of rocks in uplifted mountains produces HCO_3- into surface waters. E = cycling of CO_2 between the biosphere and atmosphere. F = burial and decay of organic carbon in surface waters (lakes,swamps and estuaries). G = storage of organic carbon in soil. H = storage of ancient organic carbon as fossil fuels. I = storage of carbon in ocean sediments and sedimentary rocks. See Chapters 9 and 10 for details on convergent plate boundaries and the processes that uplift mountains. Chapter 11 covers chemical and physical weathering and the relationship of weathering to increased surface area.

Table 12-3 Volumes of Reservoirs in the Carbon Cycle. Data from Various Sources. * Indicates Levels in 2000

Reservoir	amount (10^9 Metric Tons)
atmosphere	580-760 *
soil organics	1200-1600
ocean	38,000-40,000
marine sediments and sedimentary rocks	66,000,000 to 100,000,000
terrestrial plants	600-2200
fossil fuels & kerogen	19,000

The carbon cycle is one of the biogeochemical cycles on Earth (the others include phosphorus, nitrogen, and sulfur), so called because of the importance of cycling within the biosphere and the multiple interactions of organisms with air, water, and soil. Just as we represented the rock cycle as a cycle within the geosphere, carbon, nitrogen, and sulfur all have cycles within the biosphere. We are focusing on carbon because of the role it plays in regulating the Earth's climate. As we examine Figure 12-12 and the data in Table 12-3, it's clear that the largest reservoir for carbon consists of sediment, sedimentary rocks, and fossil fuels. These reservoirs store carbon for up to hundreds of millions of years, making it unavailable to the atmosphere. Another term for storing carbon is *sequestering*; marine

Figure 12-13 Photograph of slash-and-burn in the Amazon rainforest, Brazil, August 2010. Deforestation is a flux between the biosphere and atmosphere reservoirs in the carbon cycle. Photograph by astronauts aboard the ISS. NASA photograph ISS024-E-11941.

Image from http://earthobservatory.nasa.gov/Features/ForestCarbon/.

sediments, sedimentary rocks, and fossil fuels *sequester* carbon. Aside from mining limestone to make cement, most sedimentary rocks sequester carbon for extremely long time periods. Retrieving and burning fossil fuels, however, moves carbon that has been sequestered below the Earth's surface for millions of years into the atmosphere, where it will flux between the other reservoirs listed in Table 12-3.

Strategies to deal with the increase in atmospheric carbon include increasing its sequestration in other reservoirs. Planting

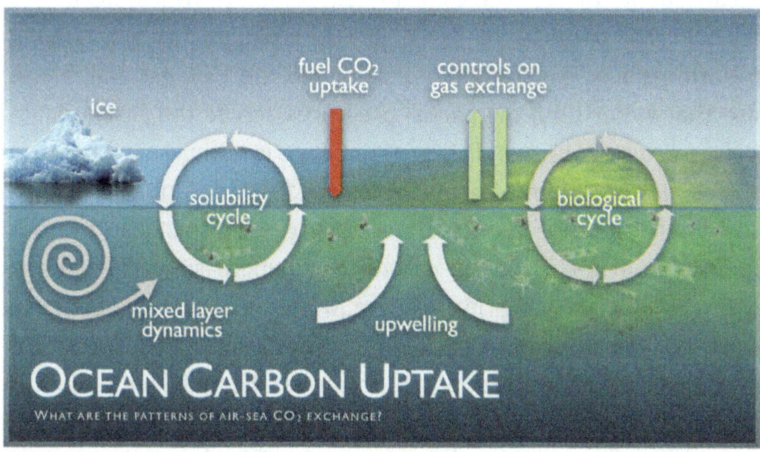

Figure 12-14 The oceanic portion of the carbon cycle. As atmospheric CO_2 levels increase, the oceans will absorb more gas. Increasing dissolution of CO_2 in seawater impacts marine ecosystems.

Image from http://pmel.noaa.gov/co2/files/pmel-research.003.jpg.

trees is an example of a carbon sequestration strategy. Can you think of the steps necessary to determine whether planting trees is an effective strategy to sequester carbon? We should have some idea of exactly how much carbon an "average tree" could store, how long that tree would live, and how many trees there are in a given area. Except for the last, these are not easy numbers to determine; the amount of carbon stored varies by tree species and by age; in other words, carbon storage changes over time. The most effective tree reservoirs are large, old trees such as "old-growth forest," and because trees in the tropics can grow year-round, forests such as the Amazon rainforest are the most effective at sequestering carbon. Unfortunately, these forests are experiencing intense deforestation activities (Figure 12-13). In general, the amount of carbon stored in a tree is equal to 50% of its dry biomass. Recent research that uses satellite-based measurements of forest cover, height, and type indicate that nearly 247 gigatons (10^9 tons) of carbon were sequestered in tropical forests. Of this, 193 gigatons are stored above ground in trunks, branches, and leaves, and 54 gigatons are stored below ground in the roots. Geographically, the forests in Central and South America account for 49% of the total, with Southeast Asia sheltering 26% and sub-Saharan Africa sheltering 25% of the carbon stored.

The oceans are considered to be the other important reservoir for sequestering carbon from the atmosphere. Chapter 11 presented evidence for ocean acidification, which is the result of ocean uptake of CO_2 as levels increased in the atmosphere. This has occurred because the partial pressure of a gas has increased in the atmosphere overlying the oceans, and the CO_2 gas will diffuse into that water until the partial pressures across the air-water interface are equilibrated. This gas exchange is complicated because of *feedback loops* in the carbon cycle. *Feedback* describes the pathway where a change in one parameter triggers a change in another parameter. In terms of the carbon cycle, increasing atmospheric levels of CO_2 warms global climate, which in turn impacts ocean thermohaline circulation. As a result of altering ocean circulation, atmospheric CO_2 uptake by the oceans is affected. The feedbacks continue as changing CO_2 uptake in the oceans alters marine ecosystems. Figure 12-14 shows more detail in the marine carbon cycle than is presented in Figure 12-12.

Because of these feedbacks, it's very difficult to predict the long-term changes in the carbon cycle, but systems theory suggests that at some point an steady state is reached.

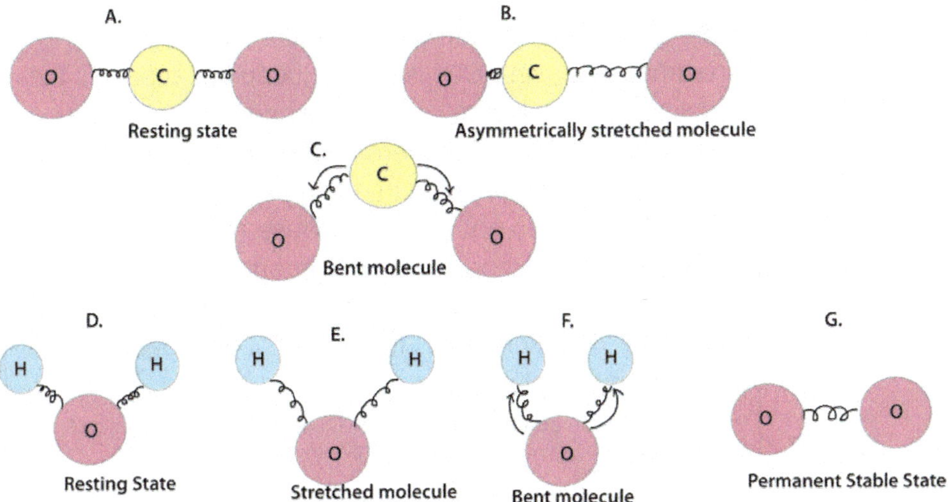

Figure 12-15 CO_2 and H_2O molecules and their behavior when interacting with infrared radiation. A–C: CO_2; D–F: H_2O. G illustrates the stable O_2 molecule, which does not interact with infrared wavelengths and is therefore not a greenhouse gas.

We will now examine why carbon molecules are considered greenhouse gases and why they impact the Earth's climate.

Greenhouse Gases

What is the "greenhouse effect," and why are some gases considered to be "greenhouse gases"? A greenhouse keeps warm by allowing solar radiation to pass through the glass during daylight, where it can be absorbed by vegetation. During the night, the absorbed heat is re-radiated back, but it is re-radiated energy at a different, nonvisible wavelength. Gases in the Earth's atmosphere act like the pane of glass in a greenhouse. They allow a range of electromagnetic spectra to pass through and be absorbed by the Earth's surface. In a greenhouse the panes of glass trap the heat inside; however, on Earth, the absorbed energy is re-radiated back into space. The most important aspect of this re-radiation is that the energy is emitted at *different wavelengths* from the incoming energy. These wavelengths interact with gases in the atmosphere and don't escape back into space.

Revisit Table 12-2 and see that CO_2 and H_2O are abundant molecules in the troposphere layer of the atmosphere. These two gases, along with methane (CH_4), chlorofluorocarbons (CFCs),

Table 12-4 Global Warming Potential (GWP) of Greenhouse Gases. CO_2 is not Included because it is the baseline gas to which all others are compared; residence time of CO_2 in the atmosphere is difficult to constrain, but is estimated to be between thirty and ninety-five years. water vapor is also not shown because of the large variability in the temperature-dependent concentrations in the atmosphere; its residence time is as short as nine days. The "time horizon" is the length of time over which the GWP is calculated.

Data from 2013 Intergovernmental Panel on Climate Change Assessment Report 5, page 714), http://www.ipcc.ch/ipccreports/tar/wg1/016.htm.

Source: Intergovernmental Panel on Climate Change (IPCC).

Gas	Lifetime (years)	GWP time	Horizon
		(20 years)	(100 years)
CH_4	12.4	86	34
HFC–134a	13.4	3,790	1,550
CFC–11	45	7,020	5,350
NO	121	268	298
CF_4	50,000	4,950	7,350

hydrofluorocarbons (HFCs), nitrous oxide (NO), and carbon tetrachloride (CF_4), are considered "greenhouse gases" because the physical structure of their molecules results in their ability to absorb a narrow range of long-wavelength electromagnetic radiation. Instead of energy being re-radiated back into space, it stays trapped in the atmosphere, absorbed by greenhouse gas molecules, where it acts as a thermal blanket for the Earth.

What Makes a Gas a Greenhouse Gas?

In order to understand why some molecules are greenhouse gases and others are not, we need to understand electromagnetic waves. Recall that there is a relationship between an electric current and the resulting magnetic field: one induces the other. If we think of an electric current flowing north-south, a magnetic field would be established in an east-west orientation, and vice versa. Imagine that we have a molecule, such as H_2O, that is dipolar (Figure 12-15). This means that one "side" of the molecule has more of a positive charge than the other does. When an electromagnetic wave interacts with this dipolar molecule, where the positive and negative charges are separated from one another, the electric field in the wave causes the bonds to deform. An analogy would be to imagine the bonds as springs that can expand and compress. In some molecules, such as CO_2 and CH_4, the springs are not symmetrically balanced, so when the electromagnetic wave passes through the molecule, the springs move around. Compounds in the atmosphere where the bonds (springs) are balanced include O_2 and N_2. Compounds where they are not balanced are the molecules we call the greenhouse gases. Energy is stored in these unbalanced molecules.

If there were no greenhouse gases in the atmosphere, the Earth would re-radiate the heat it absorbs, and the Earth's average temperature would be –18°C; in other words, it would be encased in ice. Greenhouse gases act as our thermal blanket. However, the Earth's energy budget is not currently in balance because of increasing amounts of these gases in the atmosphere. The atmospheric abundance of both CO_2 from the burning of fossil fuels and CH_4 from agriculture are both associated with human activities, so we describe it as "anthropogenically derived." As we will see in the next chapter, the abundance of greenhouse gases in the Earth's atmosphere is considered to be the primary "driver" behind current global climate change.

Do all greenhouse gases have the same heat-absorbing effect, or are some more potent thermal blankets? One way to express these relative differences is a concept called *global warming potential* (*GWP*). How much heat a greenhouse gas traps in the atmosphere is compared to that of the equivalent amount of CO_2 trapped. Values of GWP are expressed as dimensionless numbers, as they represent a relative comparison to carbon dioxide. The effectiveness of a greenhouse gas is a function of the wavelengths of infrared energy that it absorbs and the residence time of the gas in the atmosphere; a gas with a long residence time and a large band of wavelengths of infrared absorption would be a very effective greenhouse gas. Methane (CH_4) is nearly twenty-five times more potent as a heat-trapping gas than CO_2 is. Recall from Chapter 1 that methane is released from its reservoir in rocks as part of the hydrofracking retrieval process. Table 12-4 shows the global warming potential of several greenhouse gases.

If a gas has a short time horizon, it has a short residence time in the atmosphere. Thus, it may initially have a very large GWP, but it will be short-lived. A gas with a longer time horizon has a longer residence time in the atmosphere. Its initial GWP may not be as

Table 12-5 Comparison of volcanic and anthropogenic CO_2 emissions from volcanoes. Data are in gigatons (10^9, or 1 billion tons). ACM = ratio of annual anthropogenic CO_2 (35 Gt) to maximum preferred estimate for annual volcanic CO_2.

Data from US Geological Survey, http://Volcanoes.usgs.gov/hazards/gas/climate.php.

Source	Amount (Gt/yr)
Global volcanic emissions (max. estimate)	0.26
Anthropogenic CO_2 in 2010 (projected)	35
Mt. St. Helens eruptions, 1980	0.01
Mt. Pinatubo, 1991	0.05
Number of Pinatubo-equivalent eruptions equal to annual anthropogenic CO_2	700
Number of Mt. St. Helens eruptions equivalent to annual anthropogenic CO_2	3500
2010 anthropogenic CO_2 multiplier (ACM)	135
1900 ACM	18
1950 ACM	38
Number of days for anthropogenic CO_2 to equal annual global volcanism	2.7

high as that of other gases, but its persistence in the atmosphere means that over time it will have a very large cumulative effect. Examine Table 12-4 and determine which greenhouse gases are potent "up front" and which have a longer-term impact on global temperature.

Volcanic Emissions and the Earth's Atmosphere

We know that volcanoes emit both gases and particles into the atmosphere. Aren't they equally culpable in affecting the Earth's climate? After Mount Pinatubo in the Philippines erupted in 1991, the global average temperature dropped by 0.5°C, so there is evidence that this is the case. Why does this occur?

Let's review what is emitted from a volcano during an eruption (Chapter 5). Lava is a component, but pyroclastic material (pulverized rock), ash, and gases are volumetrically more important. For example, the Mount St. Helens eruption in 1980 ejected a minimum of 1.1 km^3 of uncompacted ash, which would be equivalent to 0.20 to 0.25 km^3 of magma or solid rock. In the atmosphere, this particulate material has several effects. Until it is "scrubbed" from the atmosphere through rainfall, particulate debris shields Planet Earth from incoming solar radiation by increasing albedo. This is why the global temperature dropped following Mount Pinatubo's eruption.

Gas emissions also have an impact. Table 12-1 presents data on the average gas composition from volcanoes at convergent and divergent plate boundaries as well as an intra-plate hot spot. You know that the composition of magma varies as a function of the source of the magma (what has melted). Examine the data in Table 12-1 and see if you observe a pattern to gas composition.

Three gases listed in Table 12-1 are particularly hazardous: SO_2, CO_2, HF, and HCl. We have already identified carbon dioxide as a greenhouse gas. What about sulfur dioxide, hydrogen fluoride, and hydrogen chloride? Sulfur dioxide can combine with water to produce sulfuric acid, which then dissociates to form the sulfate ion $SO_4^{=}$ (see Chapter 11). Sulfate ions condense to produce sulfate aerosols. The aerosols act as a shield against incoming solar radiation, causing a short-term decrease in incoming solar radiation. However, sulfate aerosols also interact with ultraviolet light in the stratosphere to deplete O_3 (ozone), a molecule in the stratosphere that helps block harmful UV radiation, and ozone is a greenhouse gas. Both HF and HCl are highly toxic gases that attack the respiratory systems of organisms, if ingested.

It appears that one volcanic emission (SO_2) might actually help "offset" global warming if sulfate aerosols block incoming solar radiation and lower Earth temperature. This response, however, is short lived, as the sulfates ultimately leave the atmosphere in (acidic) rain droplets. Because sulfate aerosols destroy ozone, they have the effect of increasing incoming solar radiation.

Are volcanic emissions of carbon dioxide a culprit in the current global climate change? Let's compare emissions from volcanoes to those of anthropogenic activities (Table 12-5).

Examine the data in Table 12-5 and describe the difference in the annual carbon dioxide emissions of all global volcanism versus that generated by the burning of fossil fuels. Describe the number of Mount St. Helens eruptions needed to generate the annual anthropogenic CO_2 levels emitted. Finally, a year's average volcanic eruptions equal how many days of anthropogenic production of CO_2? Based on these values, how would you respond to someone who attributes increasing levels of atmospheric CO_2 to volcanic activity?

Summary

The atmosphere, the thin gaseous layer above the Earth, originated during the same differentiation process that produced the layered solid Earth. The Earth's gravity was strong enough to hold on to these low density elements and molecules as they were degassed in volcanic eruptions, preventing their escape into space. Also like the solid Earth, the atmosphere is divisible into a series of layers, each with distinct properties. The troposphere, which supports all life, is the layer closest to the Earth's surface. The troposphere is composed primarily of nitrogen with subordinate amounts of oxygen and trace amounts of other gases such as CO_2, CO, CH_4 and SO_2. This composition has varied in the geologic past, particularly in regards to levels of CO_2 and O_2. In the earliest history of the Earth the atmosphere had approximately 30 times more carbon dioxide it in than it has now. Evolutionary trends in the biosphere, the burial of organic matter, and increased chemical weathering rates are all processes that influence the levels of CO_2 in the atmosphere. Of course today, the burning of long-buried fossil fuels is rapidly increasing atmospheric CO_2 levels. Is it possible that these rising levels are the result of increased volcanism? Analysis of the emission from a series of recent volcanic eruptions clearly indicates that this cannot be the case. It would take several hundred to several thousand large volcanic eruptions every year to equal the amount of CO_2 produced through the burning of fossil fuel in one year. Gases in the atmosphere help trap heat, keeping the Earth's surface temperatures from reaching the extremes seen on planets lacking an atmosphere, such as Mars. What makes a gas a "greenhouse gas" is the property that enhances a molecule's ability to trap some wavelengths of energy. Solar energy reaching the Earth travels through the vacuum of space in a variety of wavelengths and frequencies termed the electromagnetic spectrum. One narrow range of the EM spectrum is visible light, considered a short wavelength form of the EM spectrum. Besides supporting life, the atmosphere plays a critical role in shielding the Earth's surface from other, damaging parts of the EM spectrum, such as ultraviolet radiation. The Earth's energy budget describes the balance of "heat in" from solar radiation and "heat out." Short wave length radiation is absorbed by the Earth's surface and the atmosphere and some is not absorbed but is reflected back into space, a phenomenon termed albedo. The warmed Earth emits heat back into space, but does so as long wavelength radiation.

This form of energy interacts with the bonds of some molecules, including CO_2, CO, CH_4 and H_2O and therefore this energy does not escape into space. When this happens, "heat in" is not balanced by "heat out" and the planet warms. Anthropogenic activities drastically started increasing the levels of these gases in the atmosphere in the industrial revolution of the mid-1800's, resulting in the increase in atmospheric heat retention. Annual measurements of modern atmospheric CO_2 levels are recorded in a data set assembled by Charles Keeling, and are represented on the "Keeling Curve." Because many of the greenhouse gases are carbon molecules it is important to understand how carbon cycles through the Earth system. The largest reservoir of carbon are sediments on the sea floor and sedimentary rocks, followed by buried organic matter (fossil fuel). Much carbon is dissolved in sea water (HCO_3^-) which arrives there as a by-product of chemical weathering activities. Thus, it's clear that by burning fossil fuels we are transferring carbon from a long-term storage site below ground, into the atmosphere. The other reservoirs of carbon, for example, the biosphere, are not able to absorb these increasing levels. Ocean acidification levels (Chapter 10) suggest that the oceans have been absorbing atmospheric CO_2, but at some point they will also reach saturation.

While a water molecule has a higher heat capacity than air, atmospheric circulation plays a major role in moving heat around the Earth's surface. When water is evaporated from the oceans near the equator the energy needed to change water from a liquid to a gas is stored. This "latent heat" is released to the atmosphere when the water molecule condenses back into liquid form. Because much evaporation happens at the equator, when solar radiation is highest, when condensation occurs at higher latitudes heat has been moved poleward.

Chapter 12 Review Questions

- What are the layers of the atmosphere, their elevations above the Earth's surface and their major characteristics?
- What is the origin of the Earth's atmosphere and why is it layered?
- What are the two most abundant gases in the troposphere?
- What gases are emitted during volcanic eruptions?

- Compare the amount of CO_2 emitted from volcanic eruptions to that produced by human activities. Does this data suggest that volcanic eruptions are a reasonable cause of increasing atmospheric CO_2 levels?
- How can solar heat energy reach Earth through the vacuum of space?
- We know that solar output varies over time. Can variation in solar output explain global temperature rise over the past century?
- Define "albedo" and describe its role in the Earth's heat budget.
- Using labeled arrows, draw the fate of incoming solar energy to Earth: what percent is absorbed by the Earth's surface, what percent is reflected back into space and what percent is absorbed by clouds and the atmosphere?
- Why is the Earth considered a "black body"?
- What makes some gases "greenhouse gases"? Why is O_3 a greenhouse gas but O_2 is not?
- What is the difference between the heat received by the Earth and the heat it emits back into space?
- Because carbon molecules such as CO_2, CO and CH_4 are greenhouse gases, the pathway of carbon through the Earth system is very important. Using boxes and arrows, draw a simplified carbon cycle using only the four largest reservoirs of carbon.
- In order to reduce the amount of carbon in the atmosphere, engineers are trying to design systems to sequester carbon. What does "sequester" mean in this context?
- What is a "feedback loop"? Is melting the polar ice caps a positive or negative feedback loop for global warming?

Chapter 13

Earth's Climate

At the end of this chapter you should be able to

- Explain why it is harder to predict the weather than climate
- Define the components of the Milankovitch effect and their frequency
- Draw a concept sketch that describes how the Earth's climate system works
- Communicate the evidence on which the climate forecast for the twenty-first century is based
- Analyze and critique policy issues related to global warming

Introduction

The geosciences have been particularly useful in providing documentation of the historical record of climate change on Earth and addressing questions such as the following: How much has global climate varied in Earth's history? What is the rate at which climate changes? With what frequency does climate change? Can we distinguish between "natural" and anthropogenically induced climate change? We will explore the magnitude and rates of climate changes anticipated in the remainder of this century. This chapter will examine the data on past climates that geoscientists have retrieved and present some predictions for future changes. First, we will discuss remaining controls on the Earth's climate, including variation in the orbit of Earth around the sun.

Weather versus Climate

The difference between "weather" and "climate" is the length of time you are referring to; *weather* represents daily changes in temperature and precipitation, in other words, the state of the atmosphere at a specific place and time. Climate represents average daily weather for a location over decades to centuries. From personal experience you probably know how difficult it is to predict the weather and how often the weather forecaster is wrong! How then, can we speak about climate and climate change with any confidence? Predicting short-term events, such as weather, is extremely dependent on knowing the initial or starting conditions, as well as analyzing data that suggests how these conditions might change. Let's use a simple sports analogy. A pitcher throws a ball toward a batter. Why doesn't the batter hit the ball each time? In large part it is because the pitcher releases the ball from her hand slightly differently, depending on whether or not she is trying to throw a fastball or a curve. The same is true for predicting weather. It is very difficult to fully know all the initial weather-influencing data at a location at an instant in time because weather is so quick to change, spatially and temporally. Climate, however, because it examines longer term trends, doesn't have the "initial state" problem; daily, monthly, and seasonal variation averages out. We know that the winter ten years from now will be colder than will the summer next year. Thus, it is possible to make long-term climate predictions even though we may be unsure of next week's weather.

Variation in the Earth's Orbit around the Sun

In Chapters 11 and 12 we discussed several of the variables that control the Earth's climate, for example, variation in solar irradiance, albedo, and changes in the composition of the atmosphere. In this chapter we will discuss another control of Earth climate: variation in the Earth's orbit around the sun.

The amount of heat received at a particular location on Earth in any given year varies as a function of latitude because of the differences in the angle of incidence of solar radiation on the Earth's surface (Figure 13-1). A critical concept is that of the *subsolar point*, defined as the latitude on Earth where the sun's rays are hitting directly perpendicular to the surface. The movement of the subsolar point from the northern to southern hemispheres defines the change in seasons on Earth.

As the illustration shows, the angle at which solar rays hit the curved surface of the Earth controls the amount of surface area heated; the same amount of energy is focused on a smaller region at the equator than at the poles. The dissipation effect of the atmosphere is also a function of the angle of incidence; at shallower

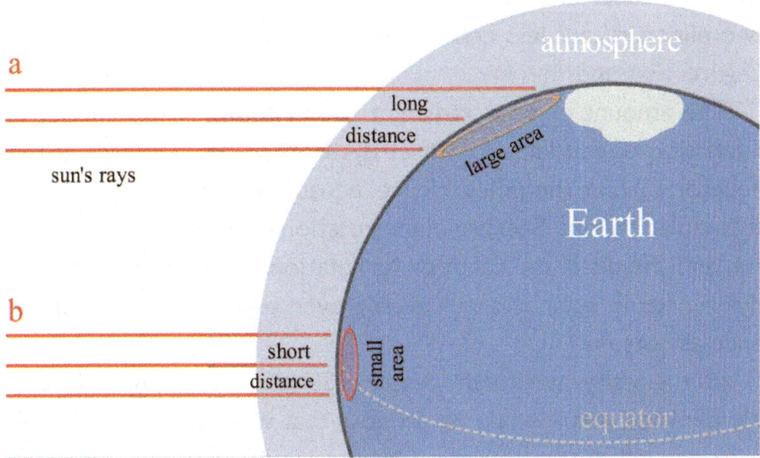

Figure 13-1 The latitudinal difference in the angle of incidence of solar radiation and its effect in heating the Earth's surface.

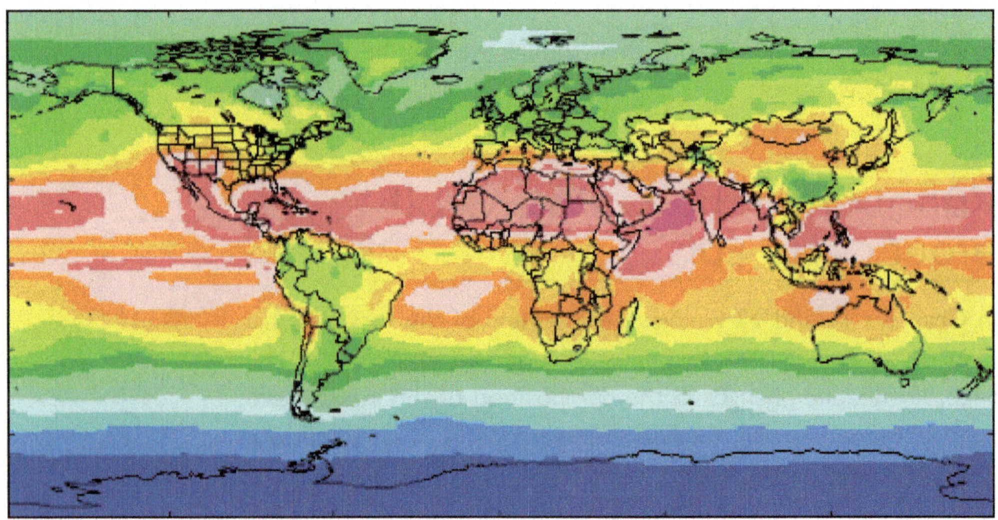

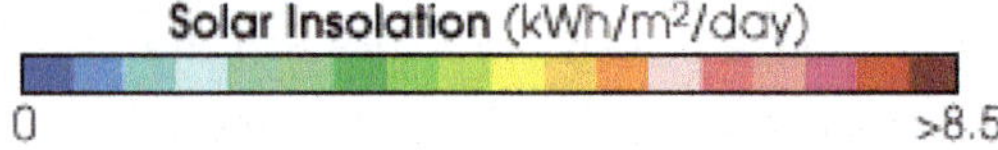

Figure 13-2 Measured values for solar radiation in April (northern hemisphere spring) over a ten-year period. Note higher solar radiation at the Earth's surface in the equatorial region, where the angle of incidence is close to perpendicular. The angle of incidence decreases toward the poles, and the amount of solar radiation received at these higher latitudes is thus substantially reduced.

Image from http://en.wikipedia.org/wiki/Insolation#mediavihttp://www.esa.int/images/07-atmosvindue.gifewer/File:Insolation.gif.

angles (higher latitudes), the atmosphere can scatter solar rays over a broader area than it can at the equator. The sum of these two effects means less heat at the poles and more at the equator (Figure 13-2 and 11-11).

The amount of solar radiation reaching the Earth's surface is a function of latitude: more heat is absorbed per unit area at the equator than at the poles. Hence, equatorial regions are warmer than polar regions. Because of the rotation of the Earth around the sun and the tilt of the Earth on its rotational axis, these latitudinal differences in solar irradiance vary over time, which we experience as seasons (Figure 13-3).

In the early twentieth century, Serbian astronomer Milutin Milankovitch was the first to propose that variations in the Earth's orbit around the sun occurred in a series of cycles. He suggested that these variations would affect the Earth's climate and might be responsible for fluctuations between glacial and interglacial periods. Milankovitch recognized three aspects to the Earth-sun orbital

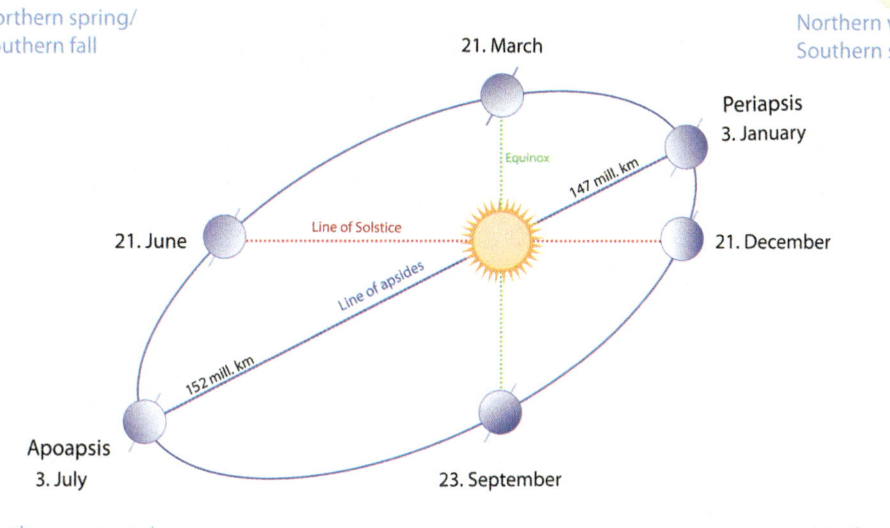

Figure 13-3 Seasons owe their origin *not* to differences in the distance of the Earth from the sun but to differences in the tilt of the Earth's axis toward or away from the sun at different times of the year. A northern hemisphere summer occurs when this region of the globe is tilted toward the sun. At this time the angle of incidence of solar radiation delivers more heat per unit area to the northern hemisphere. The ellipticity of the Earth's orbit is exaggerated,

relationship (Figure 13-4). *Obliquity* describes variation in the tilt of the Earth on its rotational axis. Currently, Earth's tilt is 23.5°, and it varies between 22° and 24°. The tilt of the northern hemisphere toward the sun causes a northern hemisphere summer as the subsolar point is in the located in these latitudes; the tilt of the northern hemisphere away from the sun characterizes a northern hemisphere winter. If there were no tilt to the Earth's rotation, there would be no seasons. Conversely, the higher the angle of inclination, the greater the differences between summer and winter temperatures. Obliquity also controls summer solar radiation differences between the high and low latitudes. When tilt is low, there are extreme temperature differences between the equator and pole because of differences in the angle of incidence of solar radiation. Because of this, heat and moisture transport by atmospheric circulation cells will be at their maximum. In turn, this promotes high rainfall in mid- to upper latitudes and, in combination with the other orbital variation, can result in ice sheet growth. When tilt is high, the solar radiation gradient between the equator and poles is lower, and there is less poleward heat and moisture transport. Ice

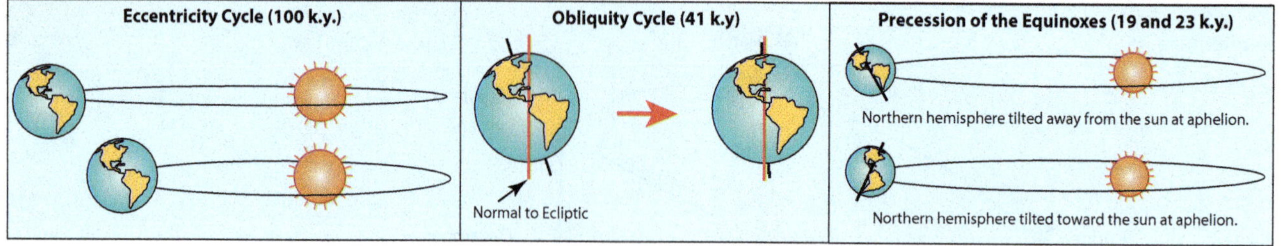

Figure 13-4 Milankovitch variables and their frequency.

sheet growth would be inhibited. Milankovitch identified a 41,000 periodicity to maxima and minima angles of tilt. We are currently in mid-cycle.

Precession describes differences between times of the year where the Earth is at aphelion (furthest distance from the sun) and perihelion (closest to the sun) in its orbit. Currently, the northern hemisphere is in summer (southern hemisphere in winter) when the Earth is far from the sun and a northern hemisphere winter (southern hemisphere summer) when the Earth is close to the sun. This describes a *precessional minima*. Approximately ten thousand years ago, we were at a precessional maxima and the current relationship was reversed (northern hemisphere summer when the Earth was close to the sun). Between the minima and maxima of precessional cycles, there is about an 8% difference in solar radiation received between the summer and winter seasons. Climatologists have also noted that precessional state influences monsoon activity in Asia and Africa (a *monsoon* is a seasonal rainy period driven by changes in wind patterns between land and sea). Monsoon activity is at its lowest during precessional minima. Milankovitch noted that precession cycles on a complicated nineteen-thousand- and twenty-three-thousand-year frequency.

Eccentricity describes the degree to which the Earth's orbit around the sun is circular to elliptical. The orbit changes between one extreme and the other on a one hundred thousand year cycle. The Earth receives more solar radiation when it is closer to the sun. The more elliptical the orbit, the greater the seasonal variation in temperature. We are currently in a state of low eccentricity (approximately 3% from circular), which produces a seasonal energy difference of about 7%. When eccentricity is at its peak (approximately 9% from circular), seasonal temperature differences can reach about 20%. Eccentricity also impacts the effects of precession. If the Earth's orbit were perfectly circular, it wouldn't matter which precessional state the orbit was in and seasonality would be controlled only by obliquity. With a more elliptical orbit, precession and obliquity both control seasonal temperature variation.

When you mathematically combine cycles that operate at different frequencies (this is termed *harmonic analysis*), you can produce a summed curve that shows the result from their interaction (Figure 13-5). Because the angle of incidence varies as a function of latitude, and because of the large amount of land area that lies at this latitude, Milankovitch chose 65°N latitude for his calculations. You can compare the summed orbital variation, termed *solar forcing*, to ancient climate data and look for correlation. Recall that "correlation doesn't mean causation," so we need to ask if there is a cause-and-effect relationship between orbital variation and climate. As our previous discussion demonstrated, there is a strong theoretical reason to propose that orbital variations of the Earth and sun would affect solar irradiance and the Earth's climate. Can it explain all climate variation?

The next section will present the record of climate change on Earth for a period of time termed the Pleistocene Epoch, spanning approximately 2.5 million years ago to approximately 11,500 years ago. There is much evidence on our landscape that during this interval of time the Earth experienced a major glaciation characterized by several advances and retreats of a large continental ice sheet. Geologists view the Pleistocene as the most recent "icebox Earth" climate extreme. You should test the Milankovitch predictions for climate variation against the data sets recording climate.

The Historical Record of Climate Change

Critical to our being able to document how climate has varied in the geologic past has been our ability to tell time. If we couldn't do this, we would not be able to determine a chronology of events. In this chapter we will examine the record of climate change as it's been recorded in glacial ice on Greenland and Antarctica for nearly half a million years.

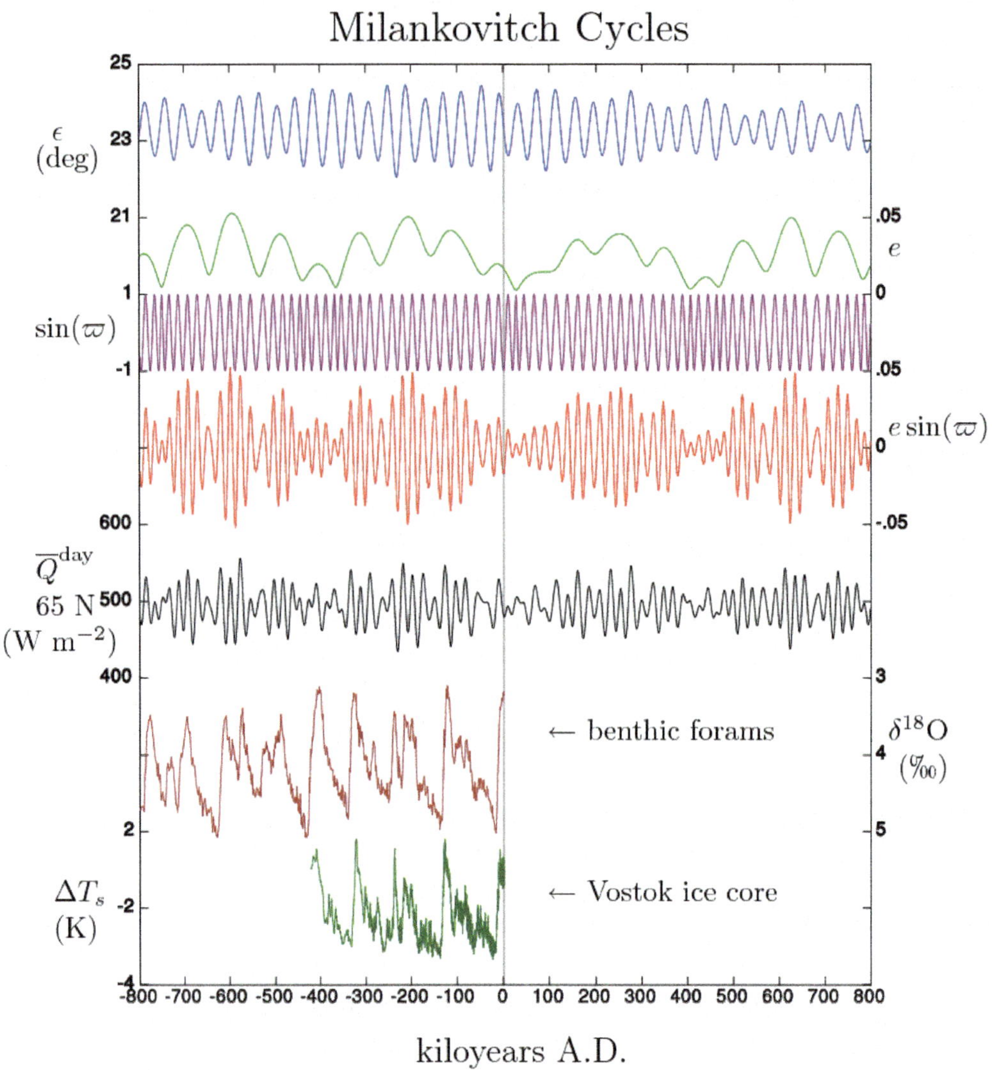

Figure 13-5 Milankovitch cycles. Epsilon (ϵ blue line) represents forty-one-thousand-year cycles of obliquity. The green line (e) represents the one-hundred-thousand-year eccentricity cycle. The purple line (ω) records longitude of perihelion; precession). The red line is the precession index, which factors precession and obliquity to represent intensity of seasonality. Q/day represents the sum of solar insolation at 65°north latitude. Proxy data on ocean temperatures are represented by benthic foraminifera oxygen isotope (brown line) and Vostok, Antarctica, ice core data (green line).

Layers in Ice

Glacial ice represents accumulated and compressed snowfall. Even in the high latitudes there are seasonal differences in the accumulation rates of ice, as well as layers of dust and volcanic ash, that make it possible to count seasonal ice layers (the ice equivalent of tree rings), date the layers, and establsih a timeline.

Figure 13-6 Photo mosaic of retrieving ice cores in Greenland.

Image from http://earthobservatory.nasa.gov/Features/Paleoclimatology_IceCores/.

It is critical that geoscientists accurately establish the chronology within any core so that their interpretations of past climates are temporally correct. Unfortunately, ice is difficult material to work with because of its physical properties. Ice compresses with age and the weight of overlying ice, so seasonal layers may become less distinct in older ice. Ice also flows, so layers may become distorted. Finally, the coring process poses challenges. A *core* is a hollow tube with a rotating drill bit on the front end. While ice is easy to drill through in terms of its softness, this softness poses problems as well, as the ice will collapse inward into the hole if a fluid isn't used to provide outward pressure to keep the hole open. Of course, the fluid must be one that won't chemically interact with the ice and contaminate the material we wish to study. Finally, the ice core needs to be stored (it must never melt!) so that it remains free of contamination from fluids and particles. In the more than thirty years of work that's been done drilling in both Greenland and Antarctica, these problems have been overcome. There is now a huge repository of ice cores and ice core data to examine.

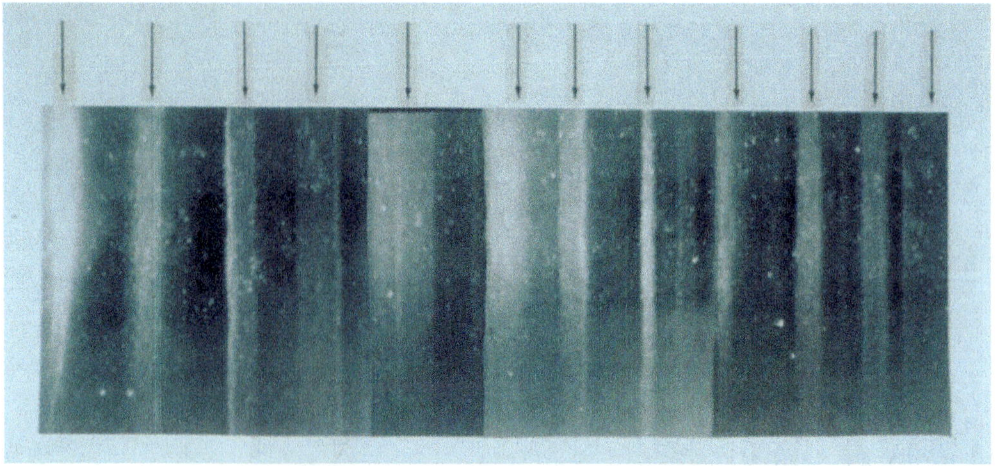

Figure 13-7 Greenland ice core segment (GISP2D1855) with seasonal layers clearly visible. This core segment is 19 centimeters long, and lighter-colored summer intervals are labeled with arrows.

Image from http://en.wikipedia.org/wiki/Ice_core#mediaviewer/File:GISP2_1855m_ice_core_layers.png.

Ice layer chronologies are synthesized with other measurements, including atmospheric deposition of volcanic debris on glaciers, which provide radiometrically datable horizons. Differences in physical properties, such as electrical conductivity of the ice from atmospheric deposition of particulate material, is another technique used to confirm ice layer ages. Still, studies by several geoscientists suggest that the ages within the Vostok Antarctic core may vary by as much as ten thousand years. This may seem like a huge source of error; however, the climate changes that we believe we see in these cores occur at a magnitude and frequency that absorbs this uncertainty.

The chemistry of ice can be analyzed to obtain climate data. Ice, of course, is H_2O, and this molecule is conveniently composed of two elements with stable isotopes that vary, or fractionate, as a function of temperature.

Climate Proxies

We know that there is a theoretical reason why levels of CO_2 in the atmosphere would be related to global warm periods, but is there a way to empirically establish the relationship? Because there are no fossil thermometers, we need to obtain data on climate indirectly,

Sidebar 13-1 Vostok, Antarctica

Established in the late 1950s by the then–Soviet Union, the Vostok research station has produced some of the most extensive and complete data on the Earth's climate. The site is located on the East Antarctic Ice Sheet, approximately 1,300 kilometers from the geographic South Pole. The sheet of ice covering Antarctica contains nearly 26.5 million km^3 of ice (Greenland's ice sheet is "only" 2.85 km^3 in volume), making it the single largest reservoir of non-marine water on Earth. Unlike the West Antarctic Ice Sheet, which overlies polar ocean water, the East Antarctic Ice Sheet is grounded on bedrock of the Antarctic continent.

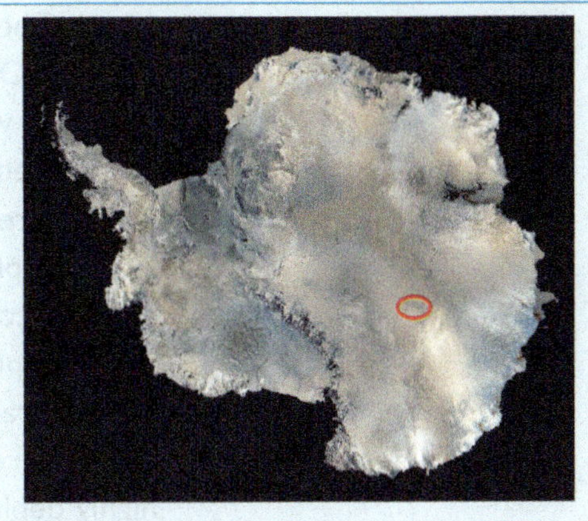

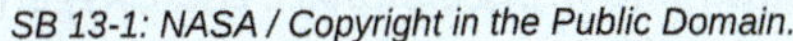

SB 13-1: NASA / Copyright in the Public Domain.

using climate proxies. A proxy is a physical archive that retains within its physical structure a permanent measurement of some variable in the environment. It can therefore substitute for actual measured data, such as temperature. There are two important climate proxies in glacial ice. The ratios of both the stable isotopes of oxygen and hydrogen vary as a function of latitude, which, because of the latitudinal gradient, is a proxy for temperature. Recall from Chapter 3 that an isotope is an element with variable atomic weight but the same atomic number. We examined unstable isotopes, such as carbon-14, which decays to nitrogen-14 over time. There are several stable isotopes, which do not decay over time. Oxygen and hydrogen isotopes are examples of these. They fractionate, or separate, as a function of temperature, not time.

Oxygen Isotopes

In nature, more than 99% of all oxygen atoms have eight protons and eight neutrons, for an atomic mass of 16. A tiny fraction have excess neutrons, to produce oxygen with an atomic mass of 18. Because of the difference in mass of the two isotopes, they behave differently in a wide range of chemical reactions. Simplistically, the lighter oxygen isotope is "more active," preferentially being taken up in the vapor phase in simple chemical reactions, such as evaporation. This preferential uptake is termed *fractionation*. The isotopes of oxygen fractionate as a function of temperature, which is why they are a useful climate proxy. Why is this the case?

The largest reservoir in the hydrosphere consists of the world's oceans. Evaporation from the ocean's surface places water vapor in the atmosphere, where it condenses and forms water droplets in clouds. The number of oxygen atoms in the H_2O are higher in the light ...^{16}O realtive to the heavier ^{18}O. This means that the ocean reservoir will be enriched in the heavier ^{18}O atoms that were not evaporated. Precipitation, surface runoff, lakes, and river waters will be preferentially enriched in light ^{16}O. Ultimately, these light ^{16}O isotopes make their way back to the ocean reservoir and the cycle is complete, but before making the "grand circuit," there are multiple events of evaporation and precipitation. Each cycle further fractionates the heavy and light oxygen isotopes. By the time you reach high latitudes and snow precipitates, the oxygen is highly depleted in the heavier ^{18}O isotope.

What happens during periods of global cooling and high rates of snowfall and glacial ice formation? If you examine the hydrologic cycle figure in Chapter 11 (Figure 11-3), you will recall that glacial ice is the second biggest current reservoir of water on Earth. The water evaporated from the oceans into the atmosphere is precipitated at the high latitudes as snow, which accumulates over the course of many years to form glacial ice. This means that the light ^{16}O atoms are temporarily stored in the glacial ice and do not return to the oceans. Repeated over thousands of years, the oceans become progressively more enriched in the heavier ^{18}O atoms that were not removed through evaporation (Figure 13-8). Thus, in times of cold climates, the ratio of ^{18}O:^{16}O increases in the ocean waters. Because marine animals make their shells from the ions dissolved in seawater, during these cold climates their shells will contain a higher percentage of ^{18}O atoms than they would in a period of warm climate. Figure 13-5 contains temperature data obtained from analysis of oxygen stable isotope ratios in foraminifera shells.

Deuterium

Hydrogen has one proton and one electron (^{1}H), but there is a stable isotope of hydrogen (^{2}H) with one proton, one neutron, and one electron (there is a third, unstable, isotope of hydrogen called tritium). The stable isotope of hydrogen is called deuterium. As is the case with oxygen, hydrogen experiences fractionation in the evaporation-precipitation cycle, such that precipitation in the polar regions contains a hydrogen isotope ratio depleted in the heavier isotope. Analysis of the heavy-to-light hydrogen isotope ratios in modern precipitation has generated a mathematical relationship between this ratio and the local temperature of this region (see Sidebar 13-2).

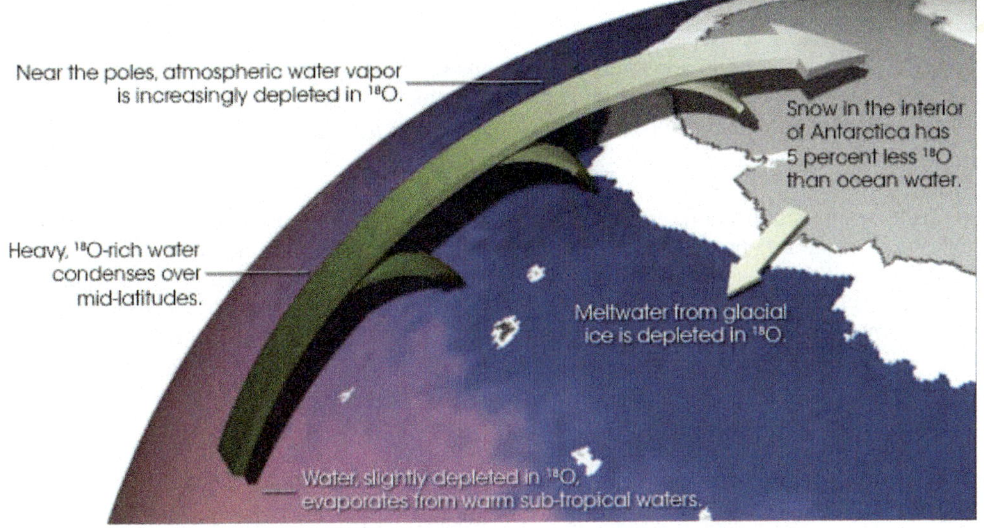

Figure 13-8 Schematic illustration of the origin of the latitudinal gradient in $\delta^{18}O$ values.

Image from http://earthobservatory.nasa.gov/Features/Paleoclimatology_OxygenBalance/oxygen_balance.php.

In addition to being able to examine the isotopic record of the ice itself, we can examine bubbles of "ancient air" trapped within the ice. Glacial ice originally starts out as snowflakes, which are loosely packed, with much air between flakes. As snowflakes compact and turn to firn, their shapes change, so that firn looks more like small grains than flakes. There is still air trapped between these grains. With further compaction the firn gets denser and becomes "solid" ice. This ice will have frozen bubbles of air within it (Figure 13-9). This air can be analyzed for its composition.

Results of Ice Core Study

The Vostok ice core analyses generated the most important record of climate change on Earth over the past four hundred thousand–plus years.

The Vostok core data, subsequently supplemented by coring data from the Greenland ice sheet, clearly shows that there have been several cycles of changing atmospheric composition and temperature over the approximately 450,000-year interval captured in the ice.

First, some basic graph interpretation. Looking at Figure 13-10, what are the maximum and minimum

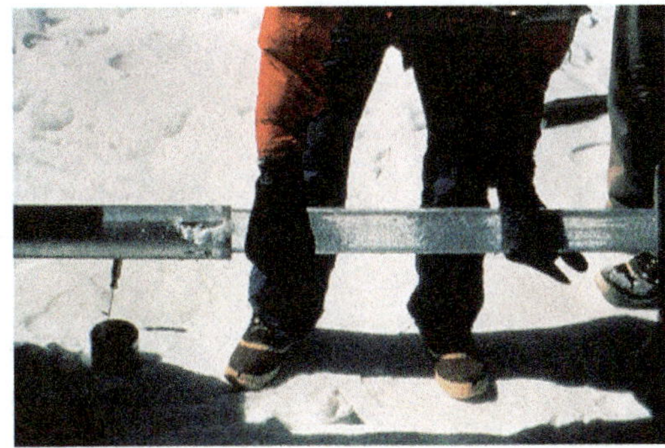

Figure 13-9 Extruding an ice core. Bubbles in the ice are clearly visible.

Image from http://icestories.exploratorium.edu/dispatches/wp-content/uploads/2008/05/ice_core.jpg.

Sidebar 13-2 Stable Isotopes

Recall from Chapter 3 that an isotope is an element with atoms of variable atomic weight. We discussed using unstable isotopes as a tool for numerical dating of rocks. Stable isotopes are the most important tool for examining climate change, and the two most valuable for this are oxygen and hydrogen.

Oxygen

There are three stable isotopes of oxygen: $^{16}O_8$, $^{17}O_8$, and $^{18}O_8$. More than 99% of the oxygen on Earth is $^{16}O_8$, and 0.2% is $^{18}O_8$ (minute amounts are $^{17}O_8$). Because of differences in their atomic weights, these isotopes behave differently (a process called *fractionation*) in chemical processes. If we examine modern ocean water, and use this as a standard, we can express the ratios of ^{16}O and ^{18}O to one another. Values are expressed as "per mil" (which equals parts per thousand) with a notation of °/oo or "delta ^{18}O":

$$\delta^{18}O\ ^{o}/oo = \frac{[(^{18}O\ /\ ^{16}O)\ \text{sample} - (^{18}O\ /\ ^{16}O)\ \text{std}] \times 1{,}000}{(^{18}O\ /\ ^{16}O)\ \text{std}}$$

As this equation illustrates, we compare the ratios of the "heavy" to "light" oxygen isotopes in a sample to that of our "standard mean ocean water" (SMOW). Delta values that are enriched in the heavy ^{18}O are positive, while those depleted in the ^{18}O are negative. An example of how these isotopes fractionate is seen in the process of evaporation of water from the largest reservoir in the hydrosphere, the ocean. This results in atmospheric water vapor (and ensuing precipitation) depleted in ^{18}O and ocean water enriched in ^{18}O. As a result of multiple evaporation-precipitation cycles between the equatorial regions and the poles, each of which fractionates ^{18}O and ^{16}O,we see a latitudinal gradient in $\delta^{18}O$, such that H_2O becomes progressively depleted in the heavier isotope with increasing latitude. Thus, large polar ice sheets are depleted in $\delta^{18}O$.

Hydrogen

Hydrogen has three isotopes, and unlike any other element, the isotopes are given names: $^{1}H_1$ (protium), $^{2}H_1$ (deuterium), and $^{3}H_1$ (tritium). Of these, more than 99.9% of all hydrogen on Earth is protium. Tritium is an unstable isotope. Protium and deuterium are present in H_2O, and, like oxygen, these isotopes fractionate during the evaporation of water from the world's equatorial oceans. By convention, delta values of $^{2}H_1$ are shown as δD. The lower vapor pressure of deuterium means that water vapor will be enriched in the lighter isotope (negative δD). When the water vapor condenses in the atmosphere to form precipitation, this material will be enriched in the heavier deuterium. Thus, as with oxygen, we see a latitudinal gradient in δD from multiple evaporation-precipitation events from equator to pole. Deuterium is a more sensitive temperature indicator than oxygen, however, because the lower temperature variation of globally mixed SMOW relative to local precipitation generates a linear numerical relationship between δD and temperature: Temperature $°C = -55.5 + (\delta D + 440) / 6$.

temperature changes in a cycle? What are the maxima and minima ppmv (parts per million per volume) of atmospheric CO_2 and CH_4 in a cycle?

The blue line represents temperature deviation from the present day. Is a temperature cycle symmetrical (does it rise and fall at the same rate)? If not, describe the pattern associated with a cool-down or warm-up. What is the frequency of the temperature deviation cycle? Perhaps you noted that there is a strong signal that temperature cycles at a frequency of approximately one hundred thousand years. Is there an orbital parameter that varies on this cycle? Do you see cycles at other frequencies?

We continue to stress that "correlation doesn't equal causation" when visually comparing data. If you wanted to test whether the cycles seen in the Vostok data were related to orbital variation, how could you do this? If you responded, "Do statistical tests," this would be a good suggestion. The researchers who generated this data set (Petit et al., 1999) conducted a type of statistical analysis that matches cycle frequencies and found good correlations for the eccentricity and obliquity signals and not as good correlation for precession, which they attributed to uncertainties in age estimates for the ice. Another suggestion to test the orbital variation-temperature correlation might be to compare the cycle frequencies seen in the Vostok data to other possible "climate drivers." For example, compare the eleven-year solar cycle record (Figure 12-11) and that of the Vostok core (Figure 13-10). Do the graphs suggest that global climate changes on a frequency of approximately one hundred thousand years are driven by variation in solar output?

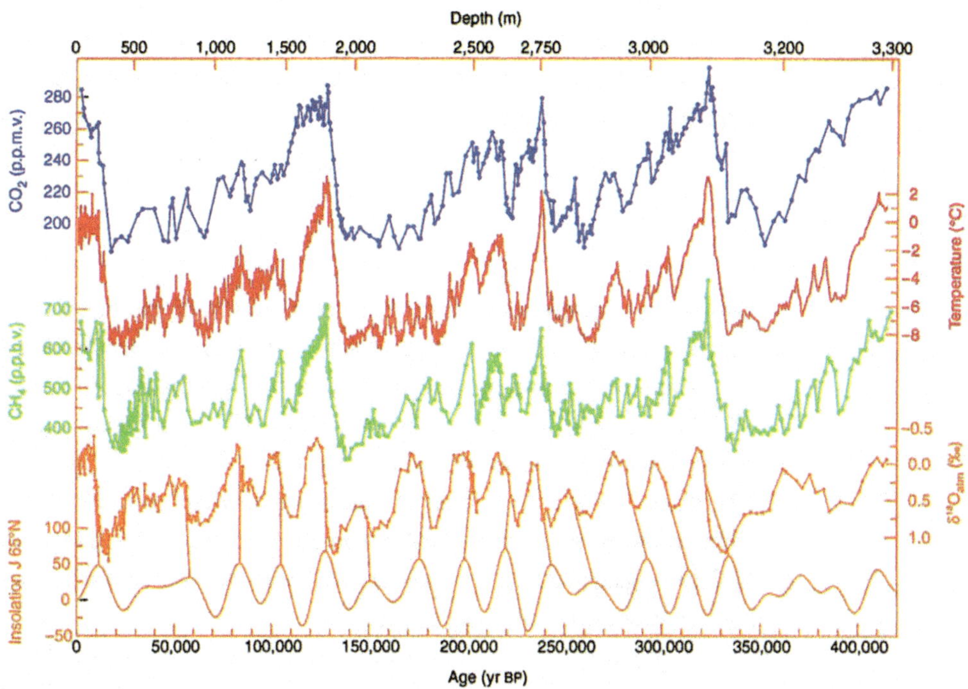

Figure 13-10 Geochemistry data from the Vostok ice core (from Petit et al., 1999) showing change in atmospheric CO_2 (blue), CH_4 (green), interpreted temperature deviation (red), measured $\delta^{18}O$ of ice (orange), and solar insolation at 65°N (brown) over the past four hundred thousand years. Temperature is recorded not as atmospheric °C but as change in °C derived from δD. Age estimates within core may vary by about ten thousand years.

Image from http://en.wikipedia.org/wiki/Vostok_Station#mediaviewer/File:Vostok_420ky_4curves_insolation.jpg.

Another control of Earth climate is the composition of the atmosphere. What cycles in atmospheric CO_2 and CH_4 are present in the Vostok core data? What are plausible explanations for these natural ("non-anthropogenic") changes in atmospheric composition? This data set should make you stop and think. We know that CO_2 and CH_4 are greenhouse gases. Their abundance in the atmosphere should trigger global warming. Do you see this relationship? Are the increases in atmospheric greenhouse gases a *cause* of temperature change or an *effect* of temperature change? How would you distinguish the two? Is there a theoretical basis to hypothesizing that these greenhouse gases would accumulate in the atmosphere as a function of increasing solar insolation? If this were the case, the abundance of CO_2 and CH_4 would be an effect, or byproduct, of global warming. Alternatively, is there a theoretical basis to hypothesizing that the increasing abundance of greenhouse gases would impact orbital variation? In order to answer the first question, you would have to identify the processes in the biosphere, atmosphere, hydrosphere, or geosphere that would increase greenhouse gas production as temperature increased, and decrease them as temperature dropped. In order to answer this, you may need to go back to Chapter 12 to review atmospheric processes.

Glacial Ice Bubble Composition

Figure 13-11 presents data on the composition of gas bubbles of "ancient air" preserved in ice bubbles within Vostok cores, plotted on a graph that shows more time resolution than the data in Figure 13-10 and also including nitrous oxide (N_2O). The deuterium curve is a proxy for temperature. What correlation do you see between interglacial periods (warmer temperatures), when continental ice sheet melting exceeded glacial advance, and values of CH_4 and N_2O? Are glacial advances (colder temperatures) characterized by high or low values of these gases?

You might have been able to hypothesize that increased global average temperatures accelerated the release of methane from frozen sub-polar environments. Perhaps increased temperatures reduced the amount of carbon dioxide gas mixing into seawater at the ocean-atmosphere interface. These interpretations can all be tested. The questions can all be framed so it is possible to determine the answer to the following: "If gas production changed

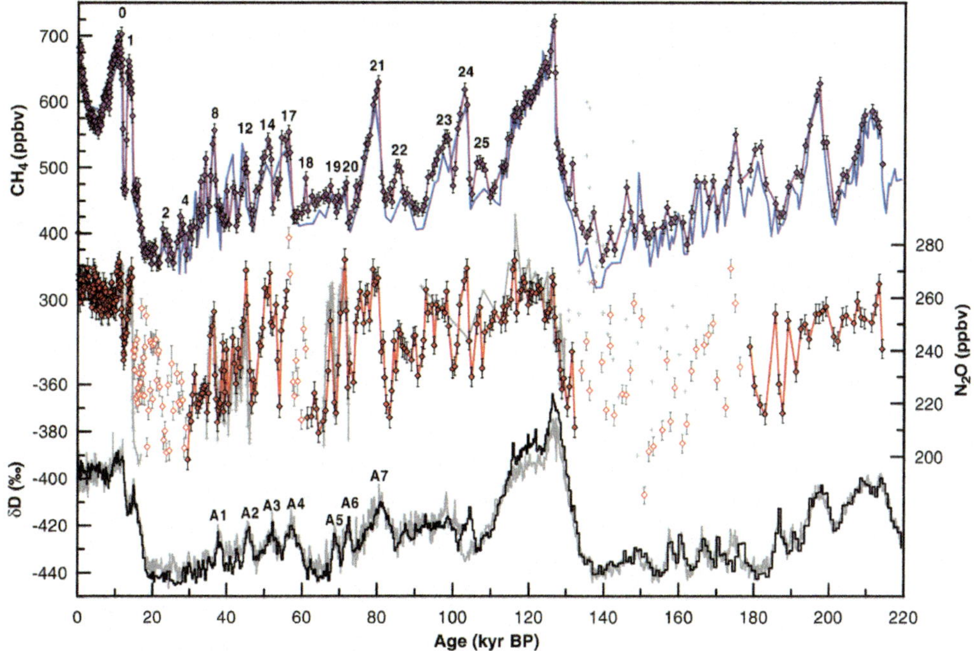

Figure 13-11 Atmospheric gas composition from Antarctic ice. CH_4 (purple line) and N_2O (red line) records along with δD data (black line), a proxy for temperature, for the interval of time between 220,000 years ago and the present. Error bars represent one standard deviation measurement of uncertainty. A1 to A7 represent Antarctic warming events. Numbers above the CH_4 curve identify Antarctic warming events.

Data from Spahni et al., 2005.

Source: http://www.whoi.edu/cms/files/Spahni05_68563.pdf.

by X amount per Y amount of temperature change, how much gas would be produced at Z temperature change?" If the gas produced was close to the observed value in the gas bubbles, this would be strong support for your hypothesis.

At the more detailed time scale shown in Figure 13-11, the shorter frequency cycles within the approximately one-hundred-thousand-year cycles are more clearly visible. The peaks labeled A1 through A7 in Figure 13-11 represent Antarctic warming events. We can compare these cycles to the Greenland ice cores (Figure 13-12), where we see that the shorter-scale climate fluctuations stand out. Are these cycles produced by orbital variation, variation in atmospheric composition, or other causes? Another factor to consider is that the cycles are more pronounced in the northern hemisphere (Greenland) than the southern. Why would this be the case?

We can expand the time scale even further. Figure 13-13 shows Greenland temperature data over a short time span of fifty thousand to thirty thousand years before the present (2000).

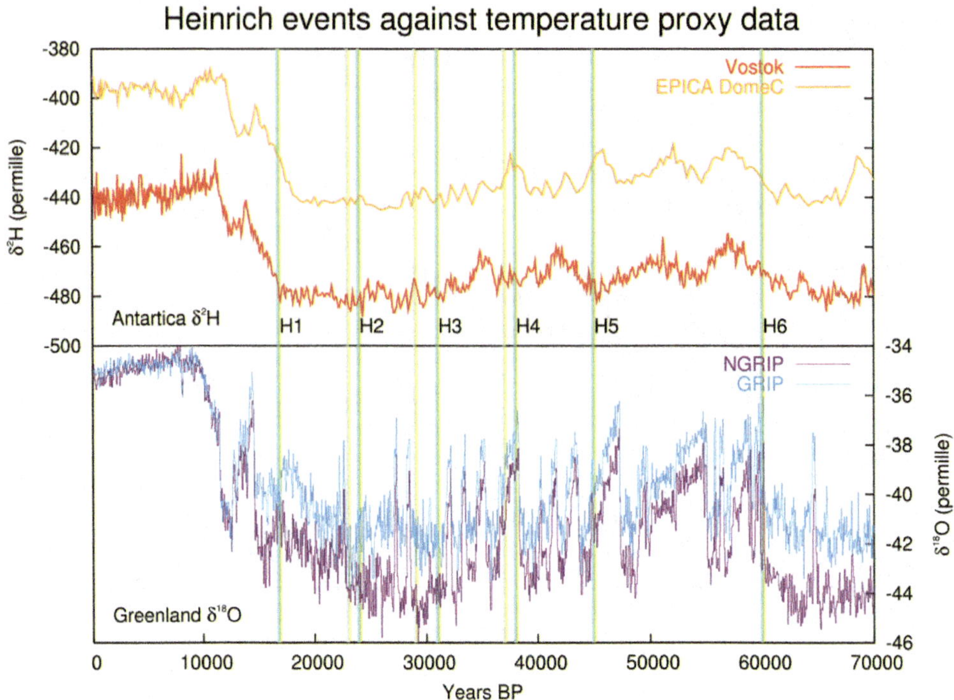

Figure 13-12 Temperature cycles in two Antarctic cores (yellow and red lines) and two Greenland ice cores (blue and purple lines) over the past 70,000 years. Less negative isotopic values indicates warmer temperatures. Vertical green lines labeled H1 to H6 are Heinrich events. See text for discussion.

When you examine this figure, you see that the cycle frequency is less than 2,000 years (approximately 1,500 years). Note the symmetry of a cycle: are the cycles symmetrical, warming and cooling at the same rate, or does it warm or cool faster? The cycles clearly seen in Figure 13-13 are termed *Dansgaard-Oeschger (D-O) cycles*. At least twenty-five D-O cycles have been identified in the Greenland ice. The origin of D-O cycles is still being explored. Why can we eliminate orbital variation as a cause, as well as solar irradiance? Many scientists attribute D-O cycles to ice sheet instability related to glacial (ice advance) and interglacial (ice sheet collapse) processes. Another possibility is that interglacial time periods result in the discharge of cold fresh-melt water into the northern Atlantic Ocean, which generates the partial collapse of thermohaline ocean circulation. Recall from Chapter 11 the role that ocean circulation plays in controlling the Earth's climate. Surface currents carry warm water poleward from the equator, transferring heat to the overlying atmosphere. In the northern Atlantic, the influx of

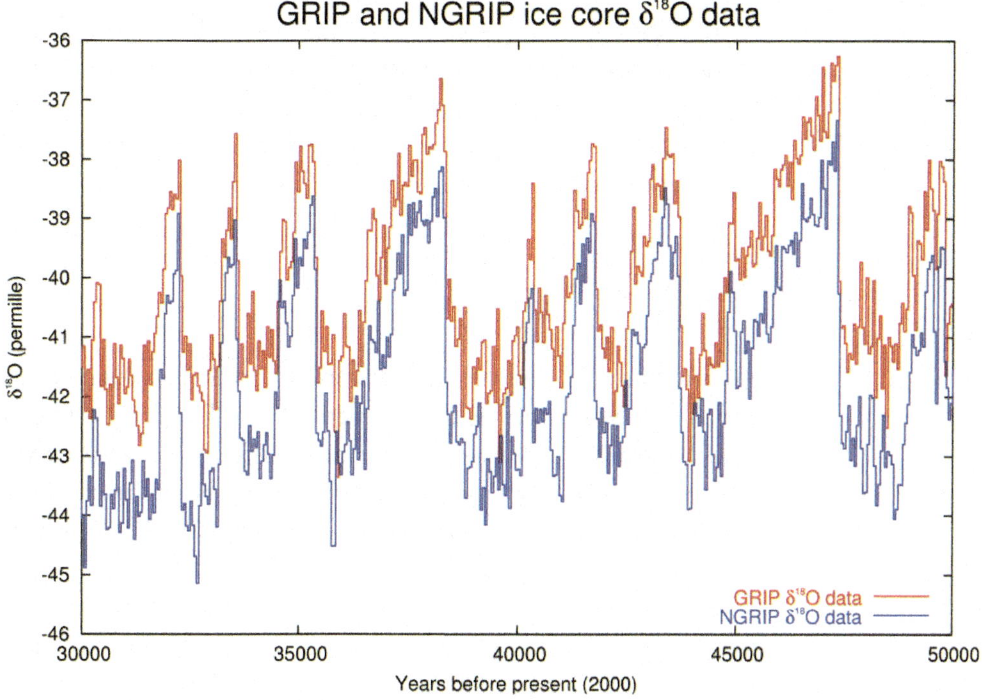

Figure 13-13 Two sets of Greenland ice core temperature data for a twenty-thousand-year period. Less negative isotopic values represent warmer temperatures.

Arctic-derived cool, lower-salinity water meets the warmer salty water and sinks, flowing southward through the Atlantic Ocean at depth, leaving warm water at the surface. If, as a result of THC collapse, more warm surface water were delivered to Antarctica, that additional heat would destabilize the production of Antarctic bottom water (ABW). Inhibited ABW water production would in turn help restore North Atlantic deep-water circulation. This scenario has been dubbed the *bi-polar seesaw* in recognition of the "back-and-forth" impacts that northern and southern hemisphere deep-water circulation have on one another.

Figure 13-12 also illustrates the timing of iceberg-deposited debris on the North Atlantic seafloor. First recognized by marine geologist Hartmut Heinrich, the timing of deposition of the debris on the seafloor is constrained by the ages of the fossils in marine sediments above and below the sand layers. While there is consensus that the Heinrich event layers represent deposition of material off the bottoms of icebergs, there is disagreement about what triggered these events. Because the climate changes are so rapid, it is difficult to distinguish whether they formed *during* warming events or immediately *prior to* them. There appear to be climate proxies that suggest both, which suggests that the speed of warming exceeds the resolution of dating techniques.

The Younger Dryas

An interval of lowered temperatures occurred between 12,800 and 11,500 years ago, right at the end of the Pleistocene Epoch, that is termed the *Younger Dryas*, named after the tundra flower *Dryas octopetala*, which was widespread throughout the northern hemisphere during this interval (Figure 13-14) and is found in circumpolar regions today.

Examine Figures 13-10, 13-11, and 13-12 and decide whether the Younger Dryas cooling event is recorded in these data sets. Figure 13-15 shows detailed temperature fluctuation from a Greenland ice core data that includes the Younger Dryas interval. This graph illustrates that during the Younger Dryas, temperature of the ice sheet dropped, as did snow accumulation, indicating a drop in precipitation. The global warming that followed the Younger Dryas

Figure 13-14 *Dryas octopetala.*

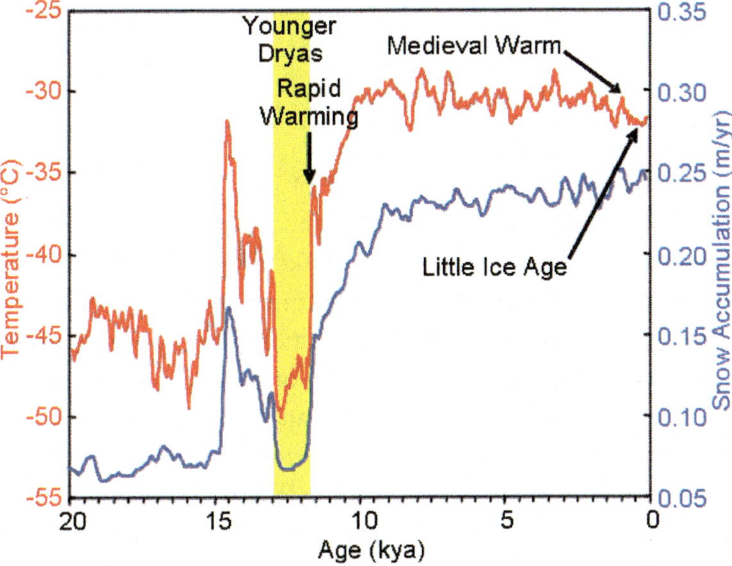

Figure 13-15 The red line represents Greenland ice sheet temperatures, and the blue line represents snow accumulation rates (Greenland). The Younger Dryas, highlighted as a yellow band, appears as the abrupt cooling event that followed a warm interval. The end of the Younger Dryas is punctuated by a rapid global warming. The Little Ice Age and Medieval Warm Period are discussed later in the chapter.

Source: Richard Alley, from The Two Mile Time Machine.

is the best analogue we have for a rapid climate change in Earth history that approaches the rate of change that we are currently experiencing. Clearly, we are interested in why the Younger Dryas cool-down occurred, as well as why it warmed so rapidly. The localization of the Younger Dryas cooling is most pronounced in Europe, which supports the theory that the origin is the collapse of thermohaline circulation in the Atlantic. If the influx of cooler, low-salinity water from the Arctic increased as a result of warming the high latitudes of Europe, oceanographers hypothesize that this large volume of water would disrupt thermohaline circulation by leaving a large mass of cool water on the Atlantic's surface west of Europe. This cold water would then trigger a climate cooling in Europe.

If the Younger Dryas records a thermohaline circulation collapse, the rapid warming that followed it would represent its reestablishment. Thus, the causal mechanism for the post-Dryas *warming* is not analogous to the present-day global climate change driven by changes in greenhouse gases in the atmosphere. However, the *rate* of post-Dryas warming might be a valuable example of the ecosystem changes that accompany rapid global warming. Examine the graph in Figure 13-15 and calculate what the rate of warming, in degrees per thousand years, might have been.

Holocene Climate Change: The Record of Global Climate Change in the Oceans

Ice cores are not the only record of past climates. Sediments accumulating on the seafloor also record global climate. Many marine animals make their shells or skeletons from $CaCO_3$, and they extract the elements to make their shells from seawater. When these organisms die, their shells fall to the seafloor, producing a long, continuous record of sediment accumulation. The chemistry of the shells can be analyzed to determine the chemistry of the seawater they utilized to make their shells. Cores can be thousands of meters in length, representing thousands to millions of years of ocean chemistry data. These data include the oxygen isotope ratio in the $CaCO_3$ of the marine shells that comprise the seafloor muds. As discussed earlier in this chapter, because of evaporation, ocean waters are enriched in the heavier ^{18}O isotope relative to ^{16}O. During glacial periods, much precipitation is stored in the polar ice caps and temporarily "stored" or removed from the evaporation-precipitation cycle. Ocean waters become even more enriched in $\delta^{18}O$. When the animal makes its shell, these heavier isotopes are present in higher levels in the $CaCO_3$. Thus, when we study cores of sediment from the seafloor, we can track glacial and interglacial periods as preserved in the chemistry of the fossil shell material. Thus, the ocean floor sediment contains a long uninterrupted record of global climate.

Single-celled planktonic organisms called foraminifera are abundant in the oceans, and the chemistry of their shells provides excellent proxies for determining sea surface temperatures. The data reflect the latitudinal differences SSTs, which owe their origin to the latitudinal control of solar radiation. The data also shows a general warming approximately one thousand years before the present, flanked by two cooler intervals. The warm event around 1000 to 1200 AD is called the Medieval Warm Event. It was followed around 1400to 1800 AD with a climate event termed the Little Ice Age. Historical records by humans confirm the occurrence of this unusually cold climate interval in northern Europe at this time. The name "Little Ice Age" conjures images of advancing ice sheets, but the cooling was confined to northern Europe and did not involve glaciation.

Do any of the global "climate drivers" we have discussed operate at this scale of hundreds of years? If not, what are the implications of this? Are there even more factors controlling climate? Before exploring what these might be, let's summarize climate change over the past 450,000 years.

Summary of Paleoclimate Data for the Late Pleistocene and Holocene Epochs

- Ice cores from both Greenland and Antarctica record numerous temperature fluctuations over the past 450,000 years.
- Over this time span, there is a clear one-hundred-thousand-year cycle frequency and a less clear forty-one-thousand-year and even *less* clear nineteen-thousand- and twenty-thousand-year frequency. While many scientists attribute the one-hundred-thousand-year cycle to orbital variation, it is puzzling why there is a less strong cycle at the shorter frequencies associated with obliquity and precession, as these parameters have a more pronounced impact on seasonal solar insolation.
- Ice bubble data reveal that there are changes in atmospheric chemistry (CO_2, CH_4, N_2O) that occur at the same frequencies as temperature cycles. This strongly suggests a causal link.
- Global climate warming is associated with elevated greenhouse gas levels in the atmosphere.
- High-resolution temperature fluctuation from Greenland cores indicates that there are glacial-interglacial fluctuations in temperature that occur on an approximately 1,500-year cycle. The origin of these cycles, termed Dansgaard-Oeschger (D-O) cycles, is not clear but is thought to be related to glacial-interglacial transitions. D-O cycles are more clear in the northern hemisphere than the southern, which suggests that these glacial-interglacial cycles influence the collapse and rebuilding of thermohaline circulation.
- The rapid warming portion of a D-O cycle is followed by a stepwise cool-down.
- The Younger Dryas represents a rapid global cooling between 12,800 and 11,500 years before the present day. Temperatures are estimated to have dropped by –5°C over

a matter of decades. Of current interest is the post–Younger Dryas warming, which was also rapid and may have been about 8°C over decades.

There are two summary takeaway points of all this: (1) climate fluctuation has been the norm on Planet Earth for the past half-million years, and (2) the origin of climate fluctuation is complex because of interactions between different climate drivers. Does this mean that *present-day* climate change is "normal" or "par for the course"? We must also consider that the detailed climate record we have reviewed is for the past half-million years only. Geologists recognize that the past 2.5 million years of Earth history has been an "icebox" world. We have been discussing the variation in climate in the unusual icebox conditions. What would Earth's climate be like if we were not in this icebox? We may find out!

The Role of Plate Tectonics in Controlling Climate

Because of the time scale over which plate tectonic processes operate, to view how plate interactions have impacted Earth climate requires taking a long time-span perspective. The paleoclimate record presented earlier in this chapter covers less than the past half-million years of Earth history. If we look back even further, we can recognize climate fluctuations that played out over millions of years.

Figure 13-16 shows a plate reconstruction for part of the globe approximately 3.5 million years ago, the Pliocene Epoch (right), and the present-day Holocene Epoch (left). The differences in the plate configuration for the Pliocene Epoch include the partial formation of Central America. The present-day convergent plate boundary along what is now the Caribbean plate had not yet uplifted volcanic arc terrain represented today in Costa Rica and Honduras. Before subduction, volcanism and crustal uplift constructed the Isthmus of Central America, and ocean circulation connected the Atlantic and Pacific oceans.

When ocean waters were able to flow from the Atlantic to the Pacific through the region now occupied by the Isthmus of Central America, surface waters were well mixed. Surface waters warmed at the equator carried heat northward into the polar

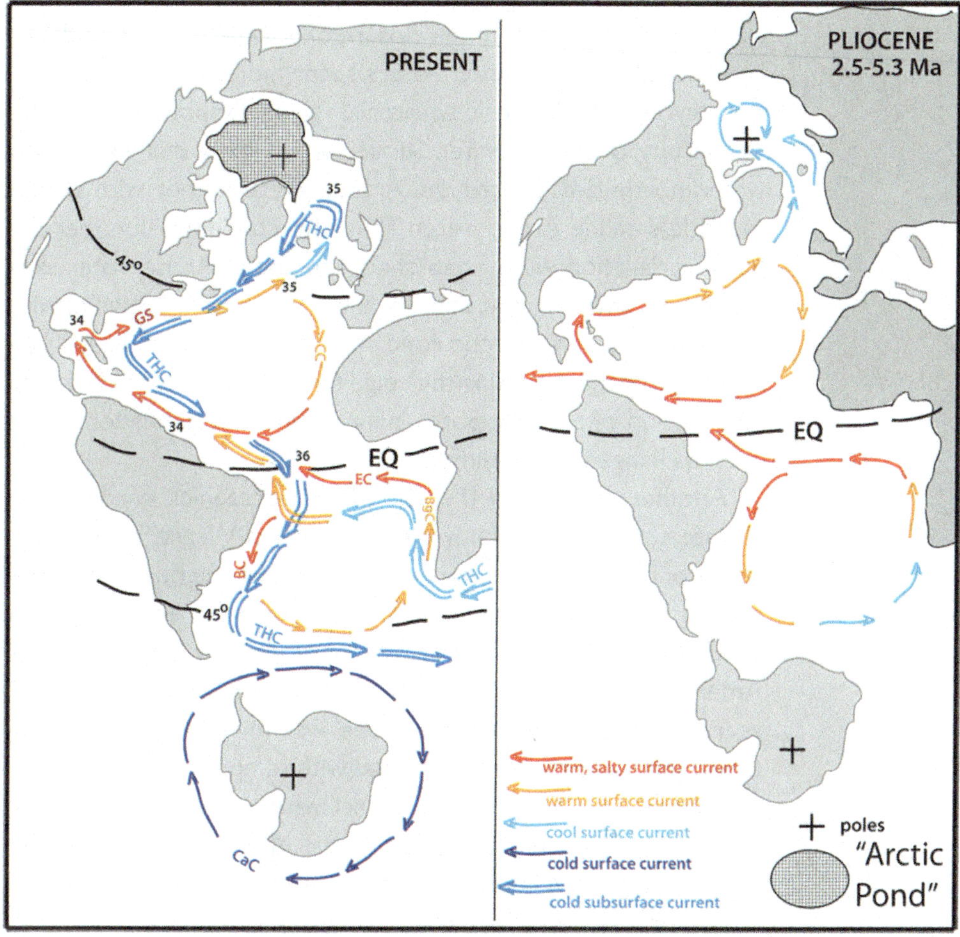

Figure 13-16 The Early Pliocene (right) map of the Atlantic region and present day (left). Numbers represent present-day salinity values for the ocean in ppt. Thin solid lines with arrows represent present-day ocean surface currents. Longer dashed lines represent near-surface thermohaline circulation, and short dashed lines represent deep ocean thermohaline circulation. The hatched region at the north pole (AP) is the cold, less dense "Arctic Pond" pool of water isolated in today's Arctic Ocean. "CG" represents the circum-Antarctic current, the source of Antarctic bottom water. In the Pliocene, thermohaline circulation was not established because of well-mixed surface waters.

oceans, warming this region. A warmer polar region means that there was not the strong latitudinal gradient in temperature that we see today. Evidence for the warmer climate at high latitudes includes accumulations of fossils of marine animals found today in subtropical settings in sediment of this age as far north as 60° latitude! It is proposed that by three million years ago, when the terrains of the convergent plate boundary accreted to form the Isthmus of Central America, the present-day ocean circulation was established. As a result of the closure of flow between the Atlantic and Pacific oceans, Atlantic waters spent more time circulating in small gyres in the Atlantic equatorial region. Evaporation elevated the salinity of this seawater. Surface winds blew this warm, salty water northward toward the Arctic, where it met with cooler, much less saline Arctic water. Thus, the conditions for thermohaline circulation were established. The cooler Arctic water sank and moved southward at depth, while warm waters ceased moving northward into the "Arctic Pond." A cool-down of the higher latitudes commenced, setting the stage for the Pleistocene glaciation. The motion of tectonic plates influenced ocean circulation, which in turn changed global climate.

Another example of the role of plate tectonics in controlling climate comes from slightly older sediments. Wegner's supercontinent of Pangaea broke apart starting around one hundred million years ago. The divergent plate boundary that had developed over thousands of kilometers ultimately separated Europe from North America and Africa from South America. For tens of millions of years the climate around the globe was mild. Even at high latitudes, animals and plants associated with temperate to subtropical conditions were widespread. Paleoclimate workers call this time period the Paleocene-Eocene Thermal Maximum in recognition of the benign conditions around the globe between about fifty-six and forty-five million years ago. Many geologists look at this time interval as our last "greenhouse world" (the alternative to the Pleistocene Epoch "icebox world"). As divergence along the mid-Atlantic Ridge continued, smaller plates also rifted apart from Africa. In the Eocene Epoch (56 to 34 Ma), South America separated from Australia, and Antarctica separated from both, so that by about thirty-five Ma, Antarctica was isolated in its present-day position at the South Pole. The isolation of Antarctic in the high-latitude global wind belt created a circumpolar current that receives no warmth from equatorial or even temperate waters. Antarctica began a long period of cool-down, which ultimately led to the onset of glaciation on that continent. The cold waters that encircle Antarctica become dense and sink, blanketing the bottom of the southern oceans with a layer of cold water. These cold waters upwell in various locations in the southern ocean, bringing cold waters to the surface. The decreasing global temperatures that started in the late Eocene continue to the present day as part of a long-term cool-down of the entire globe.

The final example of the role that plate tectonics plays in controlling the Earth's climate is the effect that the uplift of mountains has on the atmospheric carbon cycle. As discussed in Chapter 10, the uplift of mountains increases surface area available for weathering. Weathering involves the removal of CO_2 from the atmosphere as it mixes with water droplets, forms H_2CO_3, and interacts with minerals in chemical weathering (hydrolysis) reactions. The dissolved byproducts are carried by surface waters to the oceans, where the dissolved CO_2 molecules are taken up by marine organisms in precipitating their shells. Is there a record of such processes in the geologic record? Figure 13-17 shows model estimates for atmospheric CO_2 levels over geologic time.

This graph presents data generated by various models of the carbon cycle. It is based on estimates for the size of various carbon reservoirs and also on the rates of flux between them. Because of the time scales involved, such models are useful for identifying long-term trends, and as such they can possibly record

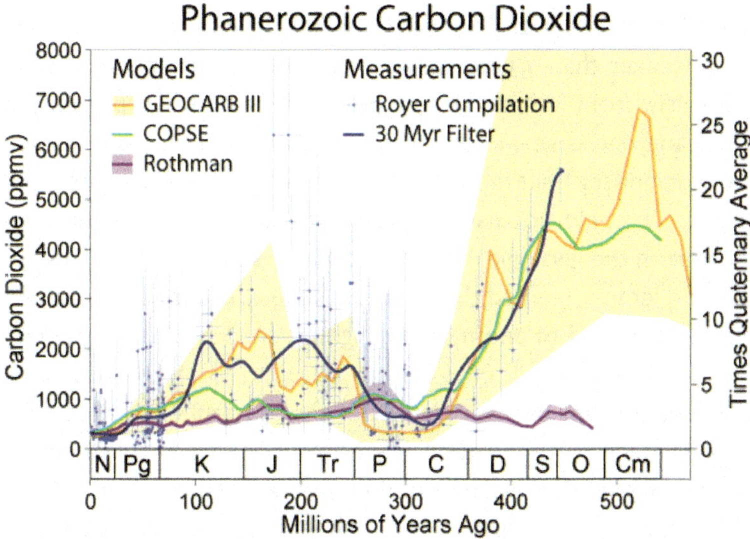

Figure 13-17 The geologic time-scale abbreviations include the following periods: Cm = Cambrian, O = Ordovician, S = Silurian, D = Devonian, C = Carboniferous, P = Permian, Tr = Triassic, Jr = Jurassic, K = Cretaceous, Pg = Paleogene, N = Neogene. Output from three models using two data sources. Note the high levels of atmospheric CO_2, estimated to be as much as twenty-five times present levels from the Cambrian through Carboniferous periods. Prior to about 350 Ma, absent or minimal levels of plant life meant that this CO_2 reservoir was not developed. The decrease in atmospheric CO_2 levels after about 400 Ma represents the diversification of terrestrial plants.

long-term processes, such as plate interactions. For example, the decrease in atmospheric CO_2 from the late Cretaceous through the Neogene (K to N on Figure 13-17) is interpreted to reflect the increase in chemical weathering rates as a result of the collision of the Indian plate with Asia and the uplift of the Himalayan mountains. To reiterate, this graph presents the results of a model that has estimated, for example, the increased surface area exposed to chemical weathering, rates of weathering the bedrock involved, and how much atmospheric CO_2 this would consume. There is supporting evidence for this increased weathering in the sediment layers that were deposited downstream and in the adjacent ocean basin, but all these values are estimated. Figure 13-17 also illustrates the increase in atmospheric CO_2 levels starting in the Jurassic and continuing into the Cretaceous, which is interpreted to reflect increasing volcanism associated with the creation of the divergent plate boundaries associated with the breakup of Pangaea.

Recall that in our discussion of solar irradiance in Chapter 12, we noted that the sun's output has increased over geologic time. When the Earth formed 4.6 billion years ago, the sun was much weaker than it is today, with nearly 70% less energy output. Obviously, from a heat budget point of view, this represents a significant decrease in "deposits" of energy to run our planet. Some scientists believe that the Earth would have been so cold that water could not have existed in a liquid state; however, we know from the geologic record that this was not the case. This apparent dilemma, termed the *Faint Sun Paradox*, is possibly explained by the estimates of anomalously high greenhouse gas concentrations at this time in Earth's history (Figure 13-17). Despite weak solar heating, the thermal blanket of the atmosphere would have trapped much solar radiation. Heat flow from the Earth's interior was also higher in this early "post-accretion" period. By about 0.5 billion years ago, the sun's output had increased to only 4% lower than it is today. This, combined with the absence of terrestrial carbon storage, created a "greenhouse world" during which much of the biosphere evolved.

In summary, plate tectonics influence global climate over long time scales associated with the geologic processes operating along divergent and convergent plate boundaries that can exists for tens to hundreds of millions of years. Some geologic processes associated with these plate boundaries produce atmospheric CO_2, while others remove it, and all work at different rates, ensuring that there will always be lag times in responses. Finally, plate configurations control the size and shape of landmasses that influence ocean currents. If plate arrangements result in the isolation of a body of water at a polar region, this sets up a condition that could result in development of thermohaline circulation. Conversely, separated plates enhance ocean water mixing and the absence of thermohaline circulation.

Atmospheric Composition and Global Warming

We are able to determine the Earth's present state of orbital variation around the sun. As discussed earlier in this chapter, we are currently at mid-point in obliquity, minima of precession, and eccentricity. The rate at which these parameters change (tens of thousands to hundreds of thousands of years) makes it clear that variation in these parameters can't explain current temperature increases. Could the eleven-year solar irradiance cycle explain the

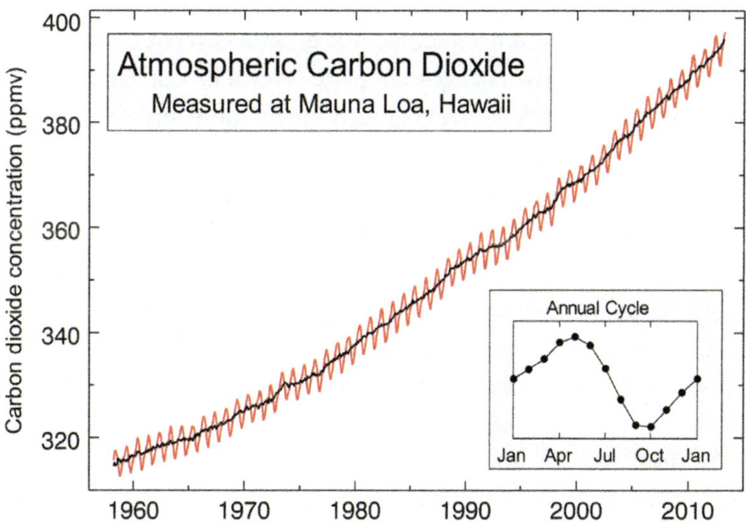

Figure 13-18 This graph, generated by Charles Keeling, shows levels of atmospheric CO_2 over the past sixty years, collected at the Mauna Loa Observatory, Hawaii. The data show the seasonal variation in CO_2 levels, which reflects seasonal plant growth and CO_2 uptake. The inset box shows details of seasonal variation. The solid black line represents the annual average, removing the seasonal effect. Atmospheric CO_2 data are in ppmv (parts per million volume).

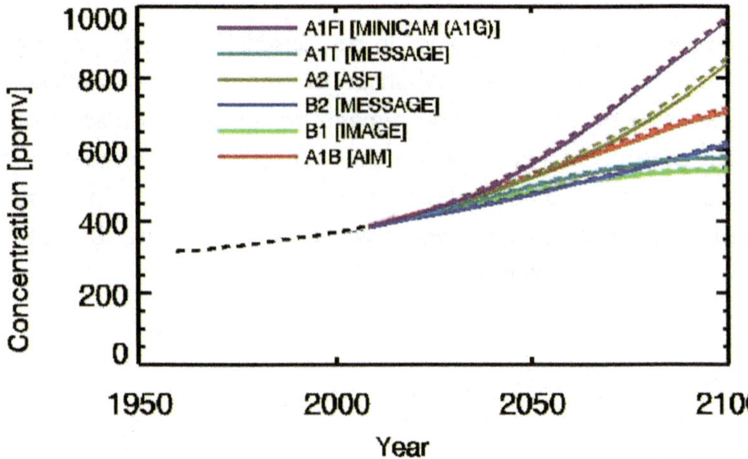

Figure 13-19 Atmospheric CO_2 levels extended back to the start of the industrial revolution. For details of the post 1960 data, see Figure 13-18; 2014 levels are 395 ppmv.

Source: Intergovernmental Panel on Climate Change (IPCC).

temperature data? Astrophysicists estimate a 0.1% difference in irradiance over the course of a cycle. If the long-term average annual solar output is 1366 watts/m^2, a 0.1% variation would produce 1.36 watts/m^2 variation and could not explain current temperature increase. Obviously, changes in plate tectonic configuration of the ocean basins occur over a much longer time scale. We therefore need to assess the potential role that changes in atmospheric composition might play in measured temperature change.

Figure 13-18 presents data on atmospheric carbon dioxide levels over the past sixty years. Compare the values for CO_2 in this graph to those in Figure 13-10. Over the time span represented on the graph, what percentage of CO_2 increase is shown here in comparison to the youngest Vostok ice data?

Figure 13-19 illustrates model predictions for atmospheric CO_2 levels to the end of this century. Model A2 represents the extrapolation of the rate of increase seen over the past 60 years continuing into the future. Figure 13-20 presents model predictions for global temperature increase as a function of increasing levels of atmospheric carbon dioxide. Because the anthropogenic output of CO_2 could vary in the future, two models were run, one assuming continued rates of production and the other a significant reduction in CO_2 emission. Results range between a 4°F and 8°F increase in global average temperatures.

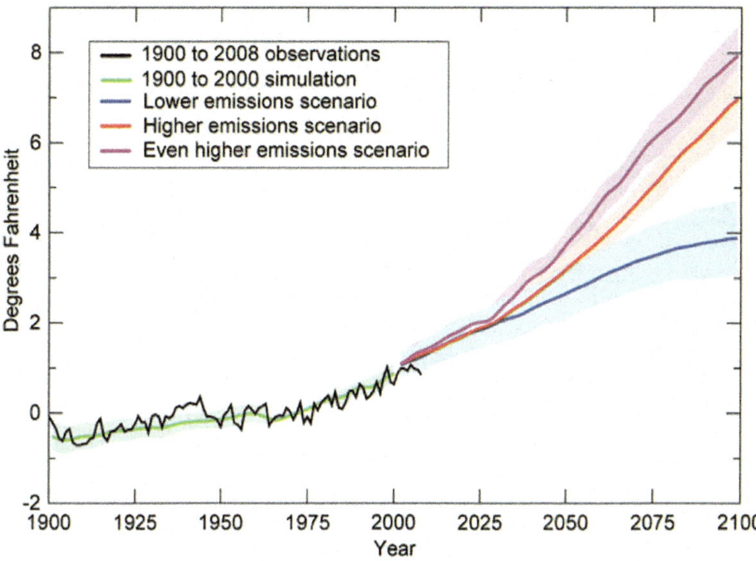

Figure 13-20 Future temperature prediction graph assuming 3 emission scenarios. The shaded areas represent the statistical ranges of values for each scenario.

Image from: http://www.epa.gov/climatechange/science/future.html

The output of models considering atmospheric CO_2 levels predicts increases in global average temperature, but are these models correct? How can you test a model to determine the reliability of its predictions? Recall from Chapter 1 that some scientists did not accept data that inferred that groundwater was contaminated from hydrofracked wells because they lacked geochemical data from *before* drilling. In other words, they could not prove that the water was not contaminated before drilling. Perhaps we should pursue a similar analysis: examine atmospheric composition data from before industrialization and fossil fuel consumption to see if there is a significant difference in the "before industrial burning of fossil fuels" and "postindustrial" levels. Such a correlation would strongly suggest an anthropogenic cause.

In the Vostok core data, we saw cyclic increases in atmospheric CO_2, CH_4, and N_2O levels that predate any possible human influence. So, how do we know that the increase in these atmospheric gas levels documented over the past fifty to three hundred years is anthropogenic? Figure 13-21 shows the results of comparing measured temperature data to predictions from global climate models. The total contributions to atmospheric CO_2 levels from all sources are shown in the pink shaded areas of the graphs (the width of the shaded areas represents the range of data from

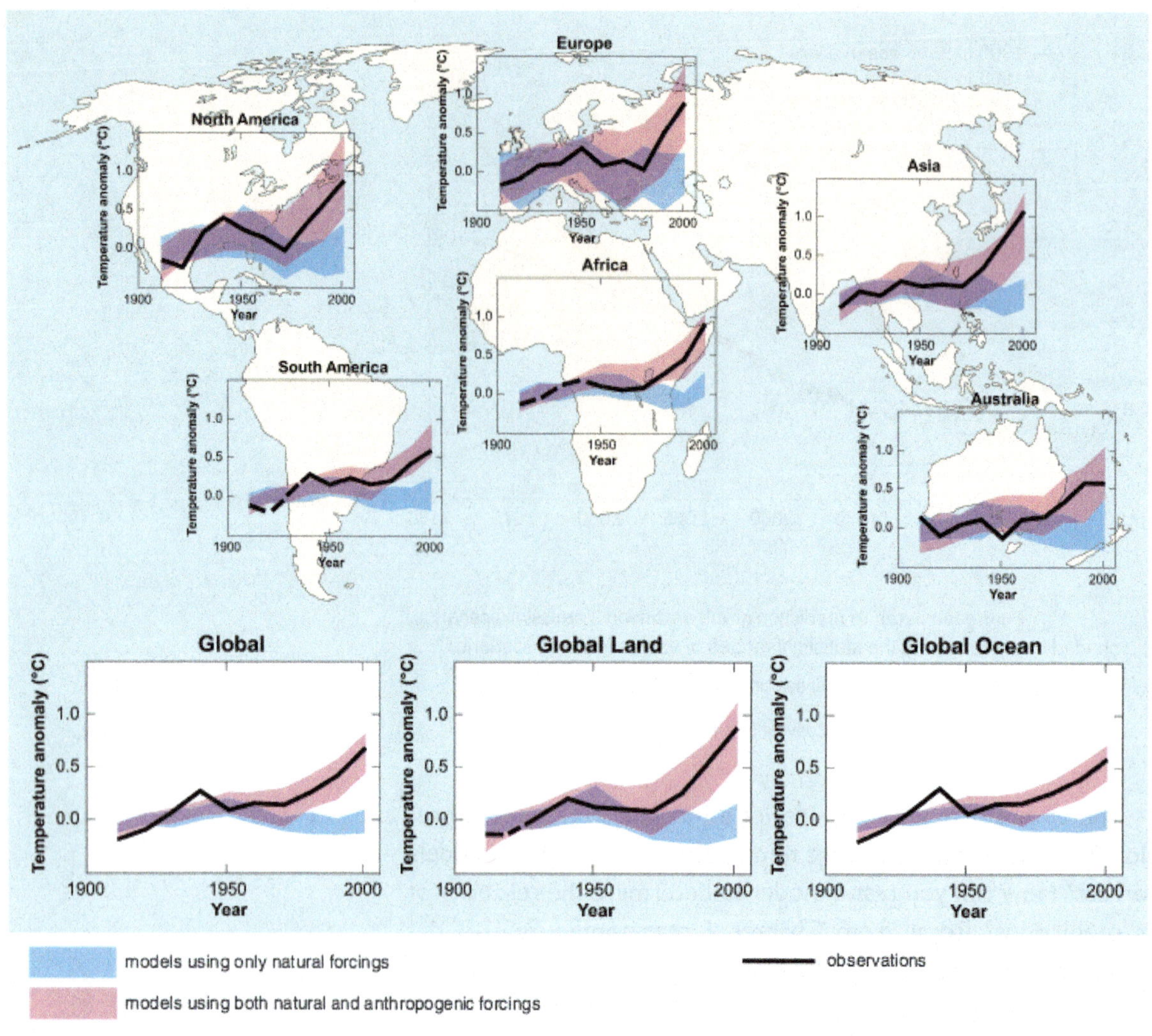

Figure 13-21 Atmospheric CO_2, CH_4, and N_2O levels from Antarctic cores between 1750 and 2000 AD. Compare the change in gas bubble composition for the past three-hundred-plus years to that of the long term (Figure 13-11). What percentage increase do you see for each gas over the long-term value and the 2000 AD value?

minimum to maximum emission levels). Note that the measured values fall in the middle of the pink shaded area, in other words, the models do an excellent job reproducing actual measured values. If we remove any human inputs the model produces results that are shown in the blue shaded area. Note that the results fall short of replicating the measured values, in other words, in order to achieve the historical temperature values measured, levels of atmospheric CO_2 must include the emissions generated by human

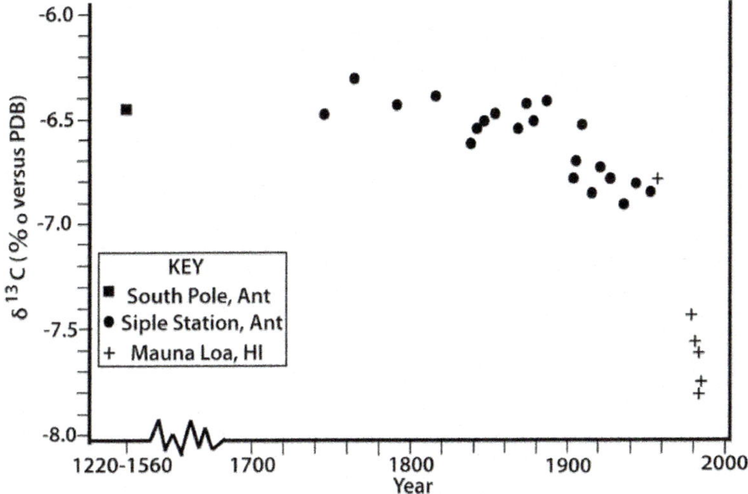

Figure 13-22 Measurements of carbon stable isotopes in CO_2 versus a standard. Data is from air bubbles in Antarctic ice (black squares and circles) and the atmosphere (Mauna Loa Observatory, Hawaii). Modified from Friedli, et al. 1986. Starting with the industrial revolution in the 1800s the atmosphere has become depleted in ^{13}C, which reflects the increasing contribution of lighter carbon stored in fossil fuels, released to the atmosphere when burned.

activities. Clearly, humans are responsible for altering atmospheric CO_2 levels and as a result, global temperatures.

The "smoking gun" linkage between atmospheric CO_2 levels and human activity was established by stable isotope analysis of carbon. Carbon has three isotopes, one of which (^{14}C) is unstable and is useful for numerical dating. The other two isotopes, ^{13}C and ^{12}C, are stable and are useful for tracking the pathway of carbon through the biosphere or geosphere. We can think of the plant fractionation of carbon, with different taxa preferentially utilizing either ^{12}C or ^{13}C as a "fingerprint" of where the carbon comes from. A study of the composition of gas bubbles in Antarctic ice cores (Friedli et al., 1986) measured the ratios of C_{13} to C_{12} in the CO_2 and CH_4 molecules. This study demonstrated that the source of the carbon was from the release of burned fossil fuels.

Multiple studies tracking the fate of carbon and oxygen through the photosynthesis cycle have demonstrated parallel trends in the anthropogenic production of CO_2 from the burning of fossil fuels and its uptake in plant material and the oceans. By understanding how a particular plant taxa fractionates carbon, and then sampling tree rings spanning many decades of growth, it is possible to track

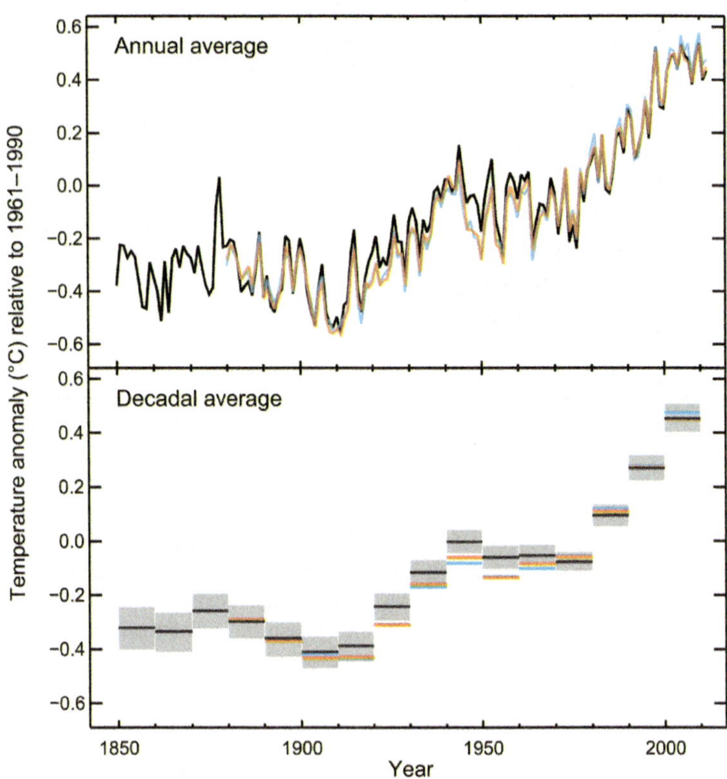

Figure 13-23 Observed global mean combined land and ocean surface temperature anomalies, from 1850 to 2012, from three data sets. Top panel: annual mean values. Bottom panel: decadal mean values, including the estimate of uncertainty for one data set (black). Anomalies are relative to the mean of 1961 through 1990.

Graph from IPCC (2013, Figure SPM1, http://www.climatechange2013.org/images/report/WG1AR5_SPM_FINAL.pdf).

Source: Intergovernmental Panel on Climate Change (IPCC).

the uptake of carbon within plant tissue as resulting from the burning of fossil fuels. These data establish without a doubt that the increase in atmospheric CO_2 levels displayed on the Keeling curve (Figure 13-18) is the result of human activities.

It is clear from Figure 13-10 that cyclically changing climate has been "the norm" for at least the past four hundred thousand years of Earth history, and other climate proxies suggest that climate fluctuation has gone on for about one billion years. How would you respond to the suggestion that the climate change that we are now experiencing is not just "more of the same"—that is, "normal"—and not something produced by human behavior?

Current and Future Temperature Trends

How has global average temperature changed over the past one hundred years, and how is it predicted to change in the future? Figure 13-23 presents temperature measurements over the past 150 years.

What has been the magnitude of annual average temperature change from 1980 to 2007? If this rate of change were extrapolated to the end of this century, what would the change in global average temperature be at this time?

Missing Heat? Another Example of the Scientific Method

While examining the annual average temperature change graph for your calculation of rate of change, you hopefully noticed that the most recent data (2000 to 2010) does not continue the upward trend that is visible in the preceding fifty years. This apparent plateauing of the temperature data is called the "global warming gap" by some scientists, or the "missing heat" by others. Both phrases refer to what appears to be a diminished warming effect that runs contrary to what the continued steep rise would predict from the observed increase in greenhouse gas emissions. Models predict climate warming of approximately 0.21°C per decade, and instead we saw temperature change of only 0.04°C over this interval. Climate change skeptics interpret this plateau to be evidence that climate models are wrong. How would you approach the problem, "Why aren't measured temperature increases matching predicted values?" To the uninformed, the simple answer is "the models are wrong." Recall that to philosopher of science Thomas Kuhn, data that are not reconcilable with existing models constitute the first step in rejecting the current paradigm and crafting a new one. Is this "heat gap" the first evidence that the global climate change paradigm is incorrect?

How might a scientist approach this problem? She might pose questions such as "Is there a problem with the climate models?" or "Is there a heat sink we haven't identified that is absorbing this heat?" Initially, scientists exploring these questions pursue research on the first question: Is there a natural variability, or cyclicity, to temperature variation that wasn't previously recognized? Earlier

Sidebar 13-3 ENSO in the Equatorial Pacific Region

The outlines of North and South America are shown on the right side of the Pacific, with Asia on the left. During El Niño conditions, wet weather is characteristic of the southern United States, and drought is common in the western Pacific (including Australia). During La Niña events, winter temperatures are above normal in the southeastern United States and below normal in the northwest.

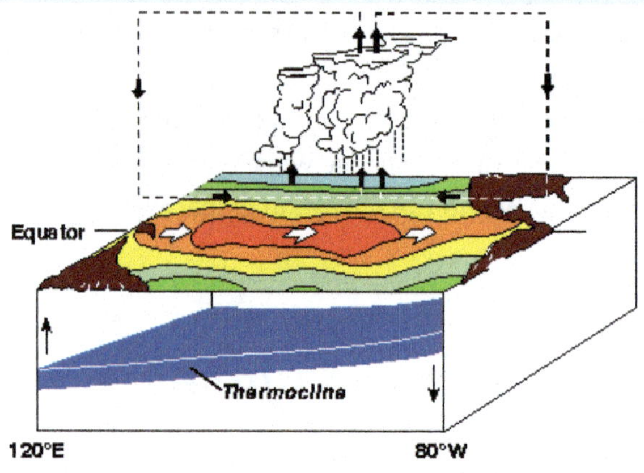

Normal Conditions

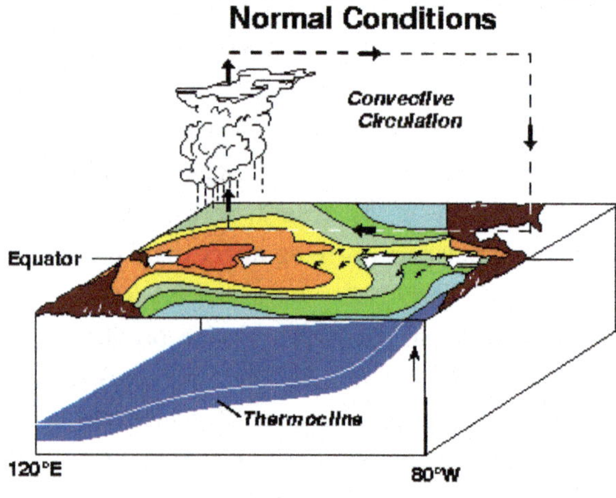

La Niña Conditions

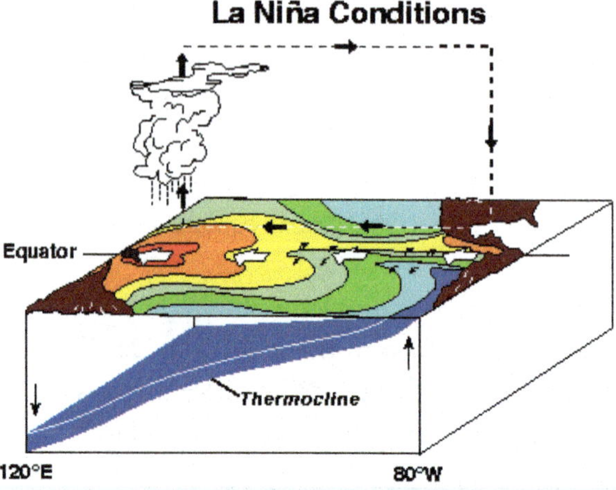

SB 13-3: NOAA / Copyright in the Public Domain.
Images from http://www.pmel.noaa.gov/tao/elnino/nino_normal.html.

in the chapter we talked about the difficulty of predicting the weather (short time-span data sets overemphasize the importance of fluctuations) versus climate (fluctuations are "averaged out" over time, so the trends show up). Are the cooler-than-expected temperatures just an anomaly within the long-term trend? Another possibility is the coincidence of the solar irradiance minima at the turn of the century (less solar heat coming in). Or, is the increasing production of aerosols, which increase albedo, from elevated levels of coal-burning in China a factor? Using satellite measurements of "heat in" and "heat out," it's been estimated that aerosols and solar irradiance variation can account for only about 20% of the missing heat. Research on the source of the remaining 80% of the anomaly is focusing on the role that the oceans play in regulating global temperatures. Recall that water has a higher heat capacity than air does. It's the obvious storage site for heat.

Stepping back from the data for a moment, let's think about what this problem means for the process of doing science. Just as Thomas Kuhn's theory of scientific revolutions predicted, an anomalous data set (in this case, less-than-predicted global average temperatures) turns out to be the one that is important for refining the existing paradigm (atmospheric CO_2 is driving current global warming). If additional data sets were to appear that also suggested that the climate models are wrong, we would be correct in entertaining the idea that we need to create a new paradigm, or theory, for how global climate and greenhouse gas emissions are linked. At this point, however, the "missing heat" problem has helped us refine the existing paradigm, not reject it for a new one.

Back to the role that the oceans play in storing heat! The surface temperature data presented in Figure 13-23 are only part of the global temperature story. Let's explore this further. How does temperature vary through the entire ocean water column? How does this temperature profile vary spatially and over time? The oceans receive heat from solar energy, which explains why the surface of the ocean is warmest, with temperature decreasing with depth. As we discussed in Chapter 11, surface water with distinct temperature and salinity properties can sink and flow at depth (thermohaline circulation). Cold bottom water can also rise to the surface in regions termed *upwelling zones*. While climate models focus on heat and moisture exchange at the ocean *surface*, we know little about heat exchange in the ocean *depths*. This is an exciting current area of exploration: How do the fluctuations

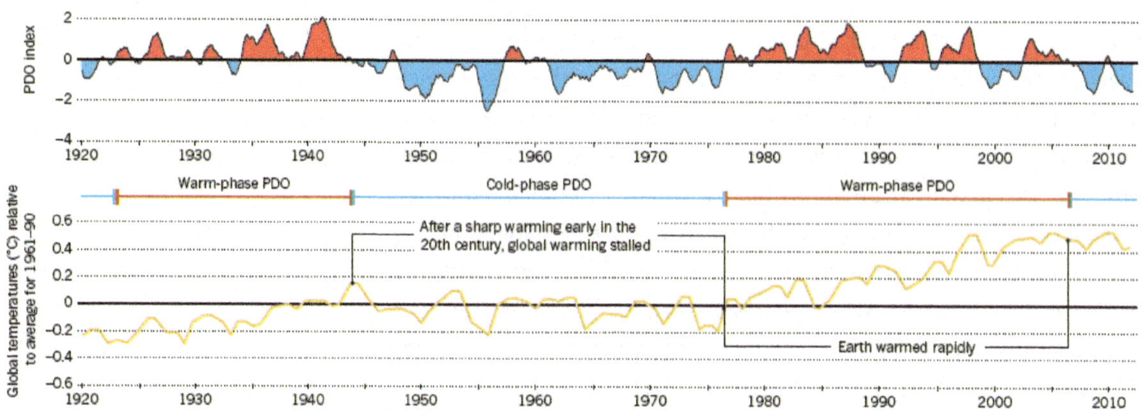

Figure 13-24 The Pacific decadal oscillation (PDO) and global temperature anomalies. When the eastern Pacific Ocean is warm (PDO positive phase), global temperatures quickly rise. When the PDO is negative (cooler Pacific Ocean), global temperature rise pauses. This is the current state of ocean circulation.

Source: Jeff Tollefson, from *Nature*, vol. 508, no. 7494, pp. 20-21.

in coupled atmosphere-ocean circulation and heat exchange influence global climate? The best example of the role of upwelling and climate is the *El Niño/Southern Oscillation (ENSO)* phenomenon, a three- to seven-year cycle in Equatorial Pacific Ocean warming (Sidebar 13-3). During *El Niño events*, a reduction in the upwelling of deep, cold water off the west coast of South America keeps surface Pacific equatorial waters very warm, which in turn heats the overlying atmosphere. During *La Niña* events, the opposite happens: deep, cold waters upwell off the South American coast, and the equatorial Pacific surface waters cool down. The overlying atmosphere cools as well. The "southern oscillation" portion of the phrase describes the effect that the altered sea surface temperatures have on high- and low-pressure regions in the atmosphere in contact with this ocean water. In the mid-1990s, an intense El Niño event created unusually warm temperatures across the equatorial Pacific, which spread globally. In addition to elevated temperatures during this period, there were several "extreme weather" events, such as severe floods, in many parts of the globe. By the end of the decade, however, circulation rapidly shifted into a La Niña event, which rapidly switched the Pacific into cool temperatures. This warm-cold cycle, which occurs on approximately ten-year interval, is termed the *Pacific decadal oscillation (PDO)*, and this fluctuation might hold the key to the "missing heat." Several studies in the early 2000s supported the idea that PDO explains why the Earth hasn't heated as much as early climate models predicted. When climate data over the past one hundred years are studied on the frequency of PDO cycles, "heat gaps" such as the one we are currently in are produced (Figure 13-24). The key finding of this research is that when the ENSO temperature effect is removed from the global average temperature data, the resulting data show a linear trend in temperature: voila, the "missing heat" signal is gone.

Another study that focuses on the role the deeper ocean waters play in storing heat focuses on subsurface flow in the Atlantic Ocean. The warm, salty water generated in equatorial regions flows north as the Gulf Stream, bringing heat northward. Elevated levels of salinity can cause the warm waters to sink, taking the heat with it. A series of buoys, termed Argo floats, have been collecting temperature and salinity data at depths up to 2,000 meters. These instruments have measured increasing water salinity and temperature at approximately 700 meters. Historical data suggest that the salinity fluctuations may occur on an approximately sixty-year oscillation. When this cycle comes to an end and less saline warm water stays near the surface for a longer time, more heat would be available to warm the atmosphere. Rates of global warming would increase as a result.

Both of these studies (ENSO and Atlantic salinity) involve subsurface ocean water storage of heat. Further work is needed to understand the ocean temperature profile and subsurface circulation. The new data suggest that the deep ocean is a larger heat sink than many older climate models considered it to be. More recent studies (Steinman, Mann and MIller, 2015) suggest that the apparent slow down in global warming is itself an artifact of how we synthesize temperature data from the Atlantic and Pacific Oceans. These scientists suggest that there is no missing heat or "pause" in the global warming trend.

There are several takeaway points from this discussion. When you examine a set of data and determine that it doesn't "fit" with what the existing model predicts, you have two options. You can "throw the baby out with the bathwater" and say, "We don't know what's going on," or you can also look at the anomaly and say, "This is interesting—what haven't we figured out yet?" When you use the anomalous data as an incentive to keep exploring, you discover new relationships and you can refine your paradigm. In this case, a portion of the data set on global temperature measurements did not agree with the predicted temperatures. New research exploring possible explanations for the anomaly led to the discovery of the previously less-understood role that subsurface ocean temperatures, and the cyclic frequency in which they interact with the ocean surface and atmosphere, play in regulating global temperature variation. What are the implications of this for future global average temperature change?

Summary

Weather represents short term (daily) changes in temperature and precipitation while climate describes average daily weather over longer term (decades and centuries). Weather is notoriously difficult to predict, in part because of the highly variable initial conditions that you need to know in order to determine the direction of change. On the other hand, we know that the winter will be colder than the summer because this has been the case for decades and decades of weather data. Despite the complexity of the Earth's climate, it is possible to make predictions about how climate is changing. In the previous chapter we discussed the role that atmospheric composition plays in controlling the Earth's climate. There is also short term (11 year cycle) variation in the amount of solar energy produced. At the other extreme, plate tectonics controls the size and shape of ocean basins and therefore controls ocean circulation and heat distribution. For example, we know that the present day ocean circulation has only been in existence for the past 2.5 million years. Prior to that, warm tropical waters did not move northward to the poles and global climate was much more latitudinally zoned. Most important in long term climate change discussions, however, is the role played by variation in the Earth's orbit around the sun. The Earth's rotation around the sun varies in several aspects. These were described and quantified by Milutin Milankovitch, a Serbian astronomer and hence these are called the Milankovitch Effects. Obliquity describes the wobble of the Earth as it rotates. Eccentricity describes how circular or elliptical the Earth's rotational path around the sun is. Precession describes whether the northern hemisphere is tilted towards or away from the sun at its furthest distance. Milankovitch determined that there are cyclic changes to each of these parameters and hence there are times when the Earth receives greater and lesser amounts of solar radiation and times when the temperature variation between summer and winter are more or less extreme. Obliquity varies on a 41,000 year frequency, eccentricity on a 100,000 year cycle and precession on two cycles of 19,000 and 23,000 years. Mathematical analysis of the cycle frequency of the Milankovitch Effects indicates that they can explain some observed climate cycles in the geologic record. The primary repository of historical climate data are ice cores from Antarctica and Greenland. Analysis of ancient climates is based on climate proxies: phenomenon that vary directly as a function of temperature. The most important climate proxy are stable isotopes of oxygen and hydrogen. By analyzing the isotope data in ice cores we are able to reconstruct temperature profiles going back nearly 5 million years. This data shows that climate has varied cyclically through this interval of time and many of the cycles align with Milankovitch cyclicity. Cycles in the abundance of CO_2 and CH_4 in the atmosphere can also be observed by analysis of tiny bubbles of "ancient air" in the glacial ice. There are cycles in climate that exist on shorter frequencies of thousands to hundreds of years. These frequencies are much shorter than can be explained by orbital variation and must have other causes. The ice core record confirms that the Earth's climate is influenced by many variables that operate at different frequencies. An important recent example of rapid climate change on Earth is termed the Younger Dryas Event. Named after a tundra-dwelling small plant, the Younger Dryas was a global cooling event which occurred between 12,800 and 11,500 years ago. The onset of this global cooling is thought to be the collapse of thermohaline circulation in the Atlantic Ocean. Following the global cooling was a very rapid warming event which might serve as a model for the rapid warming we are currently embarking on. While the cause of post-Younger Dryas warming was not burning fossil fuels, the rate at which it occurred is at the same order of magnitude as the present day and predicted change.

This chapter has presented historical data recording changes in the Earth's climate over the past 450,000 years. The data show that the climate has varied cyclically. It is possible to statistically analyze these cycles to determine their frequency. We recognize that cycles have occurred over a variety of frequencies from one hundred thousand years to less than two thousand years. The same mechanism cannot produce all these. Instead, climate on Earth appears to be driven by a range of processes operating on different scales. Most important is the combination of orbital variation and atmospheric composition, with feedback mechanisms including albedo and ocean circulation amplifying and mitigating these. The investigation into the origin of climate cycles is still ongoing. However, there is conclusive evidence, in the form of isotope geochemistry, that in the past two hundred years the role of atmospheric composition driven by anthropogenic behavior has occurred and will exert a dominant influence on the Earth's climate through the remainder of this century. The geochemistry of carbon molecules conclusively demonstrates that the source of increased levels of CO_2 and CH_4 in the atmosphere is derived from the release of previously sequestered (buried) plant-based carbon. As the climate warms from the addition of CO_2 and CH_4 to the atmosphere, various positive feedback mechanisms will kick in. For example, the melting of permafrost will release methane. The melting of glacial ice will reduce albedo. Both of these feedback mechanisms will serve to increase solar insolation. The historical climate record indicates that climate warming can be very rapid, with slightly slower, stepwise cooling.

There is an analogy for the Earth's climate that uses the concept of "tipping points" in a changing system. Imagine the Earth's climate as a canoe. There are many processes that can cause the canoe to rock from one side to the other, but it returns to a stable point, and we all stay dry. There is a point, however, where the canoe tips too far in one direction, and every attempt to bring it back to balance fails. The canoe tips over, and we all get wet. Some of the climate drivers that we've discussed in this chapter can cause the canoe to rock; volcanic emissions are one example. Because it will trigger feedback (such as decreased albedo and changes to ocean circulation), the majority of climate scientists interpret the data to indicate that the ongoing increase in atmospheric greenhouse gas levels will push our rocking canoe past its tipping point. We're all going to get wet. Research continues on *how soon* that will happen, not *whether* it will happen.

Chapter 13 Review Questions

- What is the difference between weather and climate?
- Why is the angle of incidence of solar radiation so critical to transferring heat to the Earth's surface?
- Why does the Earth have seasons?
- What are the components of the Milankovitch Effects and their frequency of cycles?
- What is a "climate proxy"
- How do the stable isotopes of oxygen and hydrogen function as climate proxies?

- Examine Figure 13-10 and calculate the average frequency of temperature cycles over the past 400,000 years. While an apparent similarity does not confirm a causal link, did your cycle frequency come close to any Milankovitch effects?
- Glacial ice frequently has small bubbles of "ancient air" trapped within it. Figure 13-11 shows some air composition data for the past 20,000 years as well as a temperature proxy. Make at least two observations about this data.
- What is a Dansgaard-Oeschger (D-O) cycle? What is the theoretical basis to thinking that these Greenland ice sheet temperature cycles might be related to ocean circulation?
- The Younger Dryas is an important climate event. When was it and what characterizes this interval of time?
- Before plate tectonic processes created the isthmus of land we call "central America" the Atlantic and Pacific Oceans were connected and water could flow east west. How was climate in the North Atlantic region different at this time? How was the climate of Antarctica different?
- What role does the uplift of mountain ranges have on the carbon cycle, and ultimately on global climate?
- In addition to triggering volcanic activity, how would you describe the effect of plate movements on the Earth's climate?
- How do the stable isotopes of carbon provide the "smoking gun" linkage between the burning of fossil fuels and levels of CO_2 in the atmosphere?
- Draw a concept map that illustrates all of the variables that control the Earth's climate and how they do so.
- Scientists recently announced that 2015 is the start of an El Nino event. What processes does "El Nino" describe and why does El Nino have an effect on global climate?

Index

A

accretionary prism 193, 195
albedo 240, 249, 259–261, 270, 272, 274, 277, 309, 315
anticline 192, 195–197, 203
aquifer 6–7, 9, 12, 207, 219, 247
Archimedes principle 197, 199, 201, 203
asthenosphere 121, 126, 129–130, 160–161, 189
atmosphere 19–23, 37, 39–40, 48, 55, 57, 60, 70, 77, 82, 87–88, 127, 135, 182, 192, 207, 209–216, 219, 222–223, 225, 228, 237–248, 249–274, 276–277, 284–285, 289–291, 294, 297, 300–302, 307, 311–316
  composition 250–256
  historical composition 281–307
  layers 252–256

B

Benioff-Wadati zone 157, 176, 180, 188, 190
Bowen's reaction series 74–77, 92, 218

C

carbon cycle 250, 263–266, 273–274, 300, 316
carbon dioxide
  atmosphere 250–253, 284, 303–307
  chemical weathering 215
  chemistry (molecule) 34
  greenhouse gas 240, 244, 267–274
  ocean acidification 20, 243–244, 266
chemical weathering 207, 215–218, 224, 230–231, 237, 246–247, 263–264, 272–273, 300–301
  dissolution reactions 263
  hydrolysis 263
  oxidation 237
cinder cone 78–79, 135, 146, 170, 172
climate proxy 285, 314, 316
closed system 18, 22–23, 30, 39, 42, 92, 210, 219, 246
compounds 28–29, 32–33, 174, 269
continental drift 157, 159–160, 188–189
convection 71, 131, 134–136, 139, 141, 150–152, 161, 174, 190, 254, 257
convergent plate boundaries 157, 162, 168, 175–182, 188, 190–191, 193–194, 202, 210, 264, 302
core, Earth's 68, 92, 121–122, 125–126, 128–130, 131, 135, 138, 141–153
Coriolis effect 225–226
correlation (statistical) 9, 10–11, 13, 288
covalent bonding 61
crust 27, 36–37, 40, 55, 71, 82, 95, 111, 120, 122, 124–130, 132–133, 138, 150–153, 158–160, 166, 180–182, 188, 190, 193–203, 236, 250–251
Curie point 131, 145–147, 150, 153

D

Dansgaard-Oeschger (D-O) cycles 291, 297, 316
decompression melting 70–71, 135, 174
deuterium and temperature 286, 288, 290
divergent plate boundaries 71, 152, 162, 168–170, 172–176, 181, 188, 190, 301

E

earthquakes 5, 7, 13–16, 109–130, 162–169
  epicenter 110–111, 114, 117–119, 122–124, 128–130, 162, 172, 177, 179
  seismic waves 110, 115–122, 141
  size 122–124
elastic rebound theory 110, 112, 129
electromagnetic spectrum 249, 257–258, 272
elements (abundance) 36–39
elements (chemical) 28–36
energy budget, Earth's 141, 249, 260, 269, 272
ENSO 310–312

F

faults 15, 78, 82, 113, 129, 149, 165–173, 177–179, 182–190, 195, 197, 203
feedback loops 20, 240, 266
focal mechanism solutions 165–168, 172, 179, 190
focus (earthquake) 110–111, 116, 122, 128–130, 163, 165–168
fold and thrust belt 193, 195
folds 195–197, 203
fractional crystallization 74, 79–81, 92–93

G

geotherm 63, 68–69, 93, 132
geothermal energy 69
geothermal gradient 68–70, 93, 132
global warming potential 268–269
greenhouse gas 250, 267–274, 289–290, 294, 297, 302, 308–309, 315
Gutenberg discontinuity 121
gyres 208, 227, 246, 298

H

half-life 95, 100–108
heat capacity of water 225, 240
heat flow 23, 112, 121, 125, 132–137, 141, 150, 157, 159, 171–174, 178–182, 186, 188–190, 302
heat flux 133–136, 150, 169, 178, 185
historical temperature data 290–295, 305, 307
hot spot 137, 139, 151–152, 251, 270
hydrogen bonds 34–35, 45, 56–61
hydrologic cycle 55, 211, 238, 241, 247, 285
hydrosphere 19–23, 55, 60, 182, 192, 207–248, 250, 263, 285, 288, 290
hydrothermal vents 171, 182

I

inner core 121, 125–126, 129
ionic bonds 33–34, 49, 56, 61, 217
isolated system 18, 21
isostasy 191–192, 194, 197–199, 201–203, 234

K

karst 219–222, 247
Keeling curve 273, 306

L

latent heat 60–62, 121, 238, 248, 261, 273
lithosphere 121, 126–130, 136–138, 160–161, 169, 173–174, 177, 180–185, 189, 194, 201, 210
low-velocity zone 120–121, 125–126, 161

M

magnetic anomalies 148–152, 171–173
magnetic field (Earth's) 131–132, 140–153
mantle 71, 74, 76–77, 90, 120–139, 141, 148,

150–151, 160–161, 174–183, 186, 193–203, 209–210, 233–236, 250
Mercalli scale 123
methane 6–10, 15, 59, 244, 267–269, 290, 315
Milankovitch effect 275, 314–316
minerals 41–62, 63–94, 104–105, 107, 113–121, 124–126, 145–146, 171, 174, 182, 197, 201, 209–210, 214–218, 221–223, 237, 245–246, 251, 263, 300
Moho 120, 125, 129–130
mountain building. *See* orogeny

N

neutron capture 28–29, 39
nucleosynthesis 28, 35–40, 127
numerical (radiometric) dating 95–96, 99–108, 147, 288, 305

O

ocean acidification 20, 243–244, 266
ocean chemistry 222, 243, 295
open system 18, 22
orogen 193–194
orogenesis 193, 202
orogeny 191, 193, 195, 197, 203
outer core 68, 92, 122, 125–126, 129–130, 141, 151, 153
oxygen isotopes 284–286, 288

P

partial melting 71–74, 92–93, 121, 125
passive margin 158, 187
phanerozoic carbon dioxide levels 247, 301, 303
physical weathering 58, 207–208, 216, 218, 230–231, 234, 246, 248, 264
plate boundaries 21, 71, 112, 134, 136, 151–152, 162–190, 191–203
plate tectonics 157–190
plate tectonics (definition) 159
plume (mantle plume) 131, 137–139, 161, 178
polarity (magnetic) 56, 144–154
polarity reversals 143–144, 146–147, 150, 152
polarity time scale 147

R

radiometric (numerical) dating. *See* numerical (radiometric) dating
rainwater composition 212–215, 246–247
rare Earth elements 42
rates of geologic processes 95, 106
refractive elements 39
relative dating 96–98, 105, 107–108
Richter scale 122–123
ridge push 157, 174
river composition 216, 218
rock cycle 63–65, 68, 92–93, 263

S

San Andreas fault 13, 112, 182–183, 185–187
sea level change 208, 239, 241
seismic tomography 137–139, 152, 178
seismic waves. *See* earthquakes
seismometers 114, 119, 122, 128, 137
shadow zones 109, 122, 129–130
shield volcanoes 78–79, 172, 174
silicates 50–51, 72, 145
silicate tetrahedral 92
slab pull 157, 181, 188
soil 88, 219, 236–238, 246
soil profile 237
solar irradiation 262–263
Steno's laws 95, 97, 105, 107–108
stratosphere 252–254, 256, 271
stratovolcanoes 78–81, 180, 194–195
strike-slip (transform) plate boundaries 152, 162, 165–167, 169, 177, 182–190, 203
subduction 176
subduction zone 157, 176–182, 188, 193–194
syncline 192, 195–197, 203

T

thermohaline circulation 227–230, 247–248, 266, 294, 297–298, 299, 302, 311, 314
time travel curves 119
topography of the Earth's surface 134, 170–173
troposphere 252–254, 256, 261, 267, 272–273

V

viscosity
  magma 64–83, 89–93, 111, 135, 135–137, 145, 148, 169, 172–174, 180–181
  mantle 177–182
volatile element 37–38, 127
volatility of magma 210
volcanic emission composition 263, 270–271
volcanoes
  global distribution 162, 164–165
  types 76–82

W

water cycle 55, 210, 216, 246
water molecule (chemistry) 34–35, 43, 49, 55–61
water table 12, 69, 219
weather versus climate 276
Wegner, Alfred 158–159, 188–189, 299

Y

Younger Dryas 293–295, 297, 314–316

CPSIA information can be obtained
at www.ICGtesting.com
Printed in the USA
LVOW02s2000100817
544542LV00003B/12/P

9 781626 618862